The Facts On File

DICTIONARY of WEATHER and CLIMATE

The Facts on File

DICTIONARY
of
WEATHER
and
CLIMATE

Edited by
Jacqueline Smith

Checkmark Books®

An imprint of Facts On File, Inc.

The Facts On File Dictionary of Weather and Climate

Copyright © 2001 by Market House Books Ltd

All rights reserved. No part of this book may be reproduced or utilized in any form or by any means, electronic or mechanical, including photocopying, recording, or by any information storage or retrieval systems, without permission in writing from the publisher. For information contact:

Facts On File, Inc.
132 West 31st Street
New York NY 10001

Library of Congress Cataloging-in-Publication Data

The Facts on File dictionary of weather and climate / edited by Jacqueline Smith.
p. cm.
ISBN 0-8160-4532-1 (hardcover : acid-free paper). — ISBN 0-8160-4533-X (pbk. : acid-free paper)
1. Meteorology—Dictionaries. 2. Weather—Dictionaries. I. Smith, Jaqueline.

QC854.F33 2001
551.5—dc21 2001040465

Facts On File books are available at special discounts when purchased in bulk quantities for businesses, associations, institutions, or sales promotions. Please call our Special Sales Department in New York at (212) 967-8800 or (800) 322-8755.

You can find Facts On File on the World Wide Web at
http://www.factsonfile.com

Compiled and typeset by Market House Books Ltd, Aylesbury, UK

Printed in the United States of America

MP 10 9 8 7 6 5 4 3 2 1
(pbk) 10 9 8 7 6 5 4 3 2 1

This book is printed on acid-free paper

CONTENTS

PREFACE

This dictionary is one of a series designed for use in schools and colleges. It is intended for students of meteorology and climatology, but we hope that it will also be helpful to other science students and to anyone interested in the subject. The other books in the series are *The Facts On File Dictionary of Astronomy*, *The Facts On File Dictionary of Biology*, *The Facts On File Dictionary of Biotechnology and Genetic Engineering*, *The Facts On File Dictionary of Chemistry*, *The Facts On File Dictionary of Computer Science*, *The Facts On File Dictionary of Earth Science*, *The Facts On File Dictionary of Environmental Science*, *The Facts On File Dictionary of Marine Science*, *The Facts On File Dictionary of Mathematics*, and *The Facts On File Dictionary of Physics*.

The dictionary contains over 2,000 headwords covering terms and concepts in modern meteorology and climatology. The Appendix contains a chronology of important events and discoveries in the subject, a bibliography, and relevant web sites, as well as a number of useful conversion tables.

We would like to thank all the people who have cooperated in producing this book. A list of contributors is given below. We are also grateful to the many people who have given additional help and advice.

ACKNOWLEDGMENTS

Contributors

Keith Boucher B.A.,.F.R.Met.S.
Clair Hanson M.Sc., Ph.D.
B.C. Harrison
R.G. Harrison Ph.D.
Ian Douglas Phillips B.Sc., M.Sc., Ph.D.
Phillip Reid B.Sc., Ph.D.
Ross Reynolds B.A., M.A.

absolute drought In the UK, a DROUGHT in which there is a period of at least 15 consecutive days, none of which have recorded 0.25 mm (0.01 in) of rain or more. This definition is inappropriate for dissimilar climates; it was originally defined in 1887 by the former British Rainfall Organisation, but is no longer much used. *See also* partial drought.

absolute humidity A measurement of the amount of moisture (WATER VAPOR) in the atmosphere, usually expressed as the mass (grams) of water vapor per cubic meter of moist air. In contrast to RELATIVE HUMIDITY, absolute humidity expresses the amount of water in the air regardless of the temperature. It is sometimes referred to as *vapor concentration*. *See also* humidity; specific humidity.

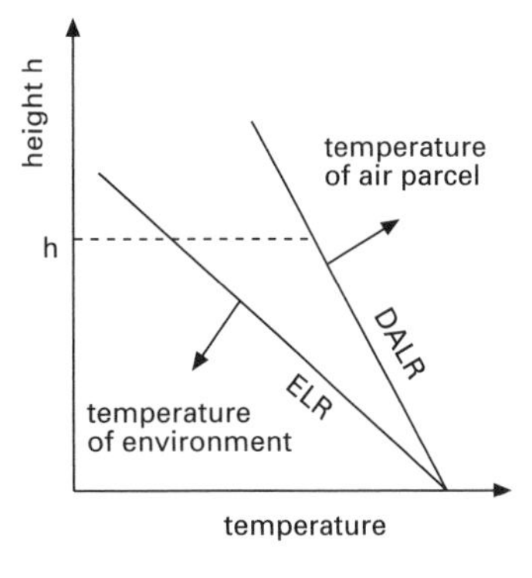

Absolute instability

absolute instability The atmospheric state in which the observed decrease in temperature with height, as defined by the ENVIRONMENTAL LAPSE RATE (ELR), is greater than that of a dry air parcel cooling adiabatically, as defined by the dry adiabatic lapse rate (9.8°C km^{-1}). Consequently, at any height *h* the air parcel will be warmer and less dense than its surroundings, and so will continue to rise, leading to *absolutely unstable air*, and often to the formation of clouds of great vertical depth.

absolutely stable air *See* absolute stability.

absolutely unstable air *See* absolute instability.

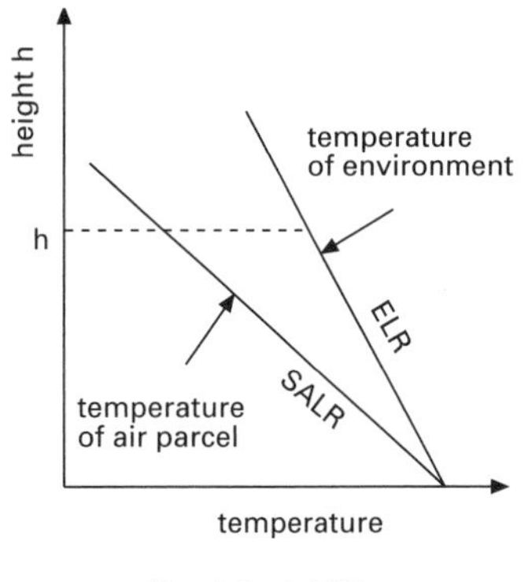

Absolute stability

absolute stability The atmospheric state in which the observed decrease in temperature with height, as defined by the ENVIRONMENTAL LAPSE RATE (ELR), is less than the SATURATED ADIABATIC LAPSE RATE (SALR). Consequently, at any height *h* the air parcel will be cooler and therefore denser than its surroundings. If an air parcel is perturbed in this situation, it will return to its original position because vertical air motion cannot be sustained. Absolutely stable air typically accompanies a summer ANTICYCLONE (high).

absolute temperature *See* thermodynamic temperature.

absolute vorticity *See* vorticity.

absolute zero The zero of THERMODYNAMIC TEMPERATURE. This is the lowest temperature theoretically possible and although it cannot be reached in practice, temperatures of 10^{-3} K are used in low-temperature physics. Absolute zero, is 0 K = –273.15°C. *See* kelvin.

absorption In the atmosphere, the conversion of electromagnetic radiation into internal energy of molecules. The absorption process depends on the energy levels of the molecule and on the wavelength of the incident radiation. For example, ozone absorbs ultraviolet radiation; carbon dioxide and water vapor both absorb infrared radiation. Absorption occurs across the whole spectrum for electromagnetic wavelengths, although there are 'atmospheric windows' in which little absorption occurs, notably in the radio and visible regions of the spectrum. Of the incoming radiation around 19% is absorbed by the atmosphere: 16% by gases and dust and 3% by clouds.

accessory cloud A smaller cloud feature that is related to a main cloud, dependent on it for development and continuance, and can either be attached or separate from it. Examples include PILEUS (cap cloud), a smooth 'cap' cloud of small horizontal extent that forms just above cumuliform clouds, PANNUS, and VELUM, which are all generally separate from the main cloud, and ARCUS (shelf and roll clouds), which are attached. *See also* cloud classification.

acclimatization The process of adaptation made by organisms, such as animals (including humans) and plants, to a change in climate and/or environment, which may result, for example, from a long-term CLIMATE CHANGE or from the organism being transferred to a new environment.

accumulated temperature The sum of departures (positive or negative) of temperature from a reference temperature for a specific period. These departures may be, for example, on a daily, monthly, or yearly mean basis. The accumulated temperature can be represented by the number of degree-hours or the number of DEGREE-DAYS. For a number of specified hours (X), the temperature exceeds the reference temperature by a number of degrees (Y), thus there are X × Y degree-hours and (XY)/24 degree-days.

accumulation The increase in the surface area of an environment by a covering of snow, firn, and ice. *Accumulation sources* include precipitation in the form of snow and freezing rain; hoar frost from radiative cooling at the surface; rime associated with supercooled water droplets in the atmosphere; and mass movement transport processes, such as avalanches and landslides.

acid rain Precipitation in which acidity is above the natural level. Normal rainfall is slightly acidic with a pH of about 5.5 as a result of chemical reaction between naturally occurring CARBON DIOXIDE and water vapor in the atmosphere forming weak carbonic acid. Precipitation with a pH value of less than 5 is considered to be acidic. The burning of such fossil fuels as coal and oil in power stations, motor vehicles, and many industrial processes releases large quantities of SULFUR DIOXIDE and NITROGEN OXIDES into the atmosphere. These pollutants can be carried considerable distances from their original source and react with water vapor in the presence of sunlight to form weak solutions of sulfuric and nitric acids.

When deposited, acid rain can cause severe environmental problems, including killing fish, damaging forests and crops, and accelerating the decay of buildings; it may also be of harm to humans. First recognized as an environmental problem in the 1950s, when increased acidity was observed in lakes in Canada and Scandinavia, it had reached serious levels by the late 1970s. Since the 1980s various measures have been introduced to reduce polluting emissions, especially by the use of catalytic converters in vehicle exhaust systems and flue-gas desulfurization plants in power stations. In the US, Phase I of the *Acid Rain Program* of the 1990 Clean Air Act Amendments set out to reduce SO_2 emissions by 10 million tons and nitrogen oxide emissions by 2 million tons by 2000. This target has been reached. Phase II, begun in 2000, tightens permitted limits and encompasses a wider range of plants. The Acid Rain Program is noteworthy for its replacement of traditional command regulations by an allowance trading system making use of market forces to reduce pollution.

actinometer An instrument used for measuring electromagnetic radiation, especially that transmitted by the Sun, scattered by the atmosphere, and reflected by the Earth's surface. There are a number of different designs including those that use bimetallic strips which, depending on their heat capacity, surface area, and reflectivity, bend differently as a result of being heated by radiation.

active remote sensing The methods in remote sensing of collecting information, for example, the measurement of atmospheric properties, using instruments that emit an artificial signal, which is then partly returned by the phenomena to the instrument. Such an instrument is said to 'active' because it actually generates the artificial beam of radiation; LIDAR and RADAR are examples of active remote sensing. *Compare* passive remote sensing.

adiabat A line drawn on a THERMODYNAMIC DIAGRAM along which an air parcel moves as it ascends or descends through the atmosphere, cooling or warming adiabatically. There are two types of adiabat: the DRY ADIABAT and the SATURATED ADIABAT.

adiabatic Denoting any process in which there is no change of heat. In meteorology it describes a thermodynamic process in which the temperature of an air parcel changes without any exchange of energy with the surroundings. By the first law of thermodynamics:

$$\Delta Q = \Delta U + \Delta W,$$

where ΔQ = heat added to the system (in this case the air parcel), ΔU = change in internal energy (a function of temperature) and ΔW = work performed against the surroundings. In an *adiabatic process*, $\Delta Q = 0$; hence $\Delta U = -\Delta W$. As an air parcel expands, its volume increases, so performing work against the surroundings. By the first law of thermodynamics, the internal energy (and therefore the temperature) of the air parcel must decrease during an *adiabatic expansion* process; conversely the air parcel's temperature increases during an *adiabatic compression* process. Two types of adiabatic processes take place: (1) the *dry adiabatic process*, when the air parcel is unsaturated; and (2) the *saturated adiabatic process*, when the air parcel is saturated. A rising saturated air parcel cools more slowly with height than a dry parcel because the release of latent heat at the change of state from gas to liquid acts to buffer the temperature decrease that occurs as a result of the adiabatic expansion process.

adiabatic change The changes in the temperature, pressure, and volume of an air parcel that occur as a result of an adiabatic process. *See* adiabatic.

adiabatic chart *See* thermodynamic diagram.

adiabatic lapse rate The rate at which a parcel of air changes temperature adiabatically as it moves vertically. In this case heat does not leave or enter the system: as an air parcel rises the reduction in surrounding pressure enables the air is to expand, which results in its temperature falling; conversely, if the air parcel sinks and is compressed its temperature rises. The SATURATED ADIABATIC LAPSE RATE (SALR) and the DRY ADIABATIC LAPSE RATE (DALR) apply to the rate of temperature change of an air parcel, depending on the moisture content of the parcel.

Advanced Interactive Weather Processing System (AWIPS) A technologically advanced interactive computer system for information processing, display, and telecommunications developed for use by the US National Weather Service (NWS). The system integrates meteorological, hydrological, satellite, and radar information and data enabling forecasts to be prepared more accurately and swiftly.

advanced very high resolution radiometer (AVHRR) The imaging scanner widely used on many meteorological satellites. It has been used on board the US National Oceanic and Atmospheric Administration's (NOAA) polar-orbiting satellites (*see* POES) since the launch of TIROS–N (1975).

advection In meteorology, the transfer of a property, such as heat or humidity, by motion within the atmosphere, usually in a predominantly horizontal direction. Thermal advection, for example, is the transport of heat by the wind. Advection is most often used to signify horizontal transport but can also apply to vertical movement, for example in a HYDRAULIC JUMP. Large-scale horizontal advection of air is a characteristic of middle-latitude zones and leads to marked changes in temperature and humidity across boundaries separating air masses of differing origins. *Compare* convection.

advection fog A fog produced when relatively warm moist air cools as it flows across a cooler surface. It occurs most commonly over the sea when water temperature is still cool in the spring and early to mid summer. Outbreaks of warm moist maritime tropical air stream poleward occasionally along the English Channel or across the Grand Banks off Newfoundland, for example. The chilly sea can cool the lowest layer of air to its dew-point temperature; further chilling will lead to condensation of the fog droplets. Advection fog occurs along the coast of California when warm moist air from over the Pacific Ocean blows across the cooler California Current. *Sea fog* is a type of advection fog, which is always characterized by windy conditions. Over land, advection fog occurs sometimes when mild moist air blows across cooler surfaces. For example, it occurs in winter over the central US when moisture-laden air from the Gulf of Mexico flows northward inland.

aeolian *See* eolian.

aeroallergen A substance or gas in the atmosphere capable of inducing an allergy in humans. For example, nitrogen dioxide is associated with allergen-induced nasal inflammation in asthmatics.

aerodynamic roughness The property of a surface, which may be solid or a layer of different density, that determines whether the flow of a fluid (e.g. air in the atmosphere) across it is turbulent or smooth. A surface is defined as aerodynamically rough when the fluid flowing over it is turbulent right down to the surface itself. In the atmosphere, this condition is very common and all surfaces are 'rough' to some degree; how much depends on the height and spacing of the roughness elements.

aerological diagram *See* thermodynamic diagram.

aerology The study of the atmosphere through its vertical extent, generally above the surface layers.

aeronomy The study of the upper atmosphere of a planet, such as the Earth, in which it is usually the study of those layers above the lower stratosphere. Dissociation and ionization are the main processes within these levels in the Earth's atmosphere; phenomena include AIRGLOW and the AURORA. Aeronomy is sometimes used for studies of all the atmospheric layers.

aerosol A dispersion of solid or liquid particles suspended in a gas. In meteorology, the gas is the atmosphere and the prime sources (both natural and artificial) of the particles are at the Earth's surface, including sea salt particles generated by breaking waves, carbon particles in smoke and soot, fine wind-blown soil and clay particles, and ash from volcanic eruptions. They are important meteorologically because some of these particles can act as condensation nuclei (*see* cloud condensation nucleus) and FREEZING NUCLEI; in significant concentration they can influence the RADIATION BUDGET of the atmosphere through ABSORPTION or SCATTERING of solar radiation.

afforestation The planting of trees to establish new forest, often for commercial purposes but also to improve the environment, for conservation purposes, or to provide a CARBON SINK. *Compare* deforestation.

African monsoon The predominantly northeastward invasion of moist air from the Gulf of Guinea during the summer months (May–September) across West Africa. The MONSOON air is accepted as being present as soon as daytime relative humidity has risen to 50%. Elsewhere over tropical Africa, complex topography and the pattern of land and sea, leads to much modified seasonal airflow patterns. This means that only some components of the seasonal reversal of winds may be detected.

afterglow **1.** A broad high arch of radiance sometimes observed in the western sky after sunset and during the deepening TWILIGHT.
2. *See* alpine glow.

ageostrophic wind A theoretical wind; the vector difference between the actual observed wind and the GEOSTROPHIC WIND at a given level. It is not, therefore, an actual wind that blows but is a very important concept associated with CONVERGENCE, DIVERGENCE, and VERTICAL MOTION, for example. These are significant in the production of cloud and, ultimately, precipitation. Surface friction is one property that leads to an ageostrophic wind.

agroclimatology The study of the effects of CLIMATE, which may include climate variability and change, on agriculture. It may include, for example, the effects of climate on crop production, the length of the growing season, the soil climate, and the location and aspects of agricultural units, such as farms in relation to climatic variables.

agrometeorology The study and use of weather and climate information in relation to agriculture. Such information may be used, for example, to expand and improve agricultural production, to assess the impact of weather on crop response and survival, and forecast the occurrence of frosts.

Agulhas Current A warm surface ocean current, originating in the Indian Ocean from the South EQUATORIAL CURRENT and also fed by the Mozambique Current. The Agulhas Current flows southwestward along the South-African coastline and joins the eastward flow between Africa and Australia; some may continue to flow westward round the Cape of Good Hope to join the Atlantic. The Agulhas Current is one of the fastest flowing currents in any of the oceans, capable of reaching a maximum velocity of 5 knots (5.8 mph, ~2.6 m/s) off the southeast coast of South Africa.

air The mechanical mixture of gases that

COMPOSITION OF AIR
(% by volume)

nitrogen	(N_2)	78.08
oxygen	(O_2)	20.95
carbon dioxide	(CO_2)	0.03
argon	(Ar)	0.93
neon	(Ne)	1.82×10^{-3}
helium	(He)	5.24×10^{-4}
methane	(CH_4)	1.5×10^{-4}
krypton	(Kr)	1.14×10^{-4}
xenon	(Xe)	8.7×10^{-5}
ozone	(O_3)	1×10^{-5}
nitrous oxide	(N_2O)	3×10^{-5}
water	(H_2O)	variable, up to 1.00
hydrogen	(H_2)	5×10^{-5}

make up the Earth's atmosphere. The two most abundant constituents are NITROGEN (78.08%) and OXYGEN (20.95%); also present are variable amounts of water vapor (the amount depending on location, temperature, and time), and argon (0.93%), while the TRACE GASES include significant amounts of CARBON DIOXIDE and METHANE.

Aircraft Meteorological Data Relay *See* AMDAR.

AIREP (aircraft report) An internationally agreed type of SYNOPTIC CODE used to transmit a message relating to dry-bulb temperature as well as wind speed and direction from a commercial aircraft in flight.

air frost The condition in which the air temperature of the lower atmosphere, measured at Stevenson screen level of 1.25 m (4 ft) above the ground surface is at or below 0°C (32°F), the freezing point of water. *See also* frost; ground frost.

airglow A faint radiant emission that occurs in the ionosphere of the upper atmosphere. It results from the ionization of atoms and molecules by radiation from the Sun.

air mass An extensive homogeneous region of air, typically a few thousand kilometers across, that possesses broadly the same temperature and humidity characteristics, with generally weak horizontal gradients of these properties. Air masses have their SOURCE REGIONS in the locations of the permanent or semi-permanent ANTICYCLONES (high-pressure zones) where winds are light and from which surface air spirals out to influence more extensive regions. The nature of an air mass depends on the location of the anticyclone with which it is associated, the season, and, for areas outside this source region, the nature of the air's track once it moves away. Air masses are classified according to temperature (i.e. tropical, polar, and Arctic or Antarctic) and on whether the surface where it originates is maritime or continental. The SUBTROPICAL ANTICYCLONES over oceans act as source regions for warm and humid air that is known as a MARITIME TROPICAL (mT) air mass. The wintertime CONTINENTAL ANTICYCLONES act as source regions for cold and dry CONTINENTAL POLAR (cP) air masses, while the highest latitude anticyclones are sources of very cold and dry ARCTIC (A) and ANTARCTIC (A or AA) air masses. CONTINENTAL TROPICAL (cT) air masses originate in the continental subtropical anticyclones. Additional classifications sometimes include equatorial (E) and Mediterranean air masses; suffixes are also sometimes used to indicate whether an air mass is colder than the surface over which it is moving (k) or whether it is warmer (w).

The air that flows away from the anticyclonic source regions is modified to varying degrees, depending on its track. That moving out from northern wintertime continents (continental polar), for example, often blows across the North Atlantic and North Pacific Oceans to be strongly heated and moistened, and transformed into MARITIME POLAR (mP) air. Where two different air masses are juxtaposed, the horizontal gradients of temperature and humidity between them are steep forming FRONTS.

air-mass thunderstorm A short-lived and generally isolated thunderstorm formed by local differential heating of an air mass, not associated with frontal or synoptic features. It forms in warm, humid, and unstable air masses, for example the maritime tropical (mT) air that extends from the Midwest to the Gulf Coast states in the US. Air mass thunderstorms are typically 1–10 km in diameter and may last for up to about 60 minutes.

air parcel A concept used most often in the context of small volumes of air that may or may not ascend or descend through their surrounding 'environmental' air. The 'parcel' is a distinct and 'isolated' bubble of air that may cool or warm (if ascending or descending respectively) at either the DRY ADIABATIC LAPSE RATE or the SATURATED ADIABATIC LAPSE RATE. Such an UNSTABLE parcel may well appear as a small CUMULUS cloud, for example.

air pollution The presence in the atmosphere of gases, aerosols, particles, or energy above background levels in such quantities and over such timescales that are likely to cause harm to all or part of the biosphere, deterioration in the fabric of buildings and structures, or changes to the way the atmospheric system operates. *See also* pollution.

air pressure *See* atmospheric pressure.

air-quality model A mathematical or conceptual attempt to simulate the pathways of pollutants through the atmosphere in order to form a first step in understanding the likely impact of air pollutants on humans and their environment. The models may incorporate components of physics, fluid dynamics, and chemistry within and above the BOUNDARY LAYER of the atmosphere.

air-quality standard The prescribed concentrations of air pollutants that should not legally be exceeded during a given period at a specified location, usually within an urban environment. Recommended guidelines for short-term exposure to pollutants are also provided. *See also* National Ambient Air Quality Standards.

air–sea interaction *See* atmosphere–ocean interaction.

Aitken nuclei The smallest and most numerous type of solid particle in the atmosphere (AEROSOL), with a radius of less than about 0.1 μm, originating mainly from deliberate combustion. They are most highly concentrated in continental air. The nuclei are named for the Scottish physicist and meteorologist, John Aitken (1839–1919).

Alaska Current A comparatively warm current that flows counterclockwise in the Gulf of Alaska; it is an example of a GYRE.

albedo The ratio, usually expressed as a percentage, of the reflected solar radiation beam to the total energy of the beam falling on to a surface as measured across the total spectrum of incoming solar radiation. The greater the albedo value, the less energy is available for other energy transformations (e.g. sensible and latent heat). Bright surfaces, such as fresh snow, reflect more than dark surfaces, such as a coniferous forest, and consequently have a higher albedo value.

ALBEDO

Surface type	*% Reflected*
fresh snow	90%
water (Sun high in sky)	2–10%
water (Sun low in sky)	10–80%
grass	15–25%
forest	5–10%

albedometer An instrument used for measuring the reflecting power, i.e. the ALBEDO, of a surface (e.g. clouds, snow, water, or crops).

Alberta clipper (Alberta low) A generally wintertime fast-moving low-pressure system that forms or redevelops to the lee of the Canadian Rocky Mountains, to the east of the mountain chain under westerly flow. These LEE DEPRESSIONS track generally eastward to often produce strong gusty winds and snowy weather in the vicinity of the Canadian/US border. Alberta clippers can travel as far as the Maritime Provinces of Canada and the northeast US, occasionally intensifying off the east coast when they encounter the relatively warm moist air over the Atlantic Ocean, and spreading heavy snow over land.

Alberta low *See* Alberta clipper.

Aleutian Current (Sub-Arctic Current) A comparatively cold ocean current that flows between latitudes 40° and 50°N in the north Pacific Ocean.

Aleutian low A semi-permanent region of low pressure that appears on average monthly MEAN-SEA-LEVEL PRESSURE charts, located over the Aleutian Islands chain that stretches westward across the North Pacific Ocean from Alaska. It is deeper in the winter than the summer and is not necessarily present from day-to-day. It is the location in which, on average, the traveling FRONTAL CYCLONES in the North Pacific reach their minimum mean-sea-level pressure value.

Alliance of Small Island States (AOSIS) A coalition of small island and low-lying countries that share concerns about their vulnerability to sea-level rise associated with global CLIMATE CHANGE. The alliance was initiated in 1990 to strengthen the states' negotiating position when voicing their opinion on greenhouse emissions within the UNITED NATIONS FRAMEWORK CONVENTION ON CLIMATE CHANGE (UNFCC). It has a membership of 43 states and observers, from all regions of the world.

almwind A warm wind that blows across the plains of southern Poland and occurs mainly during the spring. The almwind is a FÖHN-type wind originating from the forced ascent of a southerly airstream across the Tatra Mountains, which form the Polish/Slovakian border and reach heights of up to 2655 m (8711 ft). This air stream warms rapidly on its descent from the mountain range, often initiating avalanches in southern Poland.

along-slope wind system A wind system that circulates air up, along, and down the slopes of a valley. The system is comprised of ANABATIC and KATABATIC components. Such flows are generally shallow, with the maximum velocities occurring within a few meters of the slope surface.

along-valley wind system A wind system that occurs in mountainous regions. It is generated as a result of the differential heating of surfaces during the solar radiation cycle, this being induced by variations in slope aspect. An up-valley wind is most likely to develop on calm clear days. Valley wind systems may have ANABATIC or KATABATIC components, but will always be conditioned by the dynamics of the synoptic-scale GRADIENT WIND flow in the free atmosphere.

alpine glow The series of colors that appears on mountains, especially when snow-covered, as the Sun nears the horizon at sunset. The mountain tops exposed to the Sun are first yellow, then become pinkish, and finally purple. The series of colors, but in reverse order, is also visible at sunrise on mountains facing the rising Sun. A purplish glow – the *afterglow* – is sometimes visible after the Sun has gone below the horizon.

alternative energy Energy derived from nontraditional sources or sources that are renewable and do not deplete finite mineral resources. The majority of countries are heavily dependent on fossil fuels as sources of energy, but with the decline in reserves and contribution made to the enhanced GREENHOUSE EFFECT by releases of carbon dioxide from fossil fuel combustion, alternative energy sources are being increasingly used and developed. Examples include wind power, solar power, wave and tidal power, biomass energy, geothermal power, and hydroelectric power. World primary energy sources include only about 7% from renewable sources, most of which is hydroelectric.

altimeter An instrument used to measure the altitude of an object with respect to a fixed level, such as sea level. There are two main types. The *pressure altimeter*, the most common of which is the ANEROID BAROMETER, is used in aircraft; it works on the principal of measuring falls in atmospheric pressure with altitude and shows the height, in meters or feet, above sea level. The *radio altimeter* (electronic altimeter) is an absolute altimeter, providing an altitude over any surface; it measures the time it takes for a radio pulse or continuous signal to be reflected by a surface and converts this into a geometric height.

altithermal (hypsithermal) A period during which the Earth's climate was warmer than it is currently. In particular, altithermal refers to a period about 8000–4000 years ago during which, it is believed, zones of higher precipitation were shifted poleward of their present position.

altitude **1.** The vertical height of an object (e.g. an aircraft, meteorological balloon, or atmospheric layer) above a reference level, which is usually mean sea level.
2. (elevation) A horizontal coordinate; it is the angular displacement of an object or astronomical body above the horizon. It is positive if the object is above the horizon, negative if below.

alto- In meteorology, a prefix related uniquely to MIDDLE CLOUDS that have bases typically between 2000 m and 7000 m (6500 ft and 23,000 ft) above the surface. *See also* cloud classification.

altocumulus A type of MIDDLE CLOUD commonly consisting of parallel bands or small rounded masses that have angular widths between 1° and 5°. The cloud can appear as a patch or a more extensive sheet of cloud, white or gray in color, and typically partly shaded. Altocumulus clouds usually form by convection in an unstable layer aloft.

altocumulus castellanus A type of MIDDLE CLOUD; a species or special form of ALTOCUMULUS in which the cumuliform components are much taller than they are wide and so appear to be turreted or castellated. They express the presence of overturning in their layer and are often known to be useful harbingers of thundery weather.

altostratus A type of MIDDLE CLOUD, most often with a grayish or bluish tone. It is usually a sheet or layer of fairly uniform appearance with no optical phenomena, and totally or partly covers the sky. It is sometimes thin enough to view the Sun through it in which case the solar disk appears as if through ground glass. The cloud may contain water droplets only, or a mix of water droplets and ice crystals.

Ambient Air Quality Standards (AAQS) *See* National Ambient Air Quality Standards.

ambient pressure The ATMOSPHERIC PRESSURE, at any point in the atmosphere, which surrounds a physical entity, such as a cloud. Ambient pressure is most frequently referred to in discussions of air STABILITY, when the temperature and pressure of a theoretical air parcel are compared to the pressure and temperature (AMBIENT TEMPERATURE) of the surrounding air.

ambient temperature The temperature of the area surrounding a specified object, for example the temperature of the surrounding atmosphere, soil, or water.

AMDAR (Aircraft Meteorological Data Relay) The internationally agreed scheme for relaying automatically sensed and transmitted weather data from commercial aircraft in flight. The relay is handled by the telecommunications capability of GEOSTATIONARY SATELLITES.

AMIP *See* Atmospheric Model Intercomparison Project.

amorphous cloud A usually continuous layer of featureless low cloud from which rain falls.

anabatic wind A wind that blows up the slope of a valley, sometimes attaining a velocity of 12 knots (14 mph). It develops as a result of the greater insolation warming of air that is in contact with the valley slopes during the day in comparison to that of the air at the same height but vertically above the valley floor. The heated air above the slope rises in the form of a convection current, with the air of the anabatic wind moving in to take its place. An anabatic wind typically begins 30 minutes after sunrise and ceases 30 minutes before sunset. *Compare* katabatic wind.

ana-front A WARM or COLD FRONT characterized by ascent of the warm moist air over the colder drier air; it is thus active in terms of deep cloud and PRECIPITATION. *Compare* kata-front.

analog data Data represented in a continuous form. In radar, a scheme in which the radiative power emitted from a scene being imaged is converted into a proxy signal, such as an electrical current, before being transmitted for display on the ground.

analog image In remote sensing, an image in which the continuous variations in the scene being sensed is recorded as continuous variations in image tone. For example, in a photograph the variations in the scene are recorded directly as variations in tone by the grains of the photosensitive film. *Compare* digital image.

analog weather forecasting A method of weather forecasting in which past weather situations similar to the present are used to predict that the weather will develop in a similar way. The method is difficult to use and is now little used.

analysis In synoptic meteorology, an examination of the state of the atmosphere based on synoptic weather observations and used as the basis for weather forecasts. The process includes the preparation of charts, such as isobaric charts, from the data obtained from many locations at the same time, together with the interpretation of the weather elements plotted and displayed on the weather chart. *See also* isentropic analysis; isobaric analysis.

anchor ice *See* ground ice.

andhi In the northwestern region of the Indian subcontinent, a dust storm that accompanies a sudden increase in wind speed; it is caused by strong convection.

anemogram *See* anemograph.

anemograph (recording anemometer) An ANEMOMETER that continuously records wind speed and sometimes wind direction. The resulting record of wind speed is known as an *anemogram*.

anemometer An instrument for measuring the velocity of a fluid; in meteorology this is usually wind speed or force and wind direction. The most commonly used type is the CUP ANEMOMETER in which revolving cups drive an electric generator that operates an electric meter calibrated in terms of wind speed. This type of anemometer uses the kinetic energy of the wind to measure its speed. Another type is the *hot-wire anemometer*, which works on the basis of the cooling ability of a stream of air. An electrically heated wire is cooled as the air flow increases. Alternately, it is kept at a constant temperature and the electric output required to maintain that temperature is measured and then converted to an airspeed. *See also* pitot tube anemometer; pressure-plate anemometer; pressure-tube anemometer.

anemometry The science of the measurement of the speed of a fluid, especially wind speed or force.

aneroid barometer A non-liquid BAROMETER used to measure atmospheric pressure and widely used in aircraft ALTIMETERS due to its compact size and convenience. This type of barometer consists of a flexible capsule (the aneroid capsule or cell), which is usually metallic and partially exhausted of air. The capsule is held extended by an internal spring and as atmospheric pressure changes, the walls of this capsule expand and contract. These movements are transmitted either mechanically to a pointer on a dial, which registers the atmospheric pressure on a scale, or electrically to a digital read-out device.

angle of incidence The angle between a line perpendicular to a surface and the solar beam at the point of incidence. In general, it is the angle between the ZENITH and the declination of the solar beam.

Ångstrom (Å) A unit of length equal to 10^{-10} meter formerly used as a measure of the wavelength of radiation. The nearest SI unit, the nanometer (nm), equal to 10^{-9} meter, is now used, together with the micrometer (μm; 10^{-6} meter) to describe the wavelengths of electromagnetic energy.

angular momentum The momentum possessed by a body that rotates about a fixed axis. The Earth has rotational angular momentum, which is also possessed by air that is rotating with the Earth; it is equal to its linear velocity multiplied by its distance from the axis of rotation. The angular momentum for a rotating air mass is the product of the angular velocity (ω) and the radius of curvature (r), i.e. ωr.

angular velocity The rate of rotation of a body about an axis. For the Earth, the angular velocity for any point on its surface depends upon its latitude. A parcel of air at any latitude Φ shares the angular velocity of the Earth as $\Omega \sin\Phi$, where Ω is the rate at which the planet rotates about its polar axis (i.e. through 360° or 2π radians every day or every 86,400 seconds). This $\Omega \sin\Phi$ represents the angular velocity of air, not moving with respect to the surface, about the local vertical, i.e. in the local horizontal plane. It is an important component in large-scale atmospheric motion. Air that moves around a CYCLONE or an ANTICYCLONE (i.e. in a more-or-less circular path) also possesses angular velocity. It is the time rate of change of the angle subtended from a parcel of air to the pressure center as it circulates around the center.

anisotropy The property of a diffuse surface to reflect incident solar radiation in a slightly imperfect way so that the incident intensity of the solar beam does not equal the reflected intensity as the ANGLE OF INCIDENCE varies. This type of reflection is termed *anisotropically diffuse*.

anomalous propagation A type of BACKSCATTER to a RADAR in which the emitted beam is not angled upward at a shallow angle but is bent downward to intersect the Earth's surface. This occurs when there is a marked atmospheric temperature INVERSION not too far above the surface that refracts the emitted beam back down to the surface. The received signal is then an image of surface features, and of no operational use.

anomaly The deviation of a property, such as precipitation or temperature, in a given location from the normal long-term value.

Antarctic Denoting the south polar region of Antarctica, which comprises the Antarctic Ocean and the Antarctic ice-cap. It usually refers to the region to the south of and within the Antarctic Circle at 66° 32′ S, along which the Sun does not set on December 22 (the southern hemisphere summer solstice) or rise on June 21 (the southern hemisphere winter solstice).

Antarctic air mass A year-round air mass that originates from the ANTICYCLONE over the intensely cold Antarctic continent. The frigidly cold and very dry air often flows strongly down valleys toward and across the coast as KATABATIC WINDS. This air streams toward milder damper air to the north, meeting it along the ANTARCTIC FRONT that stretches as an elongated feature around the circumpolar ocean.

Antarctic bottom water A bottom-water flow of cold and very saline water found in the SOUTHERN OCEAN, which can spread as far north as 40°N. Antarctic bottom water is the main source for the bottom waters of the South Pacific, southern Indian Ocean, South Atlantic and parts of the North Atlantic. Cooling of the ocean during winter and formation of sea ice releases salt, which increases the density of the water causing a mass of cold water to sink from the surface to the bottom of the Antarctic Ocean. The cold dense water originating on the edge of the continental shelf of Antarctica spills off the shelf and flows northward along the sea floor, below other water masses, as the Antarctic bottom water. It is the most dense water of the global oceans, with salinity at around 34.65 parts per thousand and temperature originating at –0.5°C (31.1°F). *See also* Antarctic intermediate water; Arctic bottom water.

Antarctic Circumpolar Current An ocean GYRE located in the Southern Ocean that flows west to east, creating a ring around the Antarctic. The current is created from the wind stress produced by the continuously strong westerly winds found around the Antarctic. Variations in the width and path of the Antarctic Circumpolar Current are caused by the influence of landmasses, sea-floor topography, and prevailing winds; it is most restricted where if flows through the Drake Passage between South America and Antarctica. The current is alternatively known as *West Wind Drift* along its surface portion.

Antarctic Coastal Current (East Wind Drift) A narrow current that flows in a westward direction around Antarctica in response to the easterly wind blowing off the ice cap.

Antarctic Convergence (Antarctic polar front) A convergence line in the ocean surrounding Antarctica, located between latitudes 50° and 60° S. The cold Antarctic water sinks below the warmer and less dense water from the middle latitudes. The resulting water mass formed where the convergence occurs is known as the Antarctic intermediate water.

Antarctic front A zone that separates the very cold dry CONTINENTAL POLAR air blowing off Antarctica from the milder damper MARITIME POLAR air to the north. It stretches around the circumpolar ocean between roughly 55° and 65° S and is associated with the ROARING FORTIES.

Antarctic intermediate water A cold intermediate water flow within the oceans that originates where cold surface water from Antarctica sinks below warmer and less saline surface water at the Antarctic Convergence. It is less saline than ANTARCTIC BOTTOM WATER.

Antarctic Ocean *See* Southern Ocean.

Antarctic polar front *See* Antarctic Convergence.

anthropogenic source An identifiable origin of a pollutant due to human activity. In most cases this is associated with the use of energy either in industry, in transport, or the production of electricity. Sources may also be attributable to acts of sabotage and warfare.

anticyclogenesis The process of the formation of a new ANTICYCLONE (high) or the strengthening of the circulation around an existing anticyclone. The formation of a new anticyclone can be related to the processes of RADIATIVE COOLING, which is responsible for COLD ANTICYCLONES, or dynamic SUBSIDENCE, which is linked to WARM ANTICYCLONES. *Compare* anticyclolysis.

anticyclolysis The process of the dissipation of an ANTICYCLONE (high), or the weakening of the circulation around such a feature. *Compare* anticyclogenesis.

anticyclone (high) An extensive region, typically a few thousand kilometers across, in which barometric pressure is relatively higher than in the surrounding air. Anticyclones are SOURCE REGIONS for AIR MASSES from which the air at low levels spirals out clockwise in the northern and counterclockwise in the southern hemisphere. They are characterized by deep sinking motion and generally dry weather, although very widespread STRATIFORM cloud can often be trapped underneath the temperature inversion that typifies such features leading to ANTICYCLONIC GLOOM.

The world's principal anticyclones are those semi-permanently over the subtropical oceans and those that appear in the winter season deep in the interiors of the more extensive northern-hemisphere continents. *See* cold anticyclone; warm anticyclone. *See also* Siberian high; Azores high; Bermuda high; North Pacific high.

anticyclonic gloom Weather associated with stationary anticyclonic conditions in which extensive STRATUS or STRATOCUMULUS persists for many days – sometimes for more than a week – without breaking up. Conditions are therefore dull with occasional drizzle. The air in the BOUNDARY LAYER beneath the INVERSION is often polluted in such conditions of persistent gloom.

anticyclonic rotation The direction of the flow around an ANTICYCLONE (high). The air circulates out clockwise in the northern and counterclockwise in the southern hemisphere.

antitrades The winds that occur in the upper troposphere above the surface easterly TRADE WINDS. Typically occurring above 2000 m (6500 ft), the antitrades blow in the opposite direction to the surface trades. In the past, antitrades were defined as high-altitude return winds transporting, to higher latitudes, rising air from the Intertropical Convergence Zone (ITCZ). This theory has now been disproved and the earlier definition of antitrades has consequently been largely abandoned.

antitriptic wind A type of LOCAL WIND that is topographically generated; examples include the ANABATIC WIND and the KATABATIC WIND. A term initially used by the British astronomer and geophysicist Sir Harold Jeffreys (1891–1989), antritriptic winds are so called because frictional effects dictate both the wind's dynamics and its thermodynamics. Such winds are diurnal in character, and most develop on calm and clear days.

anti-wind The intermediate layer of an ALONG-VALLEY WIND SYSTEM, in which the flow is reversed from the layer below. The anti-wind acts to balance the flow of the lowest layer of the system. Since the ANABATIC and KATABATIC components account for only a few meters, the remainder of the depth of the valley is taken up by the returning anti-wind.

anvil (anvil cloud) A more-or-less horizontal sheet of icy CIRRUS and CIRROSTRATUS that flows out of the upper reaches of a deep CUMULONIMBUS cloud. It is distorted by the upper winds to stream away in an anvil-like shape.

anvil dome A large OVERSHOOTING TOP or penetrating top of a SUPERCELL THUNDERSTORM cloud that is a sign of considerable thermal activity and updraft within the cloud.

AOSIS *See* Alliance of Small Island States.

aphelion The point in a planet's orbit at which it is furthest from the Sun. In the Earth's orbit around the Sun, the planet is furthest from the Sun (about 152 million km or 1.017 astronomical units) currently on about July 4. Variations in this orbit are associated with the MILANKOVITCH CYCLES and the astronomical theory of climate change.

apogee In remote sensing, the point along a satellite's orbit at which it is furthest from the Earth or the point that it is orbiting. More generally, in astronomy any Earth satellite, including the Moon, is said to be in apogee when it is at its greatest distance from the Earth. *Compare* perigee.

apparent temperature *See* heat index.

Appleton layer *See* F-layer.

applied climatology The analysis of CLIMATE data for useful applications. Examples of potential applications include agriculture (*see* agroclimatology), urban studies, industry, aviation, water resources, disaster management, and any climate-sensitive areas of society.

applied meteorology The application of meteorological information to problems facing, for example, agriculture, commerce, and planning (e.g. the location and design of airports, factories, and other constructions).

APT *See* automatic picture transmission.

Arctic Denoting the north polar region of the Arctic. In its strictest sense it is defined as the region to the north of the Arctic Circle at 66° 32′ N, along which the Sun does not set on June 21 (the northern hemisphere summer solstice) or rise on December 22 (the northern hemisphere winter solstice). *See also* Arctic warming.

Arctic air mass (A) An air mass that resides across the Arctic Basin and can lead to anomalously cold weather to its south. In winter, when the Arctic Ocean is frozen, the continental Arctic (cA) air mass is very cold and dry. In summer, when the basin is more like a cold ocean and the air above it is cool and damp, it is defined as a maritime Arctic (mA) air mass.

Arctic bottom water A bottom-water flow found in the ARCTIC OCEAN. It originates when sea water freezes to form sea ice and releases salt, forming salt-free ice and a high-density residual brine. The increase in density causes a mass of cold water to sink to the bottom of the Arctic Ocean. This bottom water is restricted mainly to the Arctic due to submarine topographic barriers, such as the Bering Sill, which blocks flow into the Pacific Ocean. Some bottom water does escape into the western Atlantic Ocean from the region around Greenland. *See also* Antarctic bottom water.

Arctic Climate System Study (ACSYS) One of the core projects of the WORLD CLIMATE RESEARCH PROGRAM (WCRP) that seeks to understand Arctic Ocean variability and change, such as sea-ice processes, and the role that the Arctic plays in the global climate It is particularly concerned with establishing the global consequences of anthropogenic change in the Arctic climate system, and whether the climate of the Arctic is as sensitive to increased GREENHOUSE GASES as climate models predict. The organization of research was expanded into the *Climate and Cryosphere* (CliC) initiative following approval received in 2000 from the WCRP.

Arctic front A gently sloping zone separating cold Arctic air that has flowed south from the ice- and snow-covered Arctic Basin, from CONTINENTAL POLAR and MARITIME POLAR air masses. These latter two AIR MASSES have different properties over high-latitude continents and oceans in which they reside. In the winter and spring the Arctic front is often found between the south of Greenland and northern Norway as well as over central and northern Canada, for example.

Arctic Ocean The smallest of the Earth's oceans, covering an area of 14 million sq km (5.4 million sq miles). It is located to the north of North America and Asia, and entirely within the Arctic Circle, surrounding the North Pole. The average depth is approximately 1200 m (3900 ft), with a maximum depth of approximately 5500 m (18,000 ft). The Bering Strait connects the Arctic Ocean with the PACIFIC OCEAN and the Greenland Sea is its link with the ATLANTIC OCEAN. Three submarine ridges – the Alpha Ridge, Lomonosov Ridge, and Arctic Mid-Ocean Ridge, split the ocean floor. The Arctic Ocean is almost entirely covered with ice, around 0.5–4 m (1.5–13 ft) thick, for most of the year except in fringe areas. For this reason it was once also known as the Frozen Ocean. Ice cover reduces the energy exchange between the surface and the atmosphere, and reduces the amount of sunlight that can penetrate surface waters, thus reducing the ability of marine life to photosynthesize. Cold Arctic currents cool the shores of northeast North America and northeast Asia, contrasting with the coastlines of Europe and North America, which are warmed by the NORTH ATLANTIC DRIFT and the KUROSHIO CURRENT, respectively. The Arctic currents tend to be much less saline and less dense than the warmer currents and remain at the surface while Atlantic currents tend to be deeper.

Arctic sea smoke (Arctic frost smoke) A type of STEAM FOG that occurs in high latitudes when very cold air passes over relatively warm water. Water evaporates into the air, increasing the dew-point, and the air above may become saturated. The cold air directly above the warmer water is heated and rises through convection carrying the moisture, which almost immediately condenses to form thin vertical columns of condensed water vapor, upwards like small pillars of smoke. If the temperature of the air is below the freezing point it forms Arctic frost smoke.

Arctic warming The warming of the Arctic region during recent years. It is believed by many scientists that the polar regions, and in particular the Arctic, will show evidence of GLOBAL WARMING before other regions, a situation supported by many GENERAL CIRCULATION MODELS. Although climate records are sparse in this region, data, including climate reconstructions from tree-ring studies, ice cores, and lake cores, appear to support this theory. Evidence from the last 400 years indicates a gentle warming in the Arctic region but during the period 1976 to 1995 climate data show a more rapid rise. For example, winter temperatures in central Siberia rose by 6°C (11°F) during this period.

arcus A type of low ACCESSORY CLOUD formation that is produced when a thunderstorm's GUST FRONT scoops up moist air at the ground. One form of arcus is the shallowly sloping *shelf cloud*, which has an elongated smooth shelf-like appearance, sometimes layered and terraced along the leading edge, and a dark turbulent base. It is usually attached to the thunderstorm base. It is accompanied by gusty winds when it passes overhead. Another type of arcus is the ROLL CLOUD, an elongated cloud that appears to spin slowly about its horizontal axis. Roll clouds are found behind gust fronts that flow out across the surface from thunderstorms; their formation is not currently well understood. *See also* cloud classification.

area source In air pollution studies, pollutants entering the atmosphere from a zone, such as an industrial complex covering several square kilometers, in which it may be impossible to identify the exact sources of individual pollutants. It may also refer to a power park incorporating a number of power stations, or to an area in which pollution permits may be traded. *Compare* linear source; point source.

arid climate *See* dry climate.

aridity index An index that measures the degree of aridity or water deficiency below the level needed at a given location. The meteorologist C. W. Thornthwaite (1889–1963) devised the first aridity index, which was used in defining climate classifications. The aridity index = 100 d/n, where d = water deficit (the sum of the monthly differences of PRECIPITATION and POTENTIAL EVAPOTRANSPIRATION when the precipitation is less than the potential evapotranspiration), and where n is the sum of the monthly values of potential evapotranspiration for deficient months.

artificial rain Rainfall that is produced by human modification of cloud microphysical processes, mainly by CLOUD SEEDING with dry ice (frozen carbon dioxide) pellets, silver iodide, or other suitable substances. Seeding increases the number of condensation nuclei and/or freezing nuclei within the cloud. While this increases the efficiency of precipitation generation mechanisms, attempts at cloud seeding have not always been successful, leading in some cases to a reduction in precipitation, especially when the cloud has been overseeded.

Asian high (Asian anticyclone) *See* Siberian high.

Asian monsoon The invasion by the moisture-laden southwest airflow over eastern Asia, including China, from May through to October with warm humid air lying over much of central China by mid June (*See* Mai-yu season). Rainfall is mostly associated with surges in the flow over Eastern China. Mountain ranges considerably disrupt the southwest to southerly flow over western China. *See also* monsoon.

ASOS *See* Automated Surface Observation System.

aspect The direction faced by a sloping surface. It is the point of the compass (usually approximated to 'north', etc) indicated by a line projected on a horizontal plane normal to a slope. Aspect is important in microclimate studies since it is a major determinant of the receipt of solar radiation, south-facing slopes receiving most in the middle-latitude northern hemisphere.

astronomical theory of climate change *See* Milankovitch cycles.

Atlantic conveyor (Atlantic Ocean thermohaline circulation system) A thermohaline circulation system within the oceans that is important in the global transfer of heat from high to low latitudes. The downwelling of salt water takes place in the North Atlantic, near Labrador and near Greenland, when warm waters are cooled by chill winds. Together with the flow of cold water from the Norwegian Sea that spills over sills, the cold saline water sinks to the ocean floor and flows south as the North Atlantic Deep Water, eventually to the Indian and Pacific Oceans and the Southern Ocean. The cold dense water eventually warms and returns to the surface of the worlds oceans. The occurrence of a strong Atlantic conveyor has been linked by scientists to the increased probability of major storms and hurricanes during a season.

Atlantic Ocean The world's second largest ocean, covering an area of approximately 82 million sq km (31.8 million sq miles), about 20% of the Earth's surface. Its average depth is approximately 3500 m (11,500 ft) with a maximum depth of roughly 8500 m (28,000 ft) at the Puerto Rico Trench, north of Puerto Rico. The ocean extends from the Arctic to the Antarctic between North and South America in the west and Europe and Africa in the east. The ocean floor consists of a series of ridges. The Mid-Atlantic Ridge, approximately 500–1000 km (300–600 miles) wide, extends 16,000 km (10,000 miles) from Iceland to the Antarctic Circle. The ridge is produced by volcanic activity as well as sea-floor spreading; it is constantly widening causing the ocean to expand pushing apart the Americas in the west from Europe and Africa in the east. Other ridges include a shallow submarine ridge across the Strait of Gibraltar, which separates the Mediterranean basin from the Atlantic.

Atlantic period A period between about 8000 and 5000 years BP, which encompasses most of the warmest postglacial times and was characterized by oceanic climatic conditions in northwest Europe. It was preceded by the BOREAL PERIOD and followed by the Sub-Boreal period.

Atlantic polar front The gently sloping zone that separates MARITIME POLAR and MARITIME TROPICAL air masses across middle- and high-latitudes of both the North and South Atlantic. The tropical air originates from the SUBTROPICAL ANTICYCLONES of both basins while the polar air streams across the oceans from the cold CONTINENTAL ANTICYCLONES that reside over the adjacent wintertime continents. The front divides two regimes of weather associated with the tropical and polar air, and is the zone along which FRONTAL CYCLONES (frontal depressions) form and evolve.

atmometer *See* evaporimeter.

atmosphere **1.** The gaseous layer that surrounds the Earth, or another planet. The atmosphere of the Earth is composed of a mixture of gases, called the AIR, which is predominantly composed of oxygen and nitrogen. *See also* atmospheric structure.
2. (atm) A unit of pressure, equal to 101,325 Pa (1013.25 hPa). It is used in expressing pressures in excess of standard atmospheric pressure, such as the high pressures used in some chemical processes.

atmosphere–ocean interaction (air–sea interaction) The interactions occurring between the atmosphere and oceans; they include the input of energy to the ocean from wind, transfers of heat, the transfer of water through precipitation and evapora-

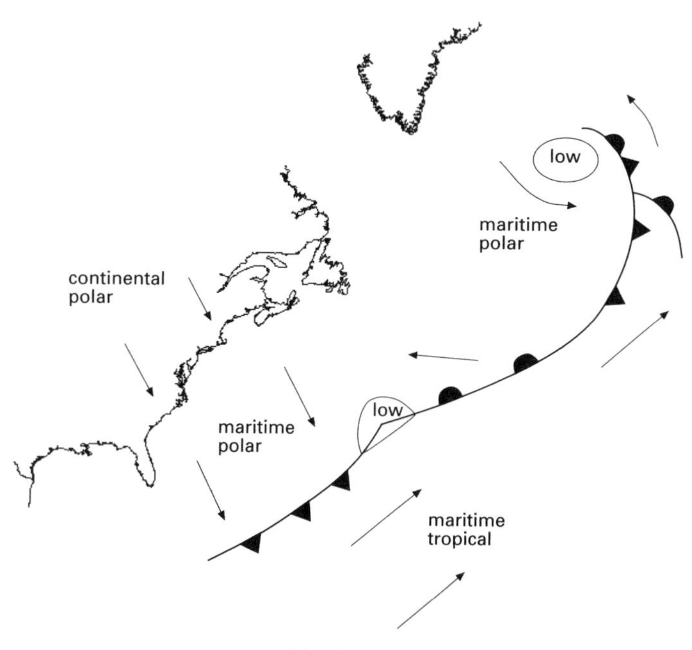

Atlantic polar front

tion, and gas exchange (e.g. carbon dioxide, dimethyl sulfide (DMS), and nitrous oxide). The interactions are of vital importance to the understanding of climate. On a global scale, changes within the ocean–atmosphere system impact on global sea level, sea-surface temperature distribution, sea-ice concentrations, the heat transferred by the ocean currents, and the distribution of heat moisture and momentum in the atmosphere. Studies of the interactions on a global scale use data gathered from satellite remote sensing, and employ large-scale coupled ocean-atmosphere models.

atmospheric boundary layer *See* planetary boundary layer.

atmospheric circulation *See* general circulation.

atmospheric energetics The study of the energy content and energy transformations, especially in the lowest 6 km (4 miles) of the troposphere. It is the study of heat, latent heat, potential energy, and kinetic energy. Analysis of the first three is carried out using THERMODYNAMIC DIAGRAMS (e.g. TEPHIGRAMS).

atmospheric heat engine A concept in which the atmosphere is regarded as a heat engine. The Sun's radiation provides the primary source of energy. Temperature is highest in the vicinity of the equatorial trough and lowest in the upper atmosphere above the poles. The engine is said to extract energy from the warm source and perform work in driving the atmospheric GENERAL CIRCULATION.

atmospheric lifetime The residence time of a pollutant in the atmosphere; this refers to the length of time that the molecules of a gas, liquid, or particle remain in the atmosphere at above its natural levels without being deposited at the surface or removed via a SINK, or altered or chemically changed into another substance in the atmosphere. The length of time depends on the source, the sinks, and the reactivity of the pollutant.

Atmospheric Model Intercomparison Project (AMIP) A project shared between many atmospheric research centers that was organized by the Working Group on Numerical Experimentation (WGNE) as a contribution to the WORLD CLIMATE RESEARCH PROGRAM. The project aimed to determine systematic errors in atmospheric MODELS using the decade 1979–88 as the test period for the simulation of climate from the participating atmospheric models, under specified conditions. The project provides a standard experimental basis for atmospheric GENERAL CIRCULATION MODELS, as well as infrastructure in support of climate model diagnosis, validation, intercomparison, documentation, and data access. This framework allows scientists to compare and analyze models with a view to model improvement and climate studies. *See also* climate model.

atmospheric moisture The WATER VAPOR in the atmosphere; all three states of water occur in the atmosphere and the state should be taken into account when referring to atmospheric moisture. Most atmospheric moisture occurs below 5000 m (16,500 ft). Water in the atmosphere may occur in its liquid state at below 0°C (32°F, the freezing point of water at sea level) in the upper atmosphere due to its lower atmospheric pressure. All forms of PRECIPITATION result from the condensation of atmospheric moisture.

atmospheric optics (atmospheric optical phenomena) The optical phenomena visible within the atmosphere at various times as a result of the processes of diffraction, reflection, refraction, and scattering.

atmospheric pollution *See* air pollution.

atmospheric pressure (air pressure, barometric pressure) The PRESSURE experienced at any point above the Earth's surface as a result of the weight of the air column, extending to the outer limit or top of the atmosphere, above that point. Consequently, pressure declines exponentially with height, the rate of decrease being a function of the temperature of the atmosphere. Atmospheric pressure is generally measured, in meteorology, either in the SI unit HECTOPASCALS (hPa) or in the c.g.s. unit of the same size, the MILLIBAR (mb) using a mercury or aneroid BAROMETER, or a BAROGRAPH. In the US surface atmospheric pressure is measured in INCHES OF MERCURY (in Hg). Global mean sea level pressure is 1013.25 hPa (or mb) = 29.29 in Hg.

atmospheric structure The layers or regions into which the Earth's ATMOSPHERE is divided. Based on differences in the rate of change of temperature with height (the LAPSE RATE) these layers, in ascending order above the Earth's surface, are the TROPOSPHERE, STRATOSPHERE, MESOSPHERE, and THERMOSPHERE. In addition, the zone beyond about 700 km (430 miles) is sometimes called the EXOSPHERE. Based on chemical composition, two divisions are recognized: the HOMOSPHERE and HETEROSPHERE. Another region, the IONOSPHERE, in which there is a high concentration of free electrons, extends through the mesosphere and thermosphere to indefinite heights. The OZONE LAYER extends up from about 10 km (6 miles), through the stratosphere.

atmospheric window The small bandwidths in the electromagnetic spectrum where the atmosphere is nearly transparent and transmittances of radiation approach unity both for incoming and outgoing radiation. The window for the visible range of solar radiation, 0.4–0.7 µm, lets most solar radiation through to the surface of the Earth; the longwave radiation window of 8–12 µm lets some terrestrial radiation escape to space The existence of such windows is vital for the Earth–atmosphere system to be maintained in near thermal equilibrium.

attenuation A reduction in energy of radiation as it passes through a medium; in meteorology it is the loss of energy of a beam of solar radiation by the solar beam SCATTERING and ABSORPTION as it passes through the atmosphere. The amount at-

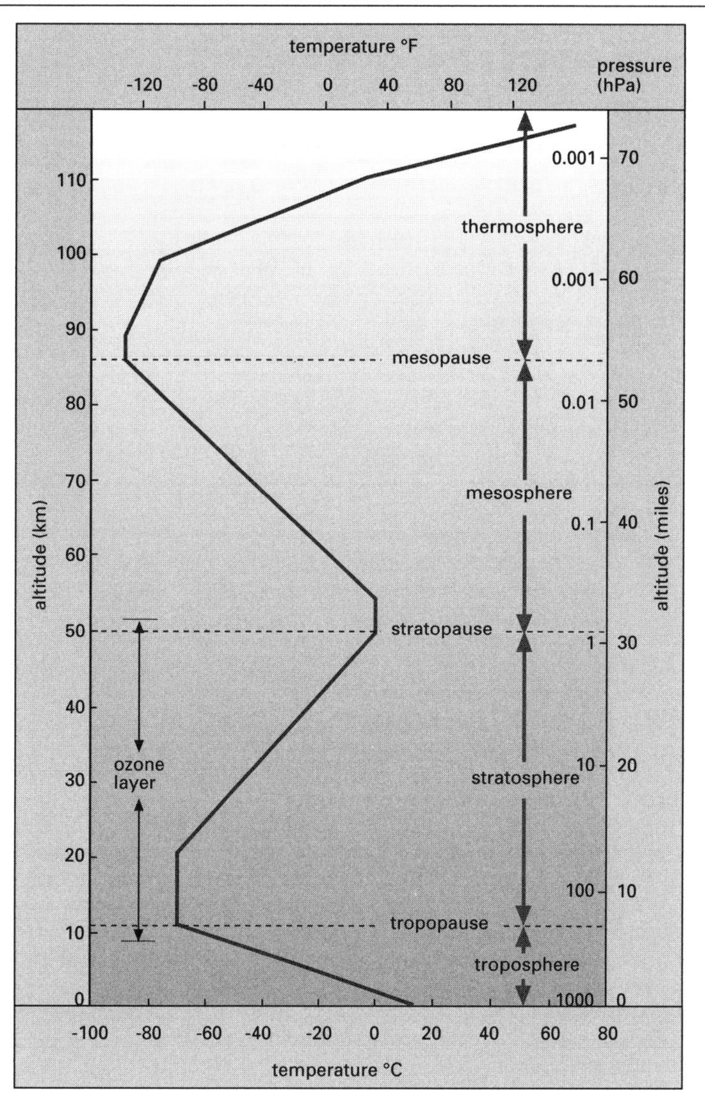

Thermal structure of the atmosphere

tenuated may be calculated by Lambert's law (Beer's Law) using extinction or attenuation coefficients, either for individual gases or for mixed gases.

aureole The luminous bluish-white disk with a brown outer ring that sometimes surrounds the Sun or Moon, and is the inner part of a CORONA.

aurora A luminous phenomenon in the sky, typically at night, that is primarily visible in high geomagnetic latitudes of both hemispheres near to the Earth's magnetic poles. At times of greater solar activity the aurora may be visible at lower latitudes. The phenomenon takes place at altitudes generally above 100 km (62 miles) in the ionosphere and may extend as high as

1000 km. It occurs when high speed charged solar particles (electrons and protons), are captured by the Earth's MAGNETOSPHERE, and are deflected downward toward the magnetic poles. In the upper atmosphere the charged particles collide with atmospheric oxygen and nitrogen atoms or molecules causing electrons to be released leaving the atom or molecule with a net positive charge. The ionized oxygen and nitrogen leads to the emission of light, especially in the red and greenish-blue colors. The display of the aurora may take the form of luminous colored rays, arcs, bands, patches, and curtains, the arc being the more persistent form of aurora, which may last for several hours.

The aurora is known as the *aurora borealis* or the *northern lights* in the northern hemisphere, and as the *aurora australis* or *southern lights* in the southern hemisphere.

automatic picture transmission (APT) The continuous transmission of ANALOG IMAGES from satellites. The system started with TIROS–8, launched in December 1963, in which cloud images from the satellite were broadcast live. This enabled any properly equipped ground station, within the field of view of the satellite, to receive and reproduce images for local use.

Automatic Surface Observation System (ASOS) In the US, an automated weather observation system jointly implemented by the Federal Aviation Administration (FAA), National Weather Service (NWS), and the Department of Defense (DoD). The first ASOS was installed in Topeka, Kansas, since when over 900 units have been planned throughout the US. The ASOS provide observations that include temperature, dew point, wind speed and direction, visibility, sky condition, and precipitation to users, such as the FAA, NWS, and DoD, and in addition, to members of the public, local authorities, air traffic control, and the media.

automatic weather station (AWS) An unmanned weather station at which all meteorological measurements are made automatically by various sensors. These stations are frequently positioned in remote places. Data acquired by the automatic weather stations (e.g. air temperature, relative humidity, wind speed and direction, rainfall, and solar radiation) are often transmitted automatically to a controlling station, either at set time intervals or continuously. Alternatively, the data may be collected from the station by an observer.

autumn (fall) The SEASON of the year between summer and winter occurring as the Sun approaches the winter solstice. It applies only in the middle-latitude areas that experience the typical four seasons. In the US, it is generally regarded as being the months September to November. It is a period of adjustment in the general circulation of the atmosphere, characterized by increasing storminess and more frequent outbreaks of air of polar or arctic origin. Declining temperatures combined with relatively high moisture content in the atmosphere, may lead to an increase in the incidence of fog. The first air frost marks the end of the GROWING SEASON.

avalanche wind A blast of air that precedes a descending avalanche. Resulting from the sudden change in air pressure, such shock waves are powerful enough to cause buildings to 'explode' before they are overwhelmed by snow and/or ice. The shock waves experienced are equivalent to air blasts of magnitude 0.5 tonnes/m^2.

AVHRR *See* advanced very high resolution radiometer.

aviation meteorology The study of meteorological phenomena that affect aircraft in flight and the use of weather data for flight planning.

AWIPS *See* Advanced Interactive Weather Processing System.

AWS *See* automatic weather station.

azimuth The directional location of an

object around the horizon, measured in degrees clockwise from true north (0°) to a point on the horizon vertically below the object.

azimuth angle The angle in the horizontal plane between north and the direction of a particular object when its location is projected vertically onto the horizon. The azimuth angle of an object that is due east of the observer has an azimuth of 90°, for example.

Azores high (Azores anticyclone) An extensive region of semi-permanent high pressure across the subtropical North Atlantic. It is a year-round feature that in the northern hemisphere winter and early spring tends to be centered over the Azores; during the summer months it migrates westward to near the island of Bermuda, when it is also known as the BERMUDA HIGH. The Azores high is the source region for MARITIME TROPICAL air flowing generally northward and of the TRADE WINDS that blow toward lower latitudes, into the INTERTROPICAL CONVERGENCE ZONE. In contrast to cold anticyclones such as the SIBERIAN HIGH, the Azores high is TROPOSPHERE deep.

back-building thunderstorm A THUNDERSTORM that appears to remain stationary or propagate backward as a result of new growth occurring on the upwind side of the storm, instead of on the more normal downwind side. In the northern hemisphere the upwind side is generally on the west or southwest side.

back-door cold front In the US, especially in New England and further south along parts of the Atlantic seaboard, a cold front that sweeps in from the east or northeast, rather than the much more common direction between west and north, and moves westward over the continental US. Its passage is associated with flow around the southeastern flank of an ANTICYCLONE to the north over Canada.

background level The concentration of a pollutant that would be present in a location without any enhancement of a particular contaminant from local sources. It may therefore also refer to concentrations of a pollutant far removed from its source. In some cases in which a pollutant also occurs naturally, it refers to the background level from natural sources only.

backing The change of the wind direction in a counterclockwise fashion. For example, suppose a northwesterly (NW) wind (315°) at a particular location at 12:00 noon is replaced by a southwesterly (SW) wind (225°) twelve hours later, then the wind is said to have backed during this 12-hour period. Backing is the opposite change of direction to VEERING.

back radiation *See* counterradiation.

backscatter In radar, the fraction of a signal emitted by an instrument that is reflected back to the radar antenna, which can therefore be compared to the power of the original signal. Precipitation radars, for example, measure the power scattered back to the antenna by raindrop-size particles in the atmosphere.

baguio The local Philippine name for a TROPICAL CYCLONE (also known in the Pacific region as a TYPHOON). Severe tropical cyclones developing annually in the western Pacific Ocean sometimes take a route across the islands of the Philippines and are particularly prevalent between July and November. *See also* tropical storm.

Bai-u season *See* Mai-yu season.

ball lightning An iridescent sphere of gas that is very occasionally observed near the Earth's surface, associated with electrically disturbed weather. There are many anecdotal reports of phenomena described as ball lightning, which vary in their reliability. Ball lightning phenomena are often reported to have a blue or orange color, and are said to float, move around, and sometimes hiss, but descriptions vary as to whether the gaseous ball is actually hot. It remains unclear where the energy arises to sustain such a plasma system for the timescales of tens of seconds commonly described: microwave energy from a parent thundercloud has been postulated as a possible source.

balloon sounding A technique employing lighter-than-air hydrogen or helium filled balloons to explore and record conditions in the Earth's atmosphere. Several types of meteorological balloon are used. The largest, a *sounding balloon*, carries meteorological instruments in a small transmitter – a RADIOSONDE – to transmit data continuously to a ground receiver. These balloons are used to examine the composition of the atmosphere, for example ozone concentrations, relative humidity, temperature, wind speeds and directions, etc., by attaching the appropriate meteorological instruments. Sounding balloons are released twice daily at 0000 and 1200 UTC from a network of sites around the world. Wind speed and direction can also be determined by using a PILOT BALLOON, which is tracked by a theodolite or radar. The smaller *ceiling balloon* is used to determine cloud-base altitude.

band In remote sensing, a limited part of the electromagnetic spectrum over which radiation is sensed, for example, by satellites. Sensors on meteorological satellites measure radiation in a number of bands, including the visible band and the thermal infrared band.

banner cloud A fixed cloud that appears to be attached to a sharp mountain peak, and streams downwind like a pennant or banner. Strong wind and relatively moist air are important ingredients. The cloud is believed to form in rising air within a large eddy that is often found to the lee of isolated mountain peaks, and may also be related to a pressure drop to their lee. The Matterhorn and Mount Everest occasionally generate these clouds.

bar (b) A c.g.s. unit of ATMOSPHERIC PRESSURE equal to 1 million dynes per square centimeter; equivalent to 750.1 mm of mercury at 0°C at latitude 45°. 1 million dynes per square centimeter is equal to 10^5 pascals (Pa), the fundamental SI unit of pressure. The bar is slightly less than 1 standard atmosphere (which equals 1.01325 million dynes per square centimeter = 760 mm of mercury = 1013.25 mb = 29.29 in Hg).

barat A LOCAL WIND experienced on the northern coast of the Indonesian island of Sulawesi (formerly Celebes). The barat is a squally and occasionally fierce northwesterly wind that blows onshore from the Celebes Sea to the north. It occurs most frequently between December and February.

baroclinic In a water body, denoting a state in which the surfaces of constant pressure intersect surfaces of constant density.

baroclinic atmosphere An atmosphere within which surfaces of constant pressure (ISOBAR) and constant density (ISOPYCNIC) intersect. The result is an atmosphere characterized by a marked horizontal temperature gradient; a frontal zone is an example of a baroclinic atmosphere. *Compare* barotropic atmosphere.

baroclinic instability A kind of dynamical instability associated with a BAROCLINIC ATMOSPHERE that often occurs in the strong westerly current associated with FRONTAL ZONES. It is linked to the development of middle- and high-latitude FRONTAL WAVES, associated with ascending warm and descending cold air. This typical vertical motion pattern means that the system's POTENTIAL ENERGY decreases, while kinetic energy is released.

baroclinic zone A confined region within the atmosphere typified by a relatively steep horizontal temperature gradient. A middle- or high-latitude FRONTAL ZONE is an example of a troposphere-deep gently sloping baroclinic zone.

barograph An instrument, usually the ANEROID BAROMETER, used to continuously record atmospheric pressure. The pressure-sensitive element of the barometer is attached mechanically by levers to a pen marker, which transfers the aneroids movements and records the changes in atmospheric pressure on a paper chart mounted around a slowly rotating drum or cylinder. This enables the production of a time series of changes in atmospheric pressure, usually for up to a week.

barometer An instrument used for measuring atmospheric pressure. There are two main types: the MERCURY BAROMETER and

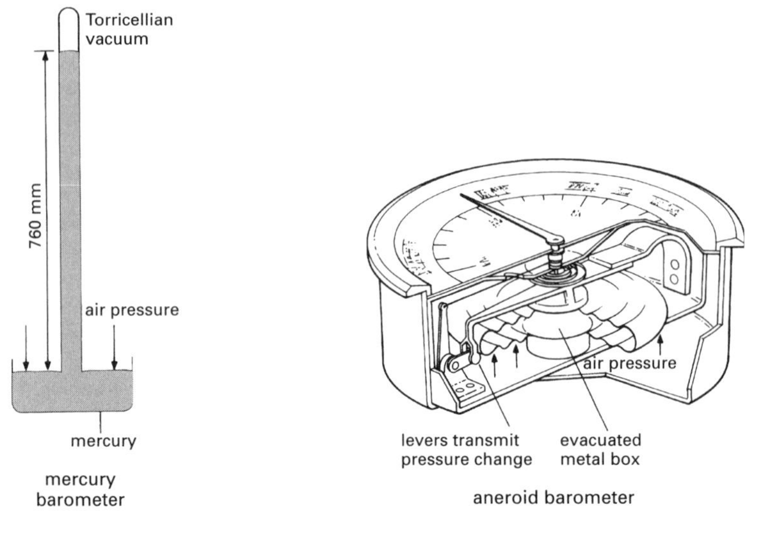

Barometer

ANEROID BAROMETER. The mercury barometer works on the basis that atmospheric pressure can support the weight of a column of mercury; as pressure alters so will the height of the column. Mercury barometers are often used to calibrate ANEROID BAROMETERS. Other forms of barometer include the use of solid-state sensors; one type measures changes in a piezoresistive transducer due to changes in barometric pressure. *See also* Fortin barometer; Kew barometer.

barometric gradient *See* pressure gradient.

barometric pressure *See* atmospheric pressure.

barometric tendency *See* pressure tendency.

barothermograph An instrument combining a BAROGRAPH and a THERMOGRAPH used to record atmospheric pressure and temperature simultaneously and continuously.

barotropic In a water body, denoting a state and type of motion in which pressure is constant or parallel on surfaces of constant density.

barotropic atmosphere An atmosphere within which isobaric and isopycnical surfaces coincide. It is an idealized atmosphere with no horizontal gradient of temperature and is therefore an approximation to conditions across an AIR MASS. *Compare* baroclinic atmosphere.

barrier jet A low-level core of higher wind speeds that sometimes occurs at a height of 1000–1500 m (3000–5000 ft) in the vicinity of a mountain range. A barrier jet is a product of the deceleration of the airflow on crossing such a topographic barrier and the associated release of latent heat changing the local thermodynamics and dynamics of the flow. For example, a south–north aligned barrier jet often develops along the west side of the Sierra Nevada mountain range in California as winter storms track eastward from the Pacific. Bands of heavy snow may develop during the evolution of a barrier jet.

baseline emission The long-term emissions, for example of greenhouse gases, that would occur if no policy intervention took place. This provides a 'business as usual' scenario, which may be used as a baseline to show the effectiveness of potential, in the case of MODEL experiments, and actual greenhouse-gas emission programs.

beaded lightning (chain lightning, pearl-necklace lightning) A rare form of lightning channel which, on its dissipation, produces separate sections of iridescence reminiscent of a beaded necklace, probably due to variations in thermal losses from the heated channel.

beam radiation *See* direct radiation.

Beaufort notation (Beaufort letters) A code consisting of a system of letters originally introduced by Admiral Sir Francis Beaufort (1774–1857) to describe present (ww) or past (WW) weather conditions. It has been much modified since; examples of the letters include c for cloudy, o for overcast, r for rain, p for passing showers, d for drizzle, s for snow, rs for sleet, h for hail, tl for thunderstorm, and f for fog. Capital letters were introduced to the system to indicate an event's intensity (e.g. S is heavy snow) and persistence (e.g. SS is prolonged heavy snow). Beaufort's alphanumeric system is now rarely used, having been replaced by an internationally recognized system of 100 geometrically based symbols (0–99) that are in current usage on official weather maps. For example, differing intensities of rain (60–69) in the ww system are represented by geometric arrangements of solid black circles.

Beaufort scale A numerical scale ranging from 0 (calm) to 12 (hurricane) that was introduced by Admiral Sir Francis Beaufort (1774–1857) in 1805 to describe variations in wind force at 10 m (30 ft) above the ground and originally based on the effect of wind on a fully rigged man-of-war. Beaufort developed his index at a time when instruments for measuring wind speed were unavailable; approximate wind speeds and effects of each *Beaufort number* on land and sea surfaces were only added at a much later date. The International Meteorological Committee adopted the scale in 1874 for use in weather telegraphs. See Table overleaf.

belt A linear feature related to a specific weather system, for example a CONVEYOR BELT is a characteristic aspect of most EXTRATROPICAL CYCLONES. There are no strict scale criteria for the feature.

Benard convection (Rayleigh–Benard convection) A type of CONVECTION that develops in a shallow fluid that is heated differentially at its bottom and top. For example, longwave emission at the top of the stratiform anvil of a mature CUMULONIMBUS cloud and absorption at the base destabilizes the anvil layer. Under such conditions, the fluid evolves into a regular array of hexagonal cells with a preferred horizontal scale. The conversion of the system's potential energy into kinetic energy becomes more efficient as the horizontal dimension of the cells decreases. This energy conversion process is eventually offset by frictional dissipation (e.g. momentum is mixed across the cell's walls), which becomes important when the horizontal dimension of the cells is comparable to the depth of the fluid. Consequently, the horizontal dimension of a typical cumulonimbus cell is comparable to the depth of the TROPOSPHERE.

Benguela Current A cold current in the South Atlantic Ocean that flows northward along the west coast of southern Africa. It is a branch of the WEST WIND DRIFT in the southern hemisphere and it merges with the westward-flowing Atlantic South EQUATORIAL CURRENT just before it reaches the Equator.

Bergen School (Norwegian School) A famous Norwegian school of meteorologists, led by Vilhelm Bjerknes (1862–1951) and including his son Jacob Bjerknes (1897–1975), Halvor Solberg, and the Swedish meteorologist Tor Bergeron (1891–1977). It produced ground-breaking work on the nature of AIR MASSES and FRONTAL CYCLONES, based on brilliant observational and theoretical work. Many famous papers appeared soon after World

THE BEAUFORT SCALE ON LAND

Beaufort number	*Descriptive term*	*Wind speed knots*	*mph*	*Visual effect*
0	calm	< 1	< 1	smoke from factories rises vertically
1	light air	1 – 4	1 – 4	wind direction shown by smoke, but not by a weather vane
2	light breeze	4 – 6	4 – 7	wind felt on face
3	gentle breeze	7 – 10	8 – 12	broken twigs from trees are in constant motion
4	moderate breeze	11 – 16	13 – 18	dust and loose paper raised from sidewalk
5	fresh breeze	17 – 21	19 – 24	swaying of small trees
6	strong breeze	22 – 27	25 – 31	large branches of trees in motion
7	near gale	27 – 33	32 – 38	whole trees vibrate
8	gale	34 – 40	39 – 46	twigs detached from trees
9	strong gale	41 – 47	47 –54	chimney pots and roof tiles fall from houses
10	storm	48 – 55	55 – 63	trees uprooted; much structural damage
11	violent storm	56 – 63	64 – 72	considerable damage. Usually experienced along the coast; very infrequent inland
12	hurricane	> 64	> 73	mostly confined to the tropics. Disastrous; loss of life, towns flattened

(As the Beaufort scale is subjective, wind speeds assigned to each Beaufort number vary; as a result some values appear in more than one wind force class.)

War I and through the 1920s from their base at the Bergen Geophysical Institute (now part of the University of Bergen).

Bergeron–Findeisen process (Bergeron process, ice-crystal theory) A theory of the process of raindrop growth in MIXED CLOUDS (composed of both ice crystals and supercooled liquid water drops), which proposes that all raindrops begin as ice crystals at upper levels. The theory is based around the differential in saturation vapor pressure (SVP) over ice and over supercooled water surfaces at cloud temperatures of between –10°C (14°F) and –30°C (–22°F). The SVP with respect to ice is less than that with respect to supercooled liquid water at the same temperature. Given this, evaporation can take place from supercooled water droplets and ice crystals can grow rapidly by direct sublimation from water vapor on to the ice crystals at the expense of the supercooled water droplets. When the ice crystals become heavy enough to overcome updrafts and fall through the cloud they may collide with supercooled water droplets which freeze on contact in a process known as ac-

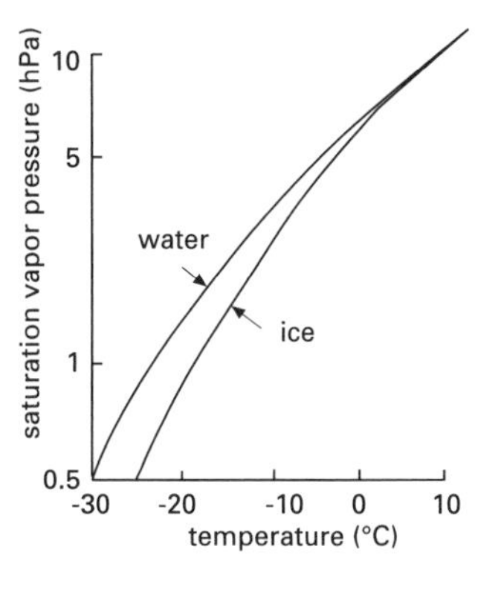

Bergeron–Findeisen process

cretion or riming. The ice particles formed, known as GRAUPEL, may collide with cloud droplets and fracture, producing more ice crystals, which can in turn grow into more graupel. During their descent the ice crystals may stick or clump together, in a process known as aggregation, to form SNOWFLAKES. This process is most common when the temperature is near 0°C (32°F). If heavy enough, the snowflakes fall through the atmosphere and, dependent on the temperatures in the lower atmosphere, fall as snow or melt to form raindrops. The theory depends on an ice/water mix in a cloud thus the mechanism does not operate in tropical clouds in which temperatures are above the freezing level.

The theory was proposed by the Swedish meteorologist, Tor Bergeron (1891–1977) and subsequently modified by Walter Findeisen. *See also* collision–coalescence process.

Berg's index of continentality *See* continentality.

berg wind A hot, dry, and occasionally dusty mountain wind of the FÖHN type that blows from South Africa's interior plateau down to the coast, generally flowing in a southerly direction. It occurs most frequently in winter when the pressure gradient is directed offshore between high pressure over the plateau and low pressure over the ocean. A berg wind may persist for several days, giving oppressive weather conditions with temperatures in excess of 35°C (95°F) and sometimes causing severe damage to crops.

Berlin mandate A mandate concluding that the commitment under the UNITED NATIONS FRAMEWORK CONVENTION ON CLIMATE CHANGE (UNFCCC) to reduce greenhouse-gas emissions to 1990 levels by the year 2000 was not adequate for the purpose of halting potential anthropogenic-induced climate change. The mandate was negotiated in Berlin, Germany, in March 1995 at the First Conference of Parties (COP 1) and included commitments to undertake action beyond the year 2000.

Bermuda high (Bermuda anticyclone) Another name for the AZORES HIGH, the semi-permanent subtropical anticyclone over the North Atlantic Ocean, especially when located in the west part of the ocean near Bermuda and near 30°N. It is also sometimes seen as a separate western high pressure cell within the broad subtropical anticyclone over the North Atlantic.

billow clouds Wave-like cloud protuberances that form on the upper surface of a CLOUD STREET line in a layer where the wind speed increases sharply with height and is just underneath an INVERSION.

bimetallic thermograph A THERMOGRAPH in which the continuous recording of temperature is based on the differing heat capacity of two metals (typically iron and brass) welded together to form a single strip. The different expansion rates of the two metals causes the strip to bend when heated. This strip is attached via gears and levers to a pen that records the changes in curvature with time on a paper chart.

bioclimatology The study of the relationships between living organisms, such as animals (including humans) and plants, and climate. This includes the effects of climate on the organism and also the effects the organism may have on climate; this includes, for example, the effects of human activities in contributing to global warming.

biogeochemical cycle (nutrient cycle) The natural processes that recycle elements between living organisms (the biotic phase) and their non-living environment (the abiotic phase). Examples of biogeochemical cycles include the CARBON CYCLE, HYDROLOGIC CYCLE, NITROGEN CYCLE, and oxygen cycle. Climatic and atmospheric changes can have implications for these cycles.

biomass 1. The total dry weight of all the living organisms of a given type (e.g. species) or of all types within a given area, such as a BIOME. It is typically measured in terms of dry mass per square meter.
2. Organic material (living and nonliving) from above and below ground, such as trees and tree litter, agricultural crops, grasses, and animals and animal waste. *Biomass energy* is produced by combusting organic material, such as wood from trees, and has been proposed as an alternative to burning fossil fuels. A current theory holds that the release of carbon dioxide from burning biomass as a fuel will not affect overall atmospheric carbon dioxide levels as long as replanting is maintained to absorb carbon dioxide. *Biomass burning* that occurs as a result of burning forests, grasslands, and cleared agricultural land through land clearance or agricultural practices, for example, releases a variety of compounds including carbon dioxide, methane, and nitrous oxide, all of which are GREENHOUSE GASES.

biome A major biotic community, extending over a large geographical area and having a characteristic type of dominant vegetation that is adapted to the climatic conditions of the area. Biomes include DESERT, grassland, SAVANNA, forest, and TUNDRA. In the mapping of biomes, areas that are now under agricultural cultivation or urbanized are mapped according to the biome that is assumed to have been present.

biometeorology The study of the relationships between living organisms, such as animals (including humans) and plants, and atmospheric conditions, such as temperature and humidity.

biosphere That part of the Earth that contains living organisms; it includes the lower part of the atmosphere in addition to the Earth's surface, seas, and outermost part of the lithosphere.

bise (bize) A cold dry wind that blows from the northwest, north, or northeast to affect mountainous regions of southern France and Switzerland during the winter. The strong outbreak of cold dry air associated with the bise is often accompanied by heavy cloud.

bishop's ring A pale whitish circle with a reddish-brown outer ring centered on the Sun or Moon. It has been observed following volcanic eruptions, the first recorded observation being after the eruption of Krakatau in 1883. The cause is believed to be the diffraction of light by fine volcanic dust or sulfur dioxide particles.

bispectral technique A remote-sensing method used in meteorology for automatically recognizing and classifying cloud type and cloud-free surface type by pairing the value of a PIXEL in both the VISIBLE and the THERMAL INFRARED images of the same scene at the same time. If, for example, the two pixels are such that there is a strong signal in the visible (very bright reflection) and a weak signal in the thermal infrared (very cold surface), it is very likely to be a CUMULONIMBUS cloud. If, in contrast, the visible displays a weak signal (poor reflection) and the thermal infrared quite a strong one (warm surface), the pixel is likely to be located over a cloud-free sea surface.

bize *See* bise.

black body A hypothetical body that absorbs and emits radiation at all the wavelengths for a given temperature, i.e. it has an emissivity of 1. The land, ocean, and most clouds are said to act as near black bodies and emit the maximum amount of radiation at their respective temperatures.

black-body radiation The radiation emitted from a surface across the full range of the electromagnetic spectrum for a given surface temperature. The rate of emission is expressed by the STEFAN–BOLTZMANN LAW.

black-bulb thermometer An instrument that is sometimes used to measure solar radiation by recording the maximum temperature reached. It is basically a MERCURY-IN-GLASS THERMOMETER with the bulb blackened and mounted in an evacuated glass sheath.

black ice A type of GLAZE that forms when rain falls on to a surface, such as a road, if the temperature is below 0°C (32°F). It is a potential hazard for highway users since the thin sheet of ice formed is dark in appearance and often hard to see.

blizzard **1.** A storm of falling and/or blowing powdery SNOW, sometimes with small ice crystals, that is accompanied by high winds of at least 30 knots (35 mph), low temperatures of –7°C (20°F) or below, visibility of less than 1/4 mile, and lasting for at least 3 hours. A *severe blizzard* is accompanied by winds in excess of 39 knots (45 mph), temperatures below –12°C (10°F), and near zero visibility. Blizzards are common in northern and central parts of the US, especially the N Great Plains, during winter. *See also* buran.
2. A general term for a severe snowstorm accompanied by strong winds.

blocking An interruption or obstruction in the normal west-to-east flow of surface CYCLONES, ANTICYCLONES, TROUGHS, and RIDGES that occurs in middle or high latitudes and may last from a few days to several weeks. Blocking occurs in both hemispheres but is more common in the northern hemisphere, especially over the eastern Pacific Ocean to the west of North America and over the eastern Atlantic Ocean to the west of northern Europe. Winter and spring are the times of greatest incidence and it is most pronounced near the tropopause. The upper-level flow changes from predominantly zonal to meridional eventually leading to the formation of CUT-OFF HIGHS and CUT-OFF LOWS. Blocking can take one of a number of forms. In the omega block, so named because the meridional flow about this block resembles the Greek letter Ω, the strongest flow is diverted to lower latitudes. A ridge often forms off the west coast of North America in winter, which can steer Pacific storms far to the north to keep British Columbia and Pacific Northwest dry and mild. *See also* meridional circulation; zonal index.

blocking anticyclone (blocking high) A large-scale stationary, or very slow-moving, ANTICYCLONE (high), often aloft, that extends down throughout the tropo sphere. This can occur when there is a split in the jet stream into two branches and the formation of a cut-off low south of the anticyclone. The mobile pressure systems are diverted either north or south (or both) to track around the flank of the block. The blocking anticyclone can persist for a week or more, diverting the usual zonal track of frontal depressions into a more meridional track. *See* blocking. *See also* meridional circulation; zonal index.

blood rain RAIN containing fine red dust particles derived from desert regions. Once entrained into the atmosphere, these particles are often transported over long distances (e.g. Saharan red dust sometimes occurs in rainfall as far north as Finland) by upper-level winds before being washed out in PRECIPITATION, which leaves a red stain on exposed surfaces.

blue Moon (blue Sun) A rare occurrence in which the Moon or Sun appears to have a bluish color. This arises when particles of a particular size, produced typically by forest fires, dust storms, or volcanic eruptions, are suspended within the atmosphere. The dust particles scatter the light leaving only the blue end of the spectrum; sometimes a green or orange color occurs. The rarity of the event is reflected in the phrase 'once in a blue moon', to mean almost never.

bomb *see* explosive cyclogenesis.

bora A regional name for a strong, cold, and often very dry north or northeasterly wind that is funneled by the valleys of the Karst and Dinaric Alps in the Balkan mountain range down to the eastern (Dalmatian) coast of the Adriatic Sea, between Trieste and Albania. It occurs mainly in winter, when air blows from a continental anticyclone (high) established over central Europe to low pressure over the Mediterranean. If a deep region of low pressure is located in the Adriatic at the time, then considerable precipitation can accompany the bora. The wind may continue for several days, sometimes with violent gusts and squalls exceeding 50 knots (57 mph). Generically, the name bora can be applied to a cold squall that moves downhill from any upland area (e.g. from the Caucasus Mountains to the Black Sea); also to a cold northeast wind from Siberia that is experienced in the Gulf of Finland.

boreal **1.** Denoting the north or the north wind.
2. Denoting the coniferous forests in the north of the northern hemisphere. *See also* boreal climate.

boreal climate The climate that extends over large areas of Eurasia and North America and is associated with the boreal forest zone. In Eurasia it extends from the Scandinavian peninsula at about 70°N eastward across Siberia to the Pacific and as far south as about 50°S. In North America it extends from Alaska across the Yukon and Northwest Territories to the Atlantic coast of Canada. Winters are long and cold and summers short, cool, and humid; there is a wide annual range of temperatures, e.g. at Yakutsk, Siberia, the range is 41°C (64°F). Much of the boreal forest zone is moist but parts of western Canada and Siberia have low annual precipitation and are cold and dry. *See also* Köppen climate classification.

Boreal period A period in the post-glacial times between 9500 and 7500 years BP during which the Earth's climate began to warm. It was preceded by the Pre-Boreal period and followed by the ATLANTIC PERIOD.

borehole In climatology and geophysics, a hole drilled into the Earth's surface, which is used for exploration purposes to gather data for studies of past climate or for geophysical purposes. Examples include ice cores drilled in the Antarctic ice sheet and sediment cores taken from ocean-floor surfaces.

boundary current An ocean current that flows either at the eastern or the western side of the subtropical GYRES in the oceanic basins, parallel to a continental margin. The *eastern boundary currents* are produced by westerly winds being deflected by the EKMAN EFFECT and the continental boundaries. This often leads to upwelling of cold water from deeper levels. The currents are broad and generally cold; the CALIFORNIA and PERU CURRENTS in the Pacific as well as the BENGUELA and CANARIES CURRENTS in the Atlantic are examples. The *western boundary currents* are faster, deeper, and narrower (less than 100 km or 62 miles wide) than the eastern boundary currents, They carry warm water along the western margins of the oceans and include the GULF STREAM in the Atlantic Ocean and the KUROSHIO in the Pacific Ocean.

boundary layer **1.** The thin layer of fluid in contact with the surface of a solid; it is stationary as a result of the adhesion between the molecules of the fluid and the surface of the solid.
2. In microclimatology, the layer of the atmosphere that is closest to the surface, such as the ocean or land, or it may refer to a very thin layer of about a millimeter covering the surfaces of vegetation through which fluxes of energy and water vapor must pass. It is the layer immediately above the active layer.
3. *See* planetary boundary layer.

boundary layer climate The integration over time of all the flows and transformations of energy and mass that affect the layer adjacent to the surface. In a restrictive sense, it may refer principally to temperature and evaporation at or below the height at which most meteorological measurements are made: the screen height, which is about 1.25 m (4 ft) above ground surface. Boundary layer climates are the result of the way in which many different surfaces receive and emit energy, and then interact with the physical processes, including precipitation and airflow, of the free air lying above the boundary layer.

bounded weak echo region (BWER) A feature of the part of the very rapidly ascending air within a THUNDERSTORM or SUPERCELL that has low RADAR reflectivity and gives a weak radar echo. The upward-rushing air moves so rapidly that raindrop-size particles form only in the upper parts of the CUMULONIMBUS cloud. Radar that views the intense updraft of this part of the cloud will see only a weak return that is often surrounded (or bounded) by stronger returns from precipitation-size particles – the bounded weak echo region.

bow echo In radar, the return signal or echo from PRECIPITATION that is associated with some SQUALL LINES that have intense rainfall and trailing STRATIFORM precipitation. As these evolve, they gradually take on a 'bow' shape that is convex in the direction of motion. A little later in their life cycle, they change into a 'comma' form - to produce a COMMA ECHO. Bow echoes are often associated with damaging surface winds near the apex of the bow; short-lived tornadoes may develop along the apex or at the north end of the bow (in the northern hemisphere).

Bowen ratio The ratio of the amount of SENSIBLE HEAT to LATENT HEAT for a surface. The major use of solar energy at the Earth's surface is either to heat the surface (H) or to evaporate water (LE). The Bowen ratio is expressed as $\beta = H/LE$, where H is the sensible heat flux and LE is the LATENT HEAT FLUX. Values rise from about 0.5 in humid middle latitudes to 5.0 in deserts where available water for evaporation is a limiting factor.

box model A computer model used in atmospheric pollution studies to calculate concentrations of pollutants. It is a black box volume in which pollutant concentrations from a number of small sources with assumed emission strengths per unit time and area are treated as a single unit source of uniform concentration within the BOUNDARY LAYER, such as an urban area. The box-model solution assumes steady state conditions and BACKGROUND LEVELS of zero.

Boyle's law A law that describes the behavior of gases: at constant temperature, the volume of a given mass of an ideal gas is inversely proportional to the pressure upon it. The law was named for the Irish physicist and chemist Robert Boyle (1627–91). *See also* Charles' law; Dalton's law; ideal gas laws.

brave west winds A nautical term for the westerly winds that occur over the oceans of the southern hemisphere between approximately 40°S and 65°S. The low percentage of land relative to ocean means that such winds blow with great force and regularity; they are associated with rough seas. *See also* roaring forties.

Brazil Current A warm current in the South Atlantic Ocean that flows southeastward along the east coast of Brazil to about 30°–40°S, where it is deflected to the east by the northward-flowing Falkland Current.

breeze 1. A wind speed of force 2, 3, 4, 5, or 6 on the BEAUFORT SCALE.
2. A local-scale air movement that is convectively forced; for example a LAND BREEZE or SEA BREEZE.
3. Any generally light wind.

brickfielder (brickfelder) A hot, dry, and often dust-laden northerly wind experienced along the southeastern coastlands of Australia (for example, in Victoria) during the summer. It is caused by a poleward advance of tropical air from the deserts of the interior. The brickfielder is usually associated with the leading edge of a trough of low pressure, and precedes the

SOUTHERLY BUSTER. Prolonged hot spells often occur when the brickfielder blows; temperatures can exceed 40°C (104°F) on a daily basis.

brocken specter An optical phenomenon in which an observer's shadow is seen against a bank of cloud, mist, or fog as greatly magnified and sometimes surrounded by colored lights. It results from the diffraction of light. The phenomenon is named after the Brocken, the highest peak in the Harz Mountains of central Germany, where it was first noted.

bromofluorocarbon *See* halon.

Brückner cycle A cycling of periods of cold and damp conditions alternating with conditions that are warm and dry in northwestern Europe. In 1890, the German geographer and climatologist Eduard Brückner (1862–1927) calculated that there is a mean period of 34.8 years between such cycles, with cycles varying from 25 to 50 years.

bubble high A mesoscale dome of high pressure that builds up from rain-cooled air in THUNDERSTORM downdrafts. Bubble highs can be large enough and persist long enough to be identified on surface charts. The edges of these high-pressure regions are marked by the GUST FRONT created by the contrast between the warm thunderstorm air and the cooler bubble-high air.

Buchan spell In the UK, a period of unseasonal weather during the year that was believed to occur with regularity. In 1869, the Scottish meteorologist, Alexander Buchan (1829–1907), attempted to construct a calendar of periods of weather that recurred with sufficient regularity at about the same time in the year to be termed 'spells' in Scottish climate. The initial listing distinguished six cold periods, with temperatures regularly below the average temperature curve, and three warm periods. Statistical analysis has subsequently cast doubt on the theory. The method was later refined by H. H. Lamb and reinterpreted in terms of atmospheric circulation types. *See* singularity.

bulk Richardson number (BRN) A dimensionless number derived from an equation containing vertical STABILITY and vertical shear terms. The BRN is generally defined as the stability term divided by the shear term. High values indicate an unstable and/or weakly sheared atmosphere; low values being indicative of weak instability and/or strong vertical shear. BRN values of 50–100 are favorable for SUPERCELL development; thunderstorms are unlikely if BRN is <10.

buoyancy force A force that acts upon an AIR PARCEL that has a different temperature (T′) from that of the surrounding environmental air (T) within which it occurs. The force is defined as positive (upward) when $T'>T$ and negative (downward) when $T'<T$.

buoy weather station A fixed or floating buoy in the ocean that carries instruments for sensing a number of meteorological parameters. The data is automatically transmitted by radio.

buran A strong cold north to northeasterly wind experienced in Russia and central Asia. Although the buran can occur in any season, it is most frequent in the winter when it is exceptionally cold, with temperatures below –30°C (–22°F) accompanied by BLIZZARDS that are dangerous to human and animal life, especially on the open steppes. The winter buran is also known as the *purga*. Occurring in the rear of a depression, the buran interrupts the comparatively calm anticyclonic conditions associated with the Siberian high. *See also* blizzard.

Bureau of Meteorology, Australia The national meteorological agency in Australia, one of the leading meteorological services of the southern hemisphere, with its headquarters in Melbourne. The Bureau commenced operations in 1908 and aims to observe and understand Australia's weather and climate, and to provide meteorological, hydrological, and oceanographic services for national use and also to support the country's international obligations. It participates in the WORLD METEOROLOGICAL ORGANIZATION's (WMO) WORLD WEATHER WATCH program with one of the three World Meteorological Centers being based in Melbourne.

burst of the monsoon The onset of the southwest MONSOON over the Indian subcontinent and southeast Asia which, by tradition, is reputed to start in a dramatic way. It marks the end of the period of great heat and DROUGHT that usually precedes the invasion of moisture-laden air over India. Indian meteorologists tend to refer to the 'timetable' of the burst to describe the progression of this change in temperature and humidity across the subcontinent. It is recorded at Cochin (far southwest) on June 1, Hyderabad (central east India) on June 5, Calcutta on June 10, New Delhi on June 15, and Lahore about July 1.

bush fire *See* fire.

butterfly effect *See* chaos theory.

Buys Ballot's law A rule, proposed in 1857 by the Dutch meteorologist Christopher Buys Ballot (1817–90), that relates the wind direction to the locations of high- and low-pressure centers. It states that if one stands (in an exposed location) with one's back to the wind, then (in the northern hemisphere) low pressure will lie to the left and high pressure to the right. The opposite is true for the southern hemisphere.

BWER *See* bounded weak echo region.

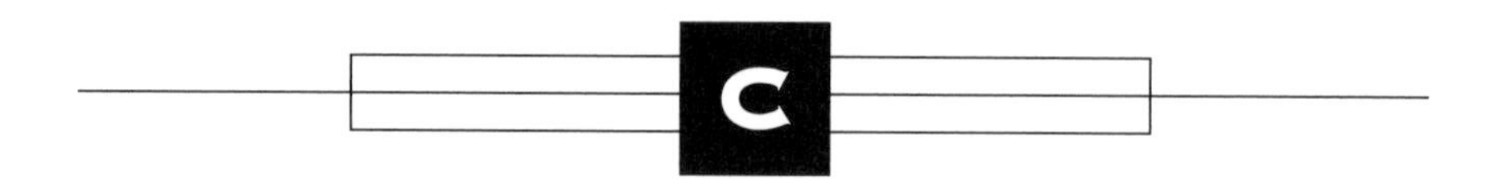

calibration The process of making and measuring instruments so that measurements can be made in specified units.

California Current A cold current in the Pacific Ocean, a southward extension of the NORTH PACIFIC CURRENT, that flows southeastward along the west coast of North America. In the summer months especially, the surface current is deflected to the west through the EKMAN EFFECT, which results in the upwelling of colder water rich in nutrients.

calm A state in the atmosphere in which there is virtually no horizontal motion, for example smoke from a factory rises vertically. It corresponds to force 0 on the BEAUFORT SCALE, with a wind speed of less than 1 knot (1.15 mph). *Calm conditions* can occur at any place and at any time, but are most common in the subtropical high-pressure belts and in the doldrums along the Equator.

calvus A species of CUMULONIMBUS cloud in which some of the protuberances across its upper surface are losing their sharp CUMULIFORM character but are not CIRRIFORM in nature. *See also* cloud classification.

Campbell–Stokes sunshine recorder A type of SUNSHINE RECORDER that uses the heating action of the Sun to record the duration of bright sunshine. A spherical glass lens focuses the Sun's rays onto a recording card graduated in time intervals, usually marked in hours. Duration and intensity of the sunshine is determined by the length and width of the burnt track on the card.

Canaries Current (Canary Current) A cold current in the North Atlantic Ocean that branches south from the NORTH ATLANTIC DRIFT to flow southward past Spain, Portugal, and northwest Africa before turning westward to join the Atlantic North EQUATORIAL CURRENT. The temperature of the current is relatively cool, a product of upwelling caused by continental offshore winds, and helps to reduce the heating effect of the Sahara to the east.

Cancer, Tropic of *See* Tropic of Cancer.

canyon wind (jet-effect wind, mountain gap wind, ravine wind) A wind generated as a result of variations in local topography. Airflow may, for instance, be funneled through a canyon, mountain pass, valley, or even a street in a city lined with high buildings, resulting in local-scale wind velocities being higher than the regionally averaged mean wind speed. Examples include the *dusenwind* (Aegean Sea, eastern Mediterranean), KOSAVA, *tehuantepecer* (Central America), and WASATCH.

cap *See* inversion.

capacity In winds, the capability of a wind current to transport material. The wind's capacity is measured by the maximum amount of detritus (e.g. silt, sand, and/or gravel) carried past a specific point in unit time. Capacity increases with wind velocity and decreases as the particle size of debris to be transported increases.

cap cloud **1.** *See* pileus.
2. A stratiform cloud above a hill or mountain and formed through orographic lift.

CAPE *See* convective available potential energy.

capillatus A species of CUMULONIMBUS cloud that is characterized in its upper reaches by CIRRIFORM strands or sheets of fibrous cloud having the form of an anvil, a plume, or a disorganized mass of hair. This type of cumulonimbus is often accompanied by a shower or by a thunderstorm. It also frequently has distinct VIRGA. *See also* cloud classification.

capping inversion *See* inversion.

Capricorn, Tropic of *See* Tropic of Capricorn.

carbon cycle The global-scale exchange of carbon substance between its different reservoirs. These include the relatively rapid continuous flow from the atmosphere (where carbon is chiefly in the form of carbon dioxide) into plant life through photosynthesis. It can then move to animals that consume plants (or other animals), returning to the atmosphere in respiration. Dead animals are decomposed by bacteria and fungi, which return the ingested carbon by respiration. Dead plants fossilize into fossil fuels, some of which return the carbon dioxide to the atmosphere on combustion (see diagram). Study of the carbon cycle involves the estimation of the magnitudes and distribution of the reservoirs, and the rates of flow of carbon between them.

carbon dating (radiocarbon dating) A method, used in paleoclimatology for example, of estimating the age of organic material; it is based on the ratio of radioactive carbon atoms to stable carbon atoms in the material. The radioactive carbon-14 (^{14}C) is formed by the interaction of cosmic rays with atmospheric nitrogen. Together with the other two carbon isotopes, it is present in living organisms; once an organism dies it ceases to absorb ^{14}C, which steadily decreases, decaying at the known rate of the half-life of 5730 ± 30 years. An estimate of the age of a material can be obtained from the residual ^{14}C present. It gives reasonably accurate results for about 40,000–50,000 years. However, since the method was originally developed in 1946–47 by Willard F. Libby (1908–80), uncertainties have arisen as to the actual ^{14}C content of the atmosphere during past ages. It is now accepted that the content varies in line with fluctuations in the cosmic ray activity that produces the ^{14}C. Corrections to dates within the recent past are now made

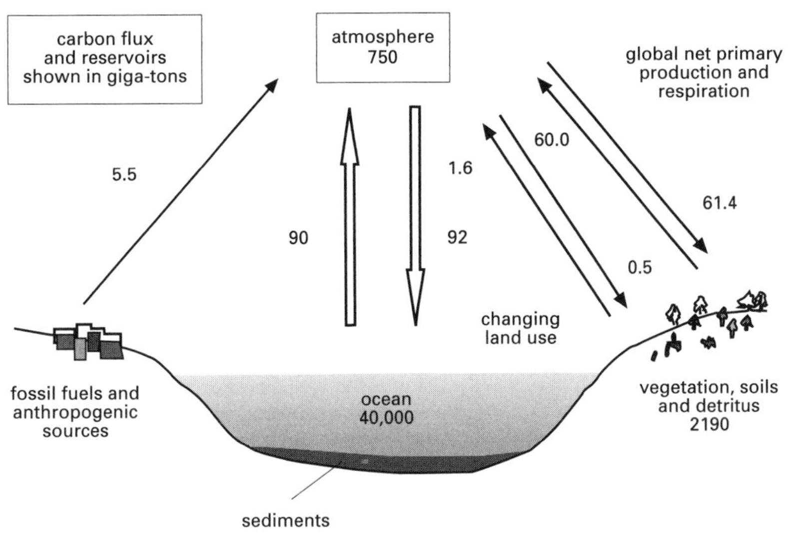

The global carbon cycle

through calibration with dates obtained through dendrochronology.

carbon dioxide (CO_2) A TRACE GAS in the Earth's atmosphere; it is colorless, odorless, and noncombustible, and is formed by the oxidization of carbon and carbon compounds. Carbon dioxide acts as an important GREENHOUSE GAS: it absorbs longwave radiation emitted from the Earth's surface and contributes to maintaining the Earth's temperature at a level at which life can be supported. Carbon dioxide concentrations vary in the range 0.3–0.4% with location and season – the northern hemisphere has higher concentrations than the southern as a result of greater number of anthropogenic sources. Levels of carbon dioxide in the atmosphere are rising in line with increased emission from human activity, especially burning fossil fuels and deforestation, which is believed to be an important cause of GLOBAL WARMING. *See also* carbon cycle.

carbon dioxide equivalent (CDE) A measure used to compare the emissions of different greenhouse gases: the amount of carbon dioxide by weight that would cause the same amount of radiative forcing as a given weight of the other greenhouse gas being measured. Carbon dioxide equivalents are generally calculated by multiplying the mass of the gas of interest (in kilograms) by its estimated GLOBAL WARMING POTENTIAL.

carbon equivalent (CE) A metric measure used to compare the emissions of different greenhouse gases. In the US, greenhouse gas emissions are usually expressed as million metric tons of carbon equivalents (MMTCE). Carbon equivalent (CE) units can be converted from CARBON DIOXIDE EQUIVALENT (CDE) units by multiplying the CDE by the ratio of the molecular weight of carbon to carbon dioxide (i.e. 12/44).

carbon sequestration The long-term storage of carbon dioxide as a means of reducing harmful levels of the gas in the atmosphere that contribute to the enhanced GREENHOUSE EFFECT and thus to global warming. The sites where it may be sequestered include plants (especially extensive forests), soils, the ocean, or underground within depleted oil and gas reservoirs.

carbon sink A reservoir that absorbs or takes up carbon that has been released elsewhere in the carbon cycle. In the general case, the atmosphere, the terrestrial biosphere, oceans, and sediments can be regarded as carbon sinks. Forests and oceans are potential large carbon sinks, which may help to offset greenhouse gas emissions.

Caribbean Current A warm surface ocean current, formed by the convergence of the North EQUATORIAL CURRENT and the Guiana Current, that flows clockwise in the Gulf of Mexico. The Caribbean Current flows through the Caribbean Sea and into the Straits of Florida via the Yucatàn Channel to form the FLORIDA CURRENT.

castellanus (castellatus) A species of clouds that have cumuliform protuberances that appear on their upper surfaces. These features appear like tall turrets of a castle and can form on, for example, CIRRUS, CIRROCUMULUS, ALTOCUMULUS, and STRATOCUMULUS. ALTOCUMULUS CASTELLANUS indicates instability in the middle TROPOSPHERE and is often the precursor of the development of thunderstorms. *See also* absolute instability; cloud classification.

CAT *See* clear-air turbulence.

CCL *See* convective condensation level.

CDE *See* carbon dioxide equivalent.

CE *See* carbon equivalent.

ceiling The height above the Earth's surface of the base of the lowest cloud layer. The cloud layer can be scattered or overcast, but must not simply be a thin layer.

ceiling balloon *See* balloon sounding.

ceilometer (cloud-base recorder, cloud-height sensor) An automated measuring and recording instrument used to determine the height of the cloud base. A beam of light is emitted vertically into the sky and the reflected light picked up by a detector, which measures the time-lag between the emitted and detected light. Trigonometry is used to determine the altitude of the cloud base. Modulating the light beam so that it can be recognized when it is reflected reduces interference from other light sources. More advanced ceilometers are a form of LIDAR in which a laser pulse is projected into the sky.

celestial sphere The imaginary sphere in the heavens of infinite radius on which the stars and other celestial bodies appear to lie. The Earth and the observer are imagined to be at the center of the sphere.

cell In meteorology, an atmospheric circulation feature that is more-or-less closed. Cells occur on different scales; for example the massive troposphere-deep HADLEY CELL that occurs in the north–south plane within the tropics, the cellular motion that characterizes open and closed cells in CELLULAR CONVECTION, or the cell formed by the UPDRAFT and/or DOWNDRAFT currents within a THUNDERSTORM.

cellular cloud The organization of CUMULUS and STRATOCUMULUS cloud across, typically, a few tens of kilometers horizontally. They can be divided into open and closed cells: the open cells consist of a ring of cumulus with a clear center, and closed cells are filled with stratocumulus with a clear surrounding rim.

cellular convection Convection that takes the form of localized convective cells, which typically develop as a result of the strong heating of the land surface, for example during the summer months in the middle latitudes. Low temperatures in the upper TROPOSPHERE, which aid the release of CONDITIONAL INSTABILITY or CONVECTIVE INSTABILITY, encourage the development of convective cells, for example as seen in tornadoes in the US. This can lead to intense precipitation events, with THUNDER, LIGHTNING, and HAIL. *See also* Benard convection.

Celsius scale A temperature scale ranging from 0 to 100 degrees in which 0°C corresponds to the freezing point of water, and 100°C to the boiling point of water. It is internationally the most widely used temperature scale although in the US, the Fahrenheit scale remains in widespread use. It is named for the Swedish astronomer Anders Celsius (1701–44) whose original scale (in which freezing point was 100° and boiling point, 0°) was proposed in 1742. The scale was known, until 1948, as the *centigrade scale*, because of the 100-degree interval between the freezing and boiling points; this name is still frequently used as a synonym for the Celsius scale. The Celsius scale can be converted from the FAHRENHEIT SCALE using the following formula:

$$°C = 5(°F - 32) \div 9.$$

centigrade scale *See* Celsius scale.

centrifugal force A fictitious force, that a body constrained to move in a circle around a central point can be said to experience. It acts outward radially from the center and has a magnitude, per unit mass, that is the body's linear velocity squared divided by the radius of curvature of its circular motion. In the atmosphere, this force acts upon a particle of air that follows a curved path around a center of low or high pressure, for example. It appears to be equal in magnitude to the CENTRIPETAL FORCE, which acts in the opposite direction. However, the use of the concept of a centrifugal force balancing a centripetal force can cause confusion because, in fact, the centrifugal force is hypothetical and the centripetal force is the gravitational force.

centripetal acceleration The radial component of acceleration of a body that is constrained to move in a circular path. It is the acceleration needed to keep an object moving in a circular path.

centripetal force The radial force experienced by a body that is required to constrain it to move in a circular or curved path; the force is directed toward the center of rotation. It appears to be balanced by the equal and opposite CENTRIFUGAL FORCE, but this latter force is hypothetical.

CFC *See* chlorofluorocarbon.

chain lightning *See* beaded lightning.

chaos theory A mathematical theory used to describe the erratic behavior of the dynamical atmospheric–oceanographic system. According to this theory a small change in the initial state leads to a rapid and unpredictable change in subsequent states. It is known as the *butterfly effect* as it has been suggested that the system is so sensitive that a butterfly flapping its wings in one part of the world could be the impetus for a tornado occurring in another part of the world. Whether or not such an effect actually exists, it is true that in the context of weather forecasting a small change in the initial conditions can have a dramatic effect on the outcome.

Chapman cycle of ozone formation A scheme proposed in the 1930s to explain the natural formation and destruction of stratospheric OZONE by a continuous process of photoassociation and photodissociation. The interaction of ultraviolet solar radiation with oxygen (O_2) molecules would split them into two atoms, some of which would combine with other molecules to form ozone (O_3). In turn, the ozone molecules are split into one oxygen molecule (O_2) and one oxygen atom (O). The reactions constitute a steady state in which the rate of ozone formation equals that of ozone destruction. *See also* ozone layer; ozone depletion.

Charles' law A law that describes the behavior of gases at constant pressure. It states that all gases expand by 1/273 of their volume at 0°C for a temperature increase of 1°C; i.e. the volume of a given mass of gas at constant pressure is directly proportional to its THERMODYNAMIC TEMPERATURE. The law is named for the French scientist Jacques Charles (1746–1823). *See also* Boyle's law; Dalton's law; ideal gas laws.

chart *See* synoptic chart.

chemistry model A mass-balance model in which most of the important chemical kinetics present in a reactive atmosphere are modeled using rate equations (speed of reaction) and controlled using observed rate parameters. The method has been applied chiefly to the modeling of photochemical SMOG.

chilled-mirror hygrometer *See* dew-point hygrometer.

chinook (snow eater) **1.** A warm dry southwesterly FÖHN-type wind experienced along the eastern side of the Rocky Mountains in parts of Canada and the US. The arrival of the wind is frequently heralded by the line of cloud known as the CHINOOK ARCH. The chinook blows on the southern side of a depression tracking eastward across the North American continent. The airflow is compressed and warmed adiabatically on its descent from the Rocky Mountains. Sometimes the warming process is so sudden that it causes the air temperature to increase by 20°C in just 15 minutes, leading to rapid snow melt during the winter and spring. Economically important in pastoral regions, frequent and strong chinooks result in mild winters; the absence of chinooks leads to severely cold winters and potentially heavy losses of livestock.
2. A southwesterly wind that brings warm and moist air to the Pacific coastal regions of Washington and Oregon, US.

chinook arch A line of cloud that overhangs the Front Range of the North American Rocky Mountains when the CHINOOK blows. A great deal of cloud develops on the upslope side of the mountains and dissipates as the air subsides along the eastern slope onto the High Plains, producing an arch-like feature that marks the line along which the cloud evaporates.

chlorofluorocarbon (CFC) Any one of a number of artificial gases composed of the elements chlorine, fluorine, and carbon, which were originally developed in the 1930s as nontoxic and nonflammable refrigerants. They include dichlorodifluoromethane (CFC-12). Chlorofluorocarbons have also been used as propellants in aerosol spray cans, as foaming agents in the production of expanded plastic foams, and as solvents for oil and grease. They contribute to the depletion of stratospheric OZONE above the Antarctic, and are powerful GREENHOUSE GASES. Their production and use is being phased out as part of the directives from the MONTREAL PROTOCOL. *See also* ozone layer; ozone depletion.

circulation In meteorology, a general term related to the organized motion of the atmosphere on a variety of space scales from global down to that, for example, of a small CUMULUS cloud. *See also* general circulation.

circulation index A numerical value that expresses the strength or magnitude of one of the large-scale atmospheric circulation components. It measures, for example, the strength of the ZONAL FLOW (east–west) or MERIDIONAL CIRCULATION (north–south) components of wind at the upper levels.

circumpolar vortex Two separate planetary-scale circulation features that are centered more or less on the North and South Poles. They affect regions at middle and higher latitudes in the middle and upper TROPOSPHERE and the STRATOSPHERE. In the troposphere the vortex consists of the complete pattern of LONG WAVES around the hemisphere, with westerly flow about an axis on the pole. In the stratosphere the vortex is westerly in the winter and easterly in the summer, in response to the reversal in the low- to high-latitude temperature gradient at that level.

cirriform Of or relating to clouds consisting entirely of ice crystals and which have a fibrous structure, such as CIRRUS, CIRROSTRATUS, and CIRROCUMULUS.

cirro- A prefix related to HIGH CLOUDS, which have bases, in the middle latitudes, above about 5000 m (16,000 ft) and are composed almost exclusively of ice crystals. *See also* cloud classification.

cirrocumulus A type of HIGH CLOUD, one of the cloud genera (*see* cloud classification), in the form of a thin white sheet made up of small elements that are ripples or patches with an apparent width of less than one degree, viewed from the surface.

cirrostratus A type of HIGH CLOUD, one of the cloud genera (*see* cloud classification), in the form of a thin sheet or whitish transparent smooth veil that normally produces optical phenomena, such as halos around the Sun and Moon. It is composed of ice crystals.

cirrus A type of HIGH CLOUD, one of the cloud genera (*see* cloud classification), with a delicate white fibrous, wispy, patchy, or banded appearance. It occurs as detached areas often with the appearance of feathers or streamers across the sky. It is composed of ice crystals. [Latin: 'wisp of hair']

CISK *See* conditional instability of the second kind.

CLAMP *See* Climate-Leaf Analysis Multivariate Program.

Clausius–Clapeyron relation The non-linear relationship between pressure and temperature when two phases of a substance (for example, liquid water and water vapor) are in equilibrium. The equation gives the SATURATION VAPOR PRESSURE over a flat surface of pure water as a function of temperature. It was named after Rudolph Clausius (1822–88), a German physicist, and Benoit-Pierre-Emile Clapeyron (1799–1864), a French engineer.

Clean Air Act 1. In the US, the two acts passed in 1965 and 1970 affirming the authority of the Federal Government to limit interstate pollution and to set national air pollution standards.

2. In the UK, the Parliamentary Act of 1956, which controlled the emission of smoke, together with the extension of that act in 1968 controlling industrial emissions.

clear-air turbulence (CAT) A layer, normally within the upper troposphere but also occurring at lower levels, in which the airflow is very choppy or turbulent, and a potential danger for jet aircraft. Much turbulent weather is made visible by the presence of CUMULUS and CUMULONIMBUS cloud with which it is associated but clear-air turbulence, in general, has no visible indicators and cannot be seen. It is often generated by the marked vertical and horizontal WIND SHEAR around the core of the POLAR FRONT JET STREAM. Such rapid changes of speed over a short distance lead to very turbulent flow. The phenomenon can also occur in clear air much lower down in the troposphere over topography, for example in lee waves.

clear ice *See* glaze.

clear slot A localized area of clearer skies or reduced cloudiness caused by the intrusion of drier air that may be visible on the west or southwest side (in the northern hemisphere) of the wall cloud below a SUPERCELL thunderstorm.

climate The average long-term prevailing WEATHER, including seasonal extremes and variations, for a particular region and time interval. It is often represented as a statistical collection of weather conditions for that region with a description of suitable boundary conditions (epoch, season, physical boundaries). Properties of interest include temperature, dynamics (wind, vertical motion, or ocean currents), thermodynamics, and hydrology (humidity, cloudiness, total column moisture, ground and surface water), and large-scale systems (pressure and density of the atmosphere, salinity of the ocean), as well as precipitation, evaporation, infrared radiation, convection, advection, and turbulence.

Climate can be seen to be the result of the redistribution of solar energy, the imbalance of the greater heating of the tropical latitudes driving the atmospheric and oceanic motions that redistribute the energy globally. The local climate for a particular location is determined by latitude, altitude, topography, and continentality or oceanicity. *See also* climate change; climatology. See illustration on page 40.

Climate Analysis Center *See* Climate Prediction Center.

Climate and Cryosphere (CliC) *See* Arctic Climate System Study.

climate change In general, the long-term fluctuations in the Earth's climate. More particularly the term is used to describe a significant change from one climatic condition to another on a global scale; for example the shifts from ice ages to warmer periods that have taken place during the Earth's geologic history. Such changes occur naturally, sometimes abruptly but at other times over considerable timescales, through CLIMATE FEEDBACKS from changes, for example, in the Earth's orbit, the energy output from the Sun, volcanic eruptions, orogeny (mountain building), meteor strikes, or lithospheric motions. Over considerably shorter timescales more recent climate changes include the periods of warmer-than-average temperatures, such as the Holocene climatic optimum and the medieval climatic optimum (*see* climatic optimum), and cooler than average periods, such as the LITTLE ICE AGE. More recently, the study of climate change has sometimes been dominated by the concept of GLOBAL WARMING and other climate changes caused by human activities. There is evidence that the increase in greenhouse gases, in particular of carbon dioxide, as a result of human acitivity is affecting climate by enhancing the natural GREENHOUSE EFFECT thereby contributing to an increase in the Earth's average surface temperature. *See also* radiative forcing.

climate classification A method to delimit and describe the major types of climate. The earliest were those of the

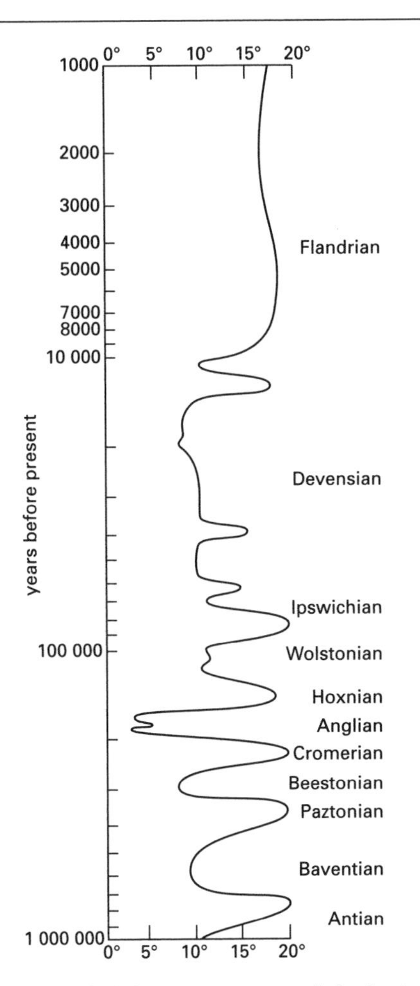

Changes in July mean temperature in lowland England

classical Greeks, divided along latitudinal lines and the length of day into the FRIGID, TEMPERATE, and TORRID ZONES. Many different modern schemes have since been developed: (1) *Generic classifications* relate climate limits to plant growth or vegetation groups and use the basic criteria of temperature and aridity. Aridity is usually expressed as effective precipitation (precipitation minus evaporation), based on the ratio of rainfall/temperature. The widely used climate classifications of Wladimir Köppen (1846–1940) are examples of this type. (2) *Energy and moisture budget classifications*, such as that of the American climatologist Charles Warren Thornthwaite (1889–1963), which is based on the moisture budget and potential evapotranspiration. (3) *Genetic classifications* are based on the causal elements of the climates, such as the atmospheric circulation in terms of wind systems and air masses. An example of this approach is H. Flohn's scheme, proposed in 1950, and based on the global wind belts and precipitation characteristics. Another genetic classification is that of the American geographer-climatologist Arthur N. Strahler (1918–) in 1951, which is based on air masses. (4) *Climatic comfort classifications*, are based on comfort indices to reflect how the climates effect human comfort. An example of such a scheme is that of the American geographer Werner H. Terjung in 1966, which was based on the parameters of temperature, relative humidity, wind speed, and solar radiation. *See also* Köppen climate classification; Flohn climate classification; Strahler climate classification; Thornthwaite climate classification.

climate feedback A feedback associated with climate processes; if the process increases the effect of climate change it is a *positive feedback*, while if it diminishes the effect it is a *negative feedback*. Generally, the interactions between the CLIMATE SYSTEMS produce multiple feedbacks, which smooth out very long-term climate change.

Climate-Leaf Analysis Multivariate Program (CLAMP) A method of reconstructing past climates, in particular the estimation of mean annual temperatures, by studying the fossil leaves of dicotyledonous plants.

climate model A MODEL that uses quantitative methods to represent dynamical and physical processes and interactions in the global system, which may include representation of the atmosphere, land, ocean, and ice. Climate models provide an important tool in the understanding and prediction of climate changes and can be used to measure anthropogenic climate change. *See also* general circulation model.

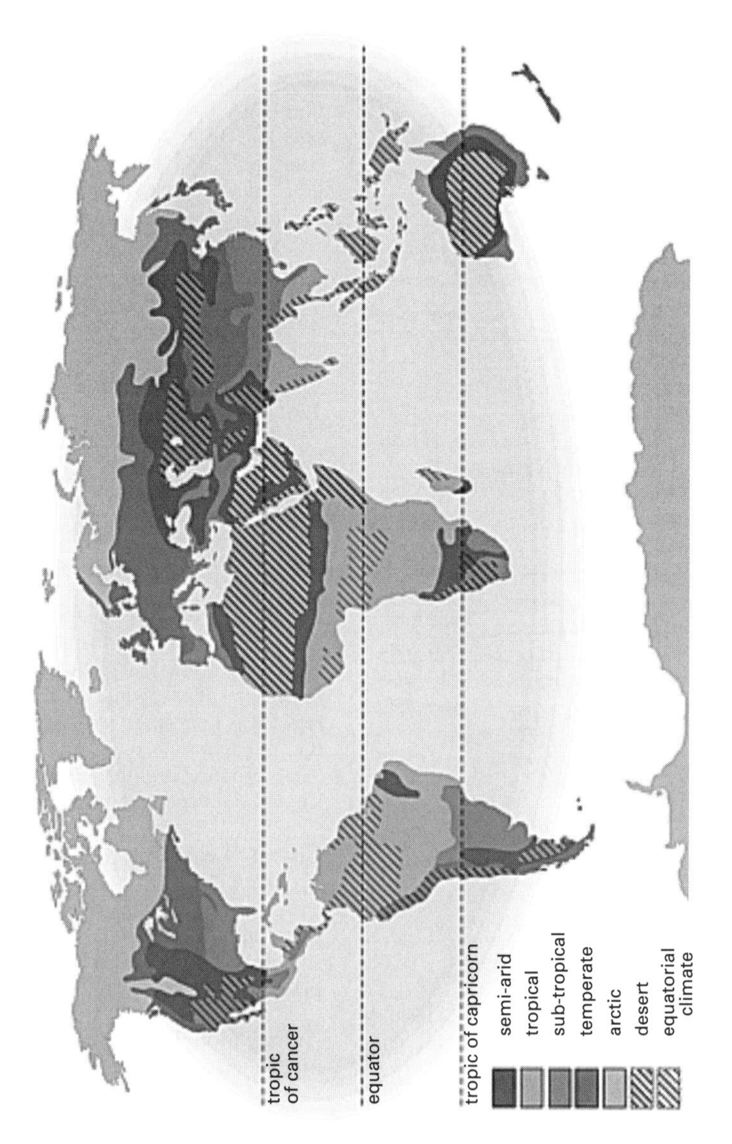
tropic
of cancer
equator
tropic of capricorn
semi-arid
tropical
sub-tropical
temperate
arctic
desert
equatorial
climate

Climate

climate modeling program A program that seeks to develop powerful computer climate MODELS of the global atmosphere and of the combined atmosphere–ocean system. These models are used to investigate climate variability, CLIMATE CHANGE associated with the enhanced GREENHOUSE EFFECT, and the likely impacts of climate change.

climate prediction Forecasting the average values of the properties of the atmosphere some years into the future. The averaging is usually for a season, a particular year, or for a group of years. Climate prediction models, for example, are used to assess the most likely impact of global warming over the coming decades and century. Their product might, for example, be a global chart of annual mean surface temperature for the year 2050, or for December–February precipitation totals. By subtracting the same fields for the current day, meteorologists can then quantify the best estimate of the impact of, for instance, increasing the atmospheric carbon dioxide concentration every year (in the model run) by 1 part per million.

Climate Prediction Center A US organization, based in Camp Springs, Maryland that fulfils the role of producing operational predictions of climate variability, the real-time monitoring of climate, and the investigation of how major climate anomalies evolve. The operational forecasts span the period from a few weeks ahead to subsequent seasons. For example, it issues extended range outlook maps for 6–10 and 8–14 days, and climate outlook maps for 1–13 months ahead.

climate system The system that connects the entire interrelated processes of the atmosphere, oceans, land, biosphere, and cryosphere. Processes within this system are divided into thermal, kinetic, and static, which are interconnected by the movement of water, radiation, and turbulence.

Climate Variability and Predictability Program (CLIVAR) An international research program, begun in 1995 as part of the World Climate Research Program (WCRP), that examines issues of natural climate variability and anthropogenic climate change, including the response of climate systems to anthropogenic forcing.

climatic geomorphology The branch of geomorphology concerned with the association of types of landform with different climates. An example would be the association of periglacial landforms (those associated with glacial conditions) in tundra climates. While there is some consistency in this association there are many instances in which this is not the case so that factors other than climate need to be considered.

climatic optimum A period during which global surface temperatures are significantly higher than the average. The period between 7000 and 4500 years BP, when global surface temperatures were significantly warmer than at present and at their highest since the last ice age, is known as the *Holocene climatic optimum*. It is believed that during this time sea levels were around three meters higher than their current level. The period of about 900–1200 AD is known as the *little climatic optimum* or *medieval climatic optimum*.

climatic region *See* climatic zone.

climatic snow line *See* snow line.

climatic zone (climatic region) An area of the Earth's surface in which the climate is generally similar. Many CLIMATE CLASSIFICATIONS have been devised to delimit climatic zones, ranging from simplistic to complex.

climatograph *See* climograph.

climatological station A station that provides observations of climatological data. The World Meteorological Organization classifications include ordinary climatological stations and principal climatological stations. At the ordinary stations observations are made of maximum and minimum temperatures and precipitation at least once a day. At principal stations readings are taken hourly or at least three times a day of a comprehensive range of climate variables (weather, wind, cloud amount, type and height of cloud base, visibility, humidity, temperature, atmospheric pressure, sunshine or solar radiation, and soil temperature).

climatology The study and description of climate, i.e. the mean state of the atmosphere and oceans, together with statistical CLIMATE variability. Climate variability may encompass those associated with seasonal variation, changes to external forcing, and internal interactions and feedbacks. Many scientists are forecasting the likelihood of climate change due to GLOBAL WARMING and therefore the study of climates is gaining greater importance.

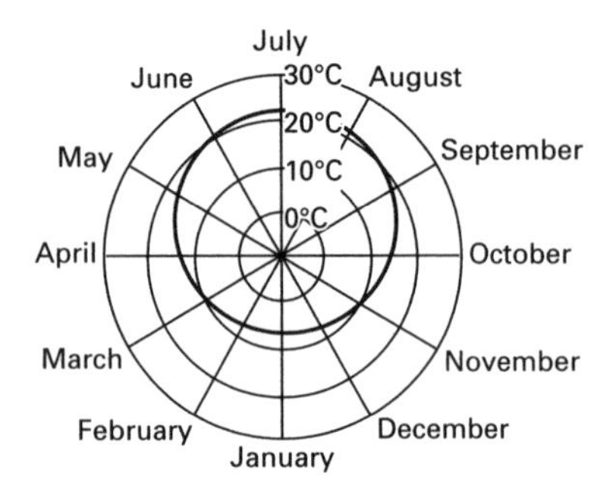

Climograph: mean monthly maximum temperatures at Kew (London)

climograph (climatograph, climatic diagram) A graph or diagram that displays two or more climate variables, for example mean temperature and mean precipitation, for a given location; the data may be displayed over different timescales but usually by month.

CLIVAR *See* Climate Variability and Predictability Program.

closed low 1. *See* cut-off low.
2. A CYCLONE (low) that is defined by more-or-less circular isobars.

cloud A visible volume of very minute water droplets or ice crystals, or a mixture of both, that occurs above the Earth's surface. The constituent droplets are no larger than 200 µm in diameter and condense out from moist air principally when it ascends and cools adiabatically in convection. Cloud can form also through orographic lift when moist air is lifted over hills and mountains or when it is forced to rise in huge volumes within the circulation of low-pressure systems. There are other forms of cloud related to smaller-scale forcing, for example the BANNER CLOUD.

Clouds are composed almost entirely of supercooled water droplets from 0°C (32°F) to –10°C (14°F) for layer clouds and –25°C (–13°F) in convective clouds. It is only when the temperature is –40°C (–40°F) or colder that a cloud is made up entirely of ice crystals. *See also* cloud classification.

cloud amount The fraction of the sky that is covered by either an individual type of cloud, or the total amount of all cloud types present. Both are reported. The fractional unit employed internationally is the OKTA, or eighth of the sky, so that zero eighths means completely cloud-free while eight oktas is overcast. There is a possibility of nine eighths, which means that the sky is obscured (e.g. by fog). The total cloud amount in oktas is represented by international convention as an increasingly blackened-in station circle that forms a component of the STATION MODEL. A slightly different scale is used in North America where it is based on tenths in place of the eighths of oktas.

cloud base The height above the Earth's surface of the base of a cloud. Technically it is the lowest zone in which the obscuration changes from that associated with clear air or haze to that linked to water droplets and/or ice crystals.

cloud-base recorder *See* ceilometer.

cloudburst A general term for brief but exceptionally heavy precipitation, usually of rain but also hail, that may be associated

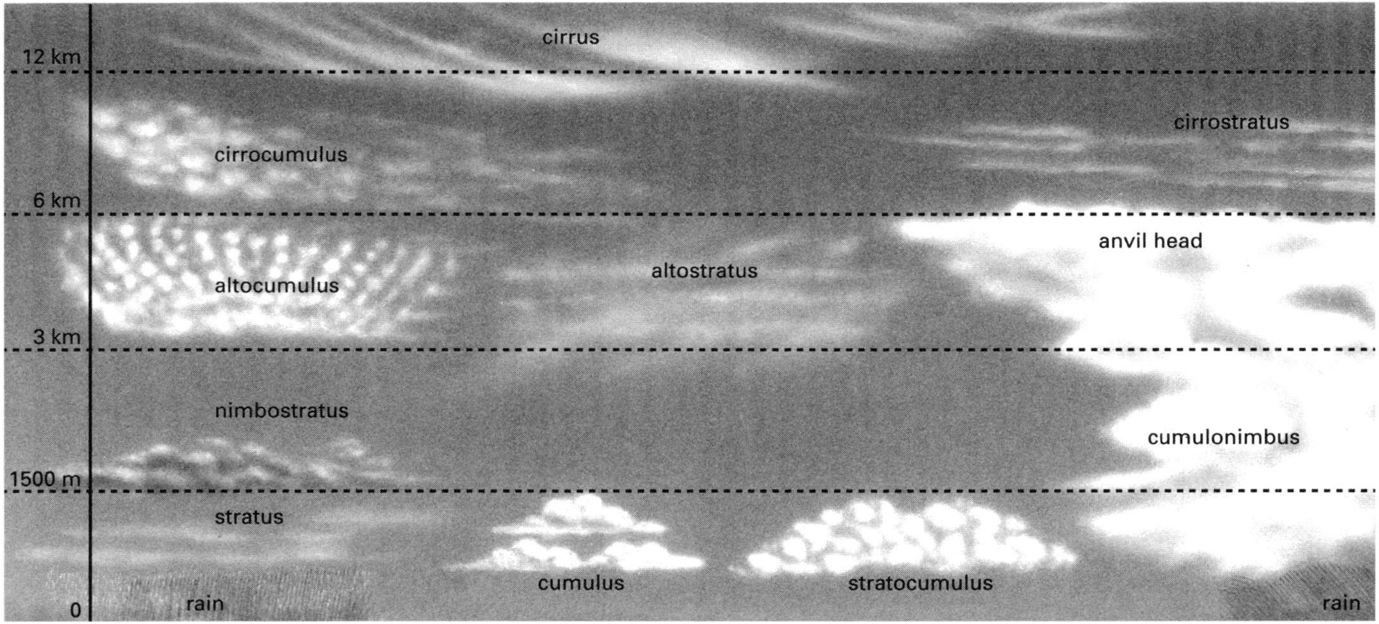

Types of cloud

CLASSIFICATION OF CLOUDS

Cloud Group	*Polar Latitudes*	*Temperate Latitudes*	*Tropical Latitudes*
High	3000 – 8000 m 10,000 – 26,000 ft	5000 – 13,000 m 16,000 – 43,000 ft	6000 – 18,000 m 20,000 – 60,000 ft
Middle	2000 – 4000 m 6500 – 13,000 ft	2000 – 7000 m 6500 – 23,000 ft	2000 – 8000 m 6500 – 26,000 ft
Low	Surface – 2000 m Surface – 6500 ft	Surface – 2000m Surface – 6500 ft	Surface – 2000 m Surface – 6500 ft

with a thunderstorm. Cloudbursts can occur anywhere but are most frequent in mountainous areas, are often associated with violent upward and downward air currents, and may cause considerable damage. *Temperate cloudbursts* are short-lived since the available supply of water vapor is soon exhausted.

cloud ceiling *See* ceiling.

cloud classification The most commonly used classification of clouds, adopted by international agreement by the World Meteorological Organization, is based on the height of a cloud's base above the Earth's surface and divides the types of clouds into four main groups: low, middle, and high, and clouds of great vertical extent. The approximate height ranges for the low, middle, and high groups is shown in the Table above.

The clouds of great vertical extent have their bases in the low cloud group. The levels are linked to the ten cloud *genera* that form the most basic cloud types: STRATUS, STRATOCUMULUS, and NIMBOSTRATUS (LOW CLOUDS); ALTOCUMULUS and ALTOSTRATUS (MIDDLE CLOUDS); CIRRUS, CIRROSTRATUS, and CIRROCUMULUS (HIGH CLOUDS); and CUMULUS and CUMULONIMBUS (clouds of great vertical extent).

Further refinements to the classification are the division of the genera into cloud *species* based on their differences in shape and internal structure, as well as the division into *varieties*, based on their arrangement and transparency. In addition, a number of *supplementary* descriptive terms and ACCESSORY CLOUDS are used. Clouds can also be classified according to air motion that led to their formation: STRATIFORM (layer) clouds that form either through ascent of the air or through turbulence; CUMULIFORM that form through convection; and OROGRAPHIC CLOUD, in which the air is forced to rise over a hill or mountain barrier. The original classification of the cloud genera was proposed in 1803 by the British pharmacist and amateur meteorologist Luke Howard (1772–1846) and, with modifications, this still forms the basis for cloud classification. The types of clouds in the classification are illustrated in the World Meteorological Organization's *International Cloud Atlas* (1956).

cloud cluster A specifically tropical phenomenon in which a region of deep vigorous cumulus and cumulonimbus clouds are clustered together over the warm ocean. Cloud clusters are very common in the humid tropics and may dominate the Intertropical Convergence Zone (ITCZ), when 10–15 clusters can occur in a month. They form through the amalgamation of areas of mesoscale convection. The clusters may persist for several days, accompanied by thunderstorms and heavy rain. These areas represent concentrations of cyclonic flow at low levels and anticyclonic flow at high levels. The release of latent heat through condensation plays a significant part in weakening the TRADE-WIND INVERSION of the tropics.

cloud condensation nucleus A hygroscopic substance (i.e. a substance that has an affinity for water) or aerosol that provides the surface on to which condensation of water or ice takes place to form a CLOUD DROPLET. The radius of the nuclei ranges from less than 0.2 μm (AITKEN NUCLEI), from 0.2 to 1.0 μm (large nuclei), and larger than 1.0 μm (giant nuclei). Condensation nuclei enter the atmosphere in a number of natural and man-made forms including volcanic dust, factory and forest-fire smoke, ocean-spray salt, and sulfate particles emitted by oceanic phytoplankton. *See also* freezing nucleus.

cloud droplet A liquid water droplet that constitutes clouds; cloud droplets are typically 20 μm diameter and so are some 100 times smaller than a typical RAINDROP. Their fall speeds range from 0.01 m s^{-1} to about 0.3 m s^{-1} depending on their size; this settling speed is usually more than balanced by the air's upward motion that forms the cloud in the first place.

cloud height The elevation of the base of a cloud above the Earth's surface.

cloud-height sensor *See* ceilometer.

cloud motion vectors (CMV) Estimates of wind speed and direction based on the successive location of particular types of cloud on half-hourly (or some other time interval) thermal infrared images from GEOSTATIONARY SATELLITES. The clouds must be of the type that act as good tracers of the wind, such as trade-wind cumulus.

cloud physics The scientific study of the formation and evolution of CLOUDS and PRECIPITATION. It includes, for example, the microphysical processes of cloud formation, nucleation, aerosol and cloud interactions, cloud electrification and lightning, and ice physics.

cloud seeding A method of distributing substances in clouds in an attempt to modify the weather in some manner, for example to induce precipitation (*see* artificial rain), or suppress hail or wind speed. A number of techniques are used to induce precipitation. Substances, such as silver iodide, which has a similar structure to that of ice, are dispersed in supercooled cloud to promote the growth of ice particles. Dry ice is used to lower temperatures to below –40°C (–40°F) to induce glaciation in clouds. Hygroscopic particles of salts (e.g. ground salt particles), which take on water, are dispersed to promote coalescence. Ice-nucleating substances such as silver iodide are released into convective clouds to suppress hail formation but results are inconclusive.

cloud street Low cloud that forms, in an unstable layer of air which is usually capped by a temperature inversion, into distinct lines or streets of CUMULUS along which the air ascends. Between the streets, the air sinks to create intervening lines of clear sky. They stream along with the wind and are normally of limited depth.

cloud tag Small distinct fragments of cloud on the outer surface of a tornado's funnel cloud that permit an observer to deduce, by tracking the 'tag' visually, that the funnel is rotating.

cloud-to-air discharge A LIGHTNING discharge in which the lightning channel propagates away from a charged region to dissipate in apparently clear air.

cloud-to-cloud discharge An electrical LIGHTNING discharge between two different clouds (*inter-cloud discharge*), in which the discharge bridges the clear air between the clouds, or between two regions of the same, extensive cloud (*intra-cloud discharge*); no charge exchange occurs with the Earth's surface. Cloud-to-cloud discharges are more common than cloud-to-ground discharges.

cloud-to-ground discharge (ground discharge) LIGHTNING that leads to the exchange of charge between an electrified cloud and the Earth's surface. Charge is most commonly transferred from the lower, negative region of a thundercloud, although the properties of individual discharges vary considerably. The STEPPED

LEADER initially facilitates the breakdown of the air, after which peak currents of 20 kA typically flow for 10 microseconds, when the main conduction channel has been established. Positive cloud-to-ground (CG) strikes are less common than negative CG strikes, but can arise from strongly sheared winter storms in which large currents flow; they also occur more often in some areas, for example the North Sea in Europe.

cloud type *See* cloud classification.

CMV *See* cloud motion vectors.

coalescence A mechanism central to the COLLISION–COALESCENCE PROCESS of raindrop growth in which large drops, together with ice crystals in frozen cloud tops, fall and grow by coalescence with smaller droplets. Factors favoring coalescence are strong upward turbulence, a high moisture content, and water cloud of great vertical extent.

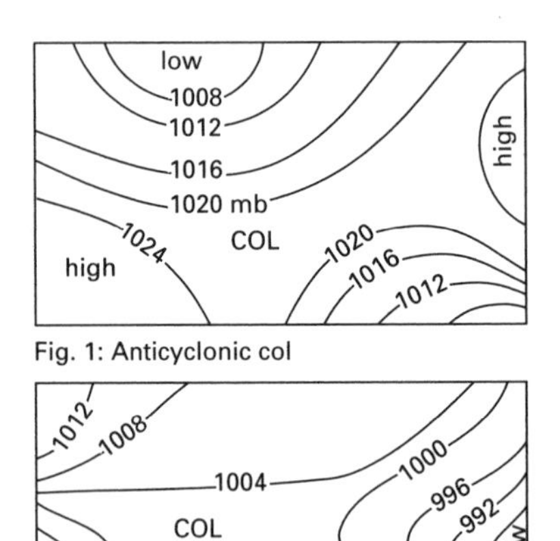

Fig. 1: Anticyclonic col

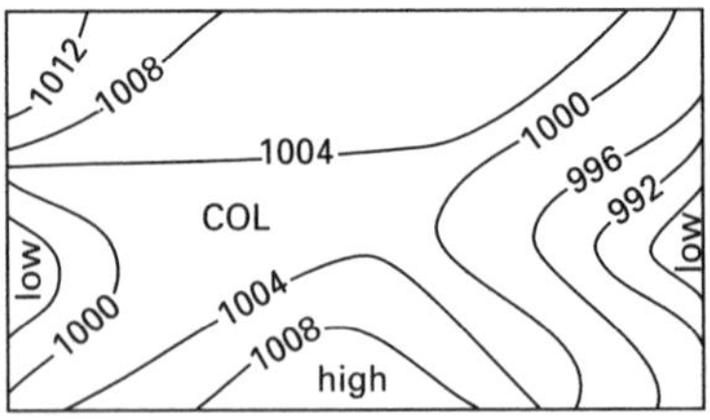

Fig. 2: Cyclonic col

Pressure patterns in a col

col A feature on an isobaric map separating two ridges of high pressure or anticyclones (highs) that are opposite one another from two troughs of low pressure or lows that are also opposite one another. It is characterized by a weak pressure gradient and is similar in pattern to a col or a saddle on a land surface topography map. In the summer, cols can be associated with thundery weather.

cold advection The horizontal movement of an air mass characterized by low temperatures and increasingly low relative humidity, to regions of greater warmth. In synoptic meteorology, cold advection is associated with the invading air following the passage of an active COLD FRONT (ANAFRONT). *See* advection.

cold air funnel A FUNNEL CLOUD or a relatively weak TORNADO formed in a convective storm that has developed in association with a pool of cold air aloft.

cold anticyclone An ANTICYCLONE (high) that forms during late fall or wintertime over middle- and high-latitude continents. Such highs are produced by intense RADIATIVE COOLING and are shallow systems that are no more than 2–3 km (1–2 miles) deep. *See* Siberian high.

cold boreal forest climate *See* Köppen climate classification.

cold conveyor belt *See* conveyor belt.

cold front The leading edge of cold air of an advancing mass of relatively colder air. It is an integral feature of a middle-latitude FRONTAL CYCLONE. As a cold front sweeps across a site, the air temperature and humidity fall. The generally overcast thick cloud of the WARM SECTOR gives way either to scattered convective cloud (over maritime areas) or cloud-free skies (particularly across extensive continents in wintertime) after a cold frontal passage. An *ana-cold front* is characterized by warm air that rides up and over it, producing a sharp line of heavy showers and/or thunderstorms slightly ahead of the surface front, and extensive precipitation behind it. In contrast, a *kata-cold front* is typified by dry air that overtakes the front in the middle troposphere so that much of the precip-

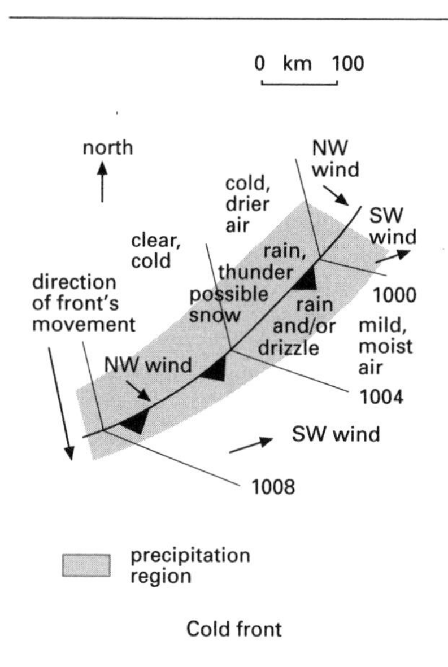

Cold front

itation is suppressed, and falls mainly in the warm sector.

cold-front clearance The improvement in visibility and clearing of cloud cover that generally follows the passage of a COLD FRONT. These conditions contrast strongly to the pre-cold frontal weather in the WARM SECTOR in which extensive low cloud and precipitation will reduce visibility.

cold low (cold pool) **1.** A low filled with air that is colder at all heights than the air surrounding it. This means that the wind circulating around the low will strengthen with height.
2. *See* cut-off low.

cold pole (pole of cold) **1.** The location in either hemisphere of the lowest mean annual temperatures of the Earth's surface. In the northern hemisphere this is in Siberia, and believed to be near Verkoyansk with a mean annual temperature of –16.3°C (2.7°F). In the southern hemisphere a mean annual temperature of –58°C (–72.4°F) has been recorded for the Pole of Inaccessibility.
2. The location in each hemisphere that holds the record for the lowest recorded temperature. The Vostok station in Antarctica holds the record for the coldest recorded temperature in the world with –89.2°C (–128.6°F) recorded on July 21, 1983. In the northern hemisphere, the lowest recorded temperature is –67.8°C (–90°F), which has been recorded at both Verkhoyansk (February 7, 1892) and Oymyakon (February 6, 1933) in Siberia.

cold pool *See* cold low.

cold sector The extensive region of colder air that follows a COLD FRONT in a developing FRONTAL CYCLONE.

cold-water desert *See* west coast desert.

cold wave An outbreak of unusually cold air bringing a rapid fall in temperature across a region. It is, for example, often associated with relatively prolonged and severe conditions across a large proportion of the US. The National Weather Service definition of a cold wave is based on the criteria of the rate of temperature fall and the minimum to which the temperature falls, depending on the location and time of year.

collar cloud A usually circular ring of cloud associated with a TORNADO; it surrounds the upper part of a WALL CLOUD. Collar cloud is sometimes used as a synonym for the wall cloud itself.

collection efficiency (E_c) A measure of the efficiency of the raindrop generation process in clouds. Higher values of E_c denote increased efficiency. It is defined as the product of the *collision efficiency* (E) and the *coalescence efficiency* (E′), i.e. $E_c = E.E'$. In the example of a larger droplet A and a smaller droplet B that is to be absorbed by A, the collision efficiency is a function of the radius (r) of droplet B (i.e. r_B); E declines as r_B decreases. The coalescence efficiency (E′) is defined as: number of droplets coalescing/number of collisions. If all colliding droplets coalesce, then $E' = 1$.

collision–coalescence process One of the currently accepted theories of the process of raindrop growth in warm clouds, based on the mechanisms of collision, coalescence, and rear capture. These mechanisms are thought to be responsible for rainfall generated from tropical convectional clouds where the BERGERON–FINDEISEN PROCESS cannot operate; they are also important in mid-latitude cloud systems.

Cloud droplets, formed by condensation of water vapor onto CLOUD CONDENSATION NUCLEI, vary in size from less than 1 μm to over 20 μm. The larger drops, with terminal velocities increasing in proportion to their diameters, begin to fall faster than the smaller droplets. During their fall the larger droplets collide with smaller droplets and often the smaller droplets are absorbed through coalescence. The larger cloud droplets thus grow into raindrops at the expense of the smaller ones. Once a raindrop reaches 5 mm it becomes unstable and breaks up into smaller drops. The collision–coalescence process is affected by the range of droplet sizes; the depth of the cloud, which determines the quantity of cloud droplets available; the strength of updrafts within the cloud; and the electrical charge of the droplets.

Colorado low A cold season low-pressure system that forms to the lee (east) of the Rocky Mountains, usually in SE Colorado, and commonly tracks northeastward, producing extensive snowfalls across the region toward the Great Lakes. A Colorado low is an example of a LEE DEPRESSION.

color display A means of portraying a remotely sensed image more clearly to the eye, by using a range of colors to stress particular features. Color coding different cloud-top temperatures for deep CONVECTIVE CLOUD can be useful to meteorologists, for example.

comfort index *See* temperature-humidity index.

comfort zone The range of temperature, humidity, and air movement within which the majority of people are most comfortable and work most effectively. *See also* temperature-humidity index.

comma cloud On the mesoscale, an extensive region of CUMULIFORM cloud organized into a distinctive comma shape. Such cloud features tend to be more common over wintertime high-latitude oceans, and are somewhat smaller features than middle-latitude FRONTAL CYCLONES. They are normally associated with heavy showers.

On the synoptic scale, comma cloud is also used more generally for the larger frontal cyclone cloud patterns that look very much like a comma when seen on satellite imagery.

comma echo The RADAR return or echo from heavy precipitation in a SQUALL LINE that takes on a comma-like pattern during the later stages of its life cycle, after it has been through the BOW ECHO phase.

concrete minimum temperature A measure of temperature recorded from a standard MINIMUM THERMOMETER that is exposed at the center of, and in contact with, a smooth slab of concrete of specified standard dimensions. In this case the bulb of the thermometer measures the temperature of the air in contact with the concrete.

condensation The process by which liquid water is formed from gaseous WATER VAPOR; it is the opposite process to EVAPORATION. The capacity of a parcel of air to hold water vapor decreases with temperature and hence, cooling of the air forces saturation (*see* saturated air) and then CONDENSATION. Condensation may occur under adiabatic conditions in which rising air expands and cools without any exchange of energy outside the air parcel (formation of clouds); by surface boundary layer mixing with cold air, particularly under inversion conditions (formation of fog); or by contact with surfaces of lower temperature (formation of dew).

condensation funnel The rapidly rotating cloud feature of a TORNADO made visible by the condensation of water droplets. These are formed in association with the extreme fall of pressure experienced as the moist ambient air is drawn into the tornadic circulation.

condensation level The level above the surface at which the dew-point is reached and CLOUD DROPLETS condense out of ascending moist air. The height of the condensation level depends on the air's surface temperature and dew-point temperature; a large difference between them will mean a high condensation level while a small difference will lead to a low condensation level or CLOUD BASE.

condensation nucleus *See* cloud condensation nucleus.

condensation trail (contrail) An initially thin trailing line of water droplets or ice crystals that condenses or freezes out of the water vapor emitted in the exhaust from jet engines. Whether a condensation trail occurs or not depends on a critical temperature at flight level. How persistent they are is largely a function of the dryness of the air at the aircraft's flight level; long-lasting condensation trails are expressions of high relative humidity where they occur and they tend to spread out on the wind. A condensation trail is also sometimes referred to as a vapor trail, which is inaccurate because vapor is invisible.

condensation-type hygrometer *See* dew-point hygrometer.

conditional instability The atmospheric state in which the magnitude of the ENVIRONMENTAL LAPSE RATE (ELR) is higher than the SATURATED ADIABATIC LAPSE RATE (SALR) but lower than the DRY ADIABATIC LAPSE RATE (DALR). A rising unsaturated air parcel will initially cool at the dry adiabatic lapse rate; as the parcel is cooler than its surroundings a condition of atmospheric STABILITY exists. With continued vertical uplift (e.g. over a mountain barrier), the air parcel will reach DEW-POINT and become saturated. Owing to the release of latent heat of condensation, the parcel will now cool at the slower saturated adiabatic lapse rate. Eventually, the air parcel will become warmer than its surroundings and hence UNSTABLE, typically resulting in the development of CUMULUS clouds and THUNDERSTORMS. The description 'conditional' is applied in this case because the instability is conditional on the saturation of the air mass.

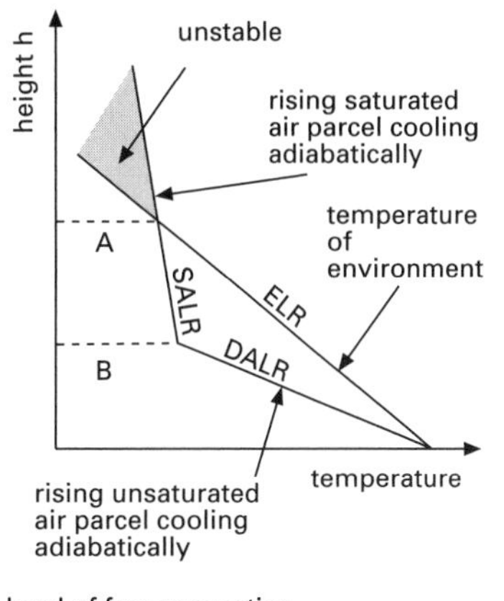

Conditional instability

conditional instability of the second kind (CISK) The process in which CONVERGENCE at low levels in a WIND FIELD produces CONVECTION and CUMULUS formation. The consequent release of latent heat enhances the degree of convergence and increases convection, so establishing a positive feedback loop that may lead ultimately to the formation of a large-scale disturbance, such as a HURRICANE.

conduction The transfer of heat by molecular diffusion from one body (such as a warm ocean current) with a relatively high surface temperature to another body (volume of air) with a lower temperature with which it is in direct contact. As air is a poor conductor of heat it is only close to the Earth's surface that conduction is significant as a mechanism of heat transfer within the atmosphere. *See also* convection; radiation.

confluence The process in which air streams from different directions flow smoothly together from either side of, and toward, an asymptotic line. It is the opposite of DIFLUENCE. *See also* convergence; divergence.

congestus (towering cumulus, convective cumulus) A species of cloud, the type of CUMULUS cloud characterized by marked sprouting of cumuliform bulges with sharp outlines in its upper reaches, which make the cloud appear like a cauliflower. It is generally bright white with a base of light to dark gray. It can grow very tall through CONVECTION, possibly extending up to the top of the troposphere, where it may spread out to form CUMULONIMBUS cloud, and can bring much precipitation in the form of showers.

Congo Air Boundary (Zaïre Air Boundary) A largely meridional boundary that may exist in the vicinity of 25°E across the Congo in central Africa. Although not universally recognized, it represents a zone of confluence between moist air from the equatorial section of the Atlantic Ocean and drier air that may move west or northwest from eastern Africa, especially if an anticyclone persists over or to the east of southern Africa. The movement eastward of the moist air is erratic and is sometimes known as the 'southwest monsoon' – despite the proximity of the Equator, which lies across the northern part of the Congo Basin. Due to sluggish air currents, the zone of active convergence may remain quasi-stationary, leading to prolonged and heavy precipitation.

coning In pollution studies, emissions from a chimney stack under atmospheric conditions of near neutral stability such that concentrations of a pollutant at a given distance downwind from the stack may be described by a normal or Gaussian distribution, being the same for both vertical and horizontal cross-sections perpendicular to the flow.

constancy of wind A statistical term referring to the variation over time in wind-flow parameters, such as velocity, direction, and vertical profile at a particular location. Constancy is a state in which these variables do not change with time. Given that the magnitudes of the forces acting on horizontal air motion are constantly changing, it is rare for such parameters to show collectively *absolute consistency* for long periods of time.

constant-pressure chart *See* isobaric chart.

constant-pressure surface *See* isobaric surface.

Contessa del vento A type of cloud formed by an eddy to the lee of an isolated peak. It has a rounded base and is topped by a protuberance that is inclined upwind; it occurs occasionally to the lee of Mount Etna, Sicily, with a west wind.

continental air mass A very extensive region of air with broadly uniform properties of temperature and humidity, that has its SOURCE REGION over a large landmass or continent and is associated with a CONTINENTAL ANTICYCLONE. Such AIR MASSES are very cold and dry in the wintertime and hot and dry in the summertime. Examples are the CONTINENTAL POLAR air masses of higher latitude North America and Eurasia in the winter, and the CONTINENTAL TROPICAL of the Saharan region of North Africa in the summer. *See also* Siberian high.

continental anticyclone (continental high) An ANTICYCLONE located over a very extensive landmass. Such highs occur across the higher latitudes of North America and Eurasia in the winter, for example, and over the subtropics of North Africa in summer.

continental climate A climatic type of the interiors of the extensive continental land masses. It is characterized by low precipitation and low humidity with a large diurnal temperature range.

continentality In climatology, a measure of the degree to which a region of the Earth's surface is influenced by a land mass. A *continentality index*, of which there are a number of variants, is a numerical method of expressing the degree of continentality of a region; it is usually based on a range of temperatures. Examples include *Berg's index of continentality*,

(K) in which the frequency of continental air masses (C) is related to that of all air masses (N): K = C/N (%). *Compare* oceanicity.

continental polar air mass (cP) A cold dry AIR MASS that forms in the high-pressure regions (anticyclones) of the northern hemisphere, especially those over central Asia (the SIBERIAN HIGH) and northern Canada, which are particularly pronounced in winter. Marked cooling occurs in the lower layers of air overlying the snow-covered land masses, producing stable air and little vertical mixing. In combination with the subsidence characteristic of high-pressure regions, a prominent temperature INVERSION typically extends up to about 850 hPa in continental polar air. During summer, continental warming over Siberia and Canada leads to considerable weakening of the source regions. No sources for continental polar air are found in the southern hemisphere as a result of the dominance of the oceans in the middle latitudes.

continental tropical air mass (cT) A warm dry AIR MASS that has its source region over an extensive arid or desert land mass in the middle or low latitudes. In the northern hemisphere summer, continental tropical air is associated with the areas of generally light winds and upper tropospheric subsidence over the major continents, such as central Asia. In the US, continental tropical air is most prevalent in summer and has its source in the desert southwest and over Northern Mexico. Such air masses are unstable at low levels, but usually stable aloft. In the southern hemisphere continental tropical air is more prevalent in the winter in Australia and southern Africa. In summer a small source of continental tropical air lies over Argentina.

contour 1. A line on a map or chart joining areas of equal elevation above, or equal depths below, sea level on topographic maps.
2. (height contour) In dynamic and synoptic meteorology, a line on an ISOBARIC CHART that connects all points at which the heights of an ISOBARIC SURFACE above sea level are equal. Conventionally these are drawn at intervals of 3, 6, or 12 decameters (1 decameter = 10 meters). The distance between contours represents the steepness of a gradient: tight contours, with little space between each contour, represent a steep gradient; contours with large gaps between them represent a slack gradient. Where the contours are close together the horizontal pressure gradient is larger.

contour current In the ocean, a bottom current that flows parallel to the slope of a continental margin, along the western side of the ocean basin.

contrail *See* condensation trail.

contrast In remote sensing, the difference in PIXEL values in an image. High contrast occurs when there is a large range of radiances over a scene, from highly reflective to dull surfaces in the visible, for example.

contrast stretching A technique used in remote sensing to enhance the CONTRAST across all or part of an image in order to highlight or exaggerate certain features that do not show up well in the original version. To enhance the contrast the range of digital number values of pixels in the image may be artificially increased.

convection The transfer and mixing of heat by mass movement through a fluid (e.g. air or water). It is one of the major mechanisms for the transfer of heat within the atmosphere, together with CONDUCTION and RADIATION. The convection process is of major importance in the troposphere, transferring sensible heat and latent heat from the Earth's surface into the BOUNDARY LAYER, and by promoting the vertical exchange of air-mass properties (e.g. heat, water vapor, and momentum) throughout the depth of the troposphere. Convection is generally accepted to be vertical circulation, whereas ADVECTION is usually horizontal.

convection current In the ocean, a circulation current that results from density differences associated with variations in temperature between water masses; it occurs when a warmer less dense mass of water rises and a mass of water that is cooler and denser sinks.

convective available potential energy (CAPE) The quantity of energy released through CONVECTION used by meteorologists to estimate the likely intensity of THUNDERSTORMS. The magnitude of CAPE can be determined from a THERMODYNAMIC DIAGRAM. In an environment conducive to storm development, the greater the area between the environment curve and the air parcel's path curve between the *level of free convection* (LFC) and the *limit of convection* (LOC), then the greater is the CAPE. Numerically:

$$\text{CAPE} = \int_{\text{LFC}}^{\text{LOC}} g(T_p - T_e)/T_e . dz,$$

where T_e is the temperature of the environment, T_p is the parcel temperature, and g is the acceleration of free fall (9.8 m s^{-2}). To estimate the thunderstorm's maximum updraft velocity, the CAPE is equated to the kinetic energy in order to obtain a measure of the storm's strength.

convective boundary layer The lower layer of the atmosphere in which dynamics and thermodynamics are affected by the presence of the underlying often aerodynamically rough land surface. This unstable boundary layer forms at the surface and grows in its depth throughout the day as the ground is heated by incoming solar radiation. CONVECTION currents transfer the heat received upward into the atmosphere.

convective cell The circulation within a fluid when warming causes some of the fluid to rise and cooling causes it to sink. For example, the distinct circulation of UPDRAFT and DOWNDRAFT that characterizes the basic cellular flow within a THUNDERSTORM. The cells can grow into much more complex patterns in an outbreak of severe CONVECTION. The HADLEY CELL is a very much larger scale example of convection within the atmosphere, although the term is not normally used in this sense.

convective cloud A cloud formed as a result of CONVECTION within the atmosphere. These clouds are typically CUMULUS, and the species are MEDIOCRIS and CONGESTUS as well as CUMULONIMBUS.

convective condensation level (CCL) The height in the atmosphere at which air that has risen from the surface by CONVECTION becomes saturated. The convective condensation level increases as the surface temperature (T°C) increases, assuming the DEW-POINT temperature (T_d) remains fairly constant. As a result, the base of CUMULUS clouds tends to occur at a higher level in the early afternoon than in the morning or early evening. The convective condensation level differs from the LIFTING CONDENSATION LEVEL (LCL) because it takes vertical mixing into account. As surface air is usually well mixed, the convective condensation level and lifting condensation level are frequently nearly identical. The lifting condensation level can be determined from a THERMODYNAMIC DIAGRAM; a numerical approximation being:

$$\text{LCL (meters)} = 120\ (T - T_d)$$

where T is the air temperature (°C), and T_d is the dew-point temperature (°C) at the surface.

convective inhibition (CIN, negative energy) The amount of energy necessary to overcome a capping INVERSION. The convective inhibition is typically much smaller than the CONVECTIVE AVAILABLE POTENTIAL ENERGY (CAPE); even a very low CIN value can effectively prevent convection. Forecasters must determine whether this cap can be broken (e.g. heating of the underlying air is one possible mechanism) and its energy released into a severe THUNDERSTORM.

convective instability (potential instability) The atmospheric condition in which a deep layer of otherwise STABLE air would become unstable if forced to rise, for example, over a mountain barrier or at a front, thereby reaching its DEW-POINT. Uplift has the effect of increasing the observed ENVIRONMENTAL LAPSE RATE (ELR) of the total thickness of the raised layer. If this new environmental lapse rate is greater than the SATURATED ADIABATIC LAPSE RATE (SALR), then this atmospheric layer becomes UNSTABLE and may overturn. Quantitatively, the EQUIVALENT POTENTIAL TEMPERATURE must decrease with height in the atmospheric layer for convective instability to occur.

convective outlook In the US, a forecast for thunderstorm activity and severity during the following 24-hour period across the country issued by the Storm Prediction Center (SPC) and National Weather Service (NWS).

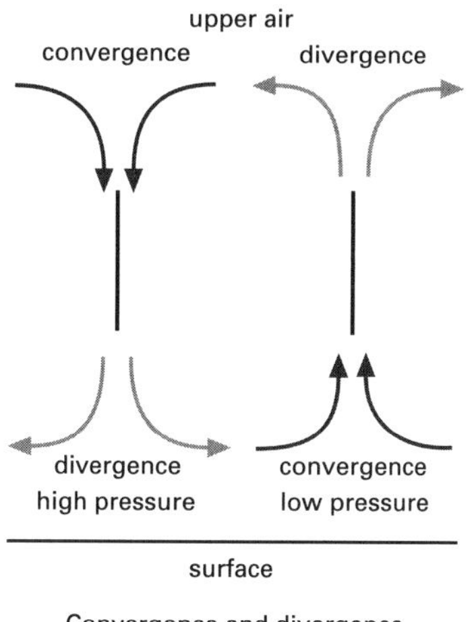

Convergence and divergence

convective precipitation (convectional rainfall, convectional precipitation) PRECIPITATION that results from convective motion (CONVECTION) in the atmosphere. When moist air that lies above a heated land surface is warmed by conduction it rises, expands, and then cools adiabatically until its temperature falls below the DEW-POINT. The water vapor within the air condenses to form clouds, typically of the CUMULONIMBUS type, from which heavy intense rain may fall. Convective rainfall is typically associated with equatorial climates, the cold front of an unstable polar air mass, and thundery rain that occurs in temperate regions on a hot summer afternoon.

convective temperature The temperature to which air near the ground must be heated in order to generate surface-based CONVECTION; one indicator of the likelihood of THUNDERSTORM development. Its calculation rests on a number of assumptions: for example, storms can sometimes develop long before or well after the convective temperature has been reached. Nonetheless, the convective temperature is a useful parameter for forecasting the onset of convection in certain flow conditions.

convergence **1.** In the atmosphere, the process in which there is an inflow of air into an area in association with a piling up of mass, which is compensated by vertical motion, and a decrease in speed along the lines of motion of the air streams. In the lower atmosphere (generally below 550 hPa) the compensatory vertical air motion is upward. With altitude the rate of inflow of air gradually drops away to zero until it changes to one of outflow, with DIVERGENCE (the opposite of convergence) in the higher levels above the lower-level convergence. When convergence is occurring above about 550 hPa the air motion is downward, with divergence taking place below at the surface. An example of a major area of convergence is the INTERTROPICAL CONVERGENCE ZONE, where the northeast and southeast trade winds slow down quite dramatically as they flow into the zone from either side. *See also* confluence.

2. In the ocean, the meeting of oceanic water masses or surface currents at a point or along a line. The denser water mass on one side sinks below the lighter less dense water on the other side. *See also* Antarctic Convergence.

convergence zone A region within which CONVERGENCE is marked. It can be elongated, as in the INTERTROPICAL CONVERGENCE ZONE, or roughly circular as in the air flow toward a mid-latitude CYCLONE.

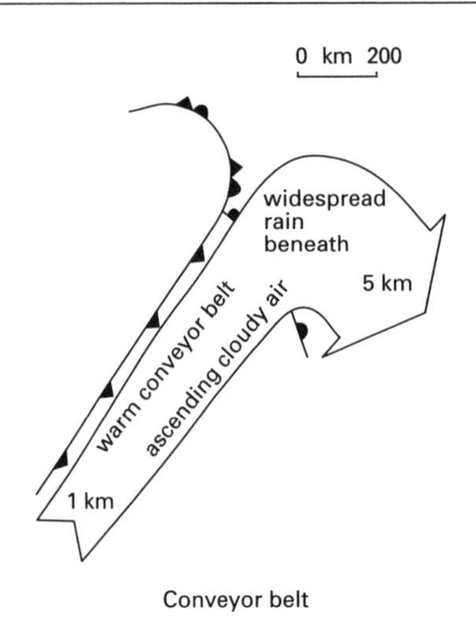

Conveyor belt

conveyor belt In meteorology, an air flow characteristic of most middle-latitude FRONTAL CYCLONES (frontal depressions). The *warm conveyor belt* (WCB) is a major air current that runs relative to the moving frontal cyclone at about 1 kilometer above the surface. It has its root in the WARM SECTOR and flows ahead of and parallel to the COLD FRONT, gradually ascending to turn anticyclonically over the WARM FRONT. It is the air stream that conveys the major proportion of a frontal cyclone's heat and moisture poleward. It can, in the case of an ana-cold front, turn rearward to flow up and over such a front. The *cold conveyor belt* (CCB) is a compensating cold air air flow in frontal cyclones.

Coordinated Universal Time (UTC) The timescale in which weather observations and forecasts are usually reported, based on the local time at the 0° meridian at Greenwich, England. The timescale known as International Atomic Time is computed by the Bureau International des Poids et Mesures (BIPM), near Paris, France, from a number of atomic clocks located internationally at various observatories; from this the Coordinated Universal Time is derived. *See also* Greenwich Mean Time; Universal Time; Zulu Time.

Coriolis effect (Coriolis force) The apparent deflection, due to the rotation of the Earth, experienced by an object or by air as it moves across the surface of the Earth. It occurs when this motion is referred to the coordinate system that is fixed to the Earth and rotates with it. The size or magnitude of the effect depends directly on the rotation rate of the planet (which is a constant), the latitude, and the speed of the air (i.e. the wind speed). The Coriolis effect is zero at the Equator and increases to a maximum at the poles. It works in such a way that an air parcel is deflected to the right in the northern hemisphere, and to the left in the southern hemisphere. It is responsible, for example, for the eastward drift of the Gulf Stream. The *Coriolis force* is strictly a fictitious force that has to be taken into account when air (or a projectile) moves across the surface of the rotating Earth. It is quantified as the acceleration per unit mass experienced by air and is equal to the product of twice the ANGULAR VELOCITY of the Earth, the sine of the air parcel's latitude, and the speed at which the air is traveling (i.e. the wind speed). The *geostrophic balance* is defined as the equality between this Coriolis force and the horizontal PRESSURE GRADIENT FORCE. The Coriolis force acts to the right of the observed wind in the northern hemisphere, and to the left in the southern hemisphere.

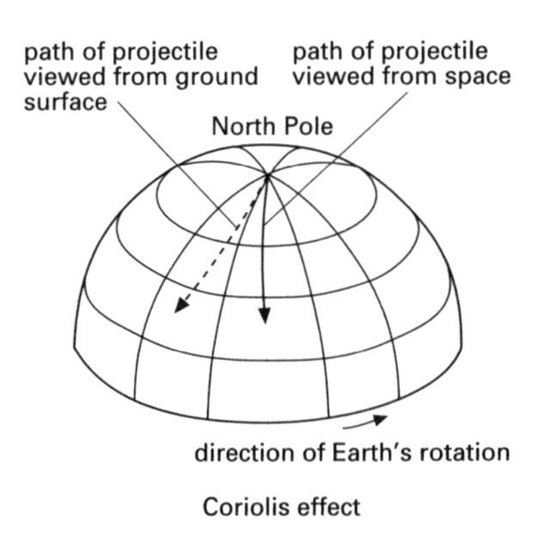

Coriolis effect

corona 1. An optical phenomenon in which one or more colored circles of light appear to surround the Sun or Moon. The

colors range from blue on the inner circle through red on the outside and are produced by the DIFFRACTION of light by water droplets in a thin layer of cloud, usually altostratus, in the atmosphere. In a full corona a series of rings surround the inner AUREOLE. *Compare* halo.
2. The outermost layer of the Sun's atmosphere, which extends into space for over two million km from the visible surface of the Sun (the photosphere). During a solar eclipse the corona is apparent as radiant light forming a faint halo around the dark rim of the Sun.
[Latin: crown]

cosmic radiation Very high energy radiation, some of which may be of solar origin, which may penetrate the Earth's atmosphere where most is absorbed or transformed into secondary radiation. Monitoring is providing valuable information about meteorological phenomena, especially lunar tidal motions in the upper atmosphere.

counterradiation (back radiation, incoming longwave radiation) The emission of longwave radiation by clouds and gases in the troposphere back to the Earth's surface. The amount depends upon the emissivity of dense clouds, which is high (90%) and of the GREENHOUSE GASES (variable, but about 60%).

coupled ocean–atmosphere model *See* general circulation model.

crachin In the region of the coast of southern China and Gulf of Tonkins, low stratus cloud and persistent drizzle that occurs during the early part of the year. It is associated with the replacement of dry continental air by moist maritime air.

crepuscular rays The rays of light interspersed with darker rays of shadow that appear to diverge from the Sun when it is partially obscured, by clouds, behind hills or mountains, or just below the horizon. The apparent divergence is a result of perspective as the rays are in reality parallel. Without the effect of SCATTERING, produced by particles, such as dust and water vapor within the atmosphere, the rays would not be visible. When the Sun is relatively high the crepuscular rays shining downward are known by a number of popular names, for example, *Jacob's ladder* and the *Sun drawing water*. When the Sun is low the rays of light appear to diverge upward and may be yellowish to reddish as a result of the differential scattering effect. Occasionally the rays of light and shadow appear to converge at the antisolar point (the point on the horizon opposite to that of the setting Sun) and are known as *anticrepuscular rays*. The illusion that the rays are converging is also produced by perspective.

critical temperature 1. (critical point) That temperature above which a gas cannot be liquefied by an increase in pressure alone.
2. A minimum temperature level below which a particular plant cannot grow or below which it cannot survive; for example temperatures below freezing point, 0°C (32°F), can kill or damage some plants.

Cromwell Current *See* Equatorial Undercurrent.

cross-valley wind system A thermally induced wind system that blows during the day across the axis of a valley toward the heated valley sideslope.

crown fire A fire that spreads across the tops of trees or bushes more or less independently of any surface fire. A crown fire can be classed as either running or dependent as a result of the degree of association it has with a surface fire. Wind is important in spreading the fire and once started it is difficult to bring under control.

cryosphere The parts of the Earth in which water is in its frozen state. This includes snow cover, ice sheets, ice caps, and glaciers, floating ice, seasonally frozen ground, and perenially frozen ground (permafrost). Satellite-borne instruments have provided a method of studying the extensive and often inaccessible cryosphere to observe and measure how it is responding to changes in climate. Changes in the cryosphere have relevance to sea levels.

cryotic Denoting temperatures below 0°C.

cumuliform Of or relating to heaped clouds, such as CUMULUS and CUMULONIMBUS, that form as a result of convection. *Compare* stratiform. *See also* cloud classification.

cumulonimbus A large low cloud of very great vertical extent, one of the cloud genera (*see* cloud classification), consisting of huge towers whose upper reaches appear fibrous, indicating streamers of ice crystals being stretched out in the wind. Additionally, it is often topped by smooth ice crystal cloud often in the form of an anvil or extensive horizontal plume where the glaciated top spreads out at the top of the troposphere. Cumulonimbus contains deep and very powerful updrafts than can sometimes exceed 50 knots (25 m s^{-1}). If convection and updrafts are especially strong the momentum of the rising air may push upward into the stable stratosphere and form an OVERSHOOTING TOP. Beneath the cloud base, which is often very dark, there are ragged clouds, frequently VIRGA, and rain shafts. The cumulonimbus base may be as low as 300 m (1000 ft) above the surface and top up to 13,000 m (43,000 ft) or more in low latitudes. The cloud is composed of water droplets in its lower levels rising to ice crystals in the upper levels in the glaciated top; strong updrafts and downdrafts associated with the convection lead to mixing of both water droplets and ice crystals in the middle levels. Cumulonimbus clouds are often associated with heavy showers of hail, rain, or snow, as well as lightning and thunder. Cumulonimbus may appear as a THUNDERSTORM and a group may merge into a MESOSCALE CONVECTIVE COMPLEX; the SUPERCELL is an especially turbulent long-lasting cumulonimbus that can produce TORNADOES.

cumulus A type of LOW CLOUD, one of the cloud genera (*see* cloud classification), with a flat base and 'bubbling' upper surface. They are usually scattered as individual small 'units' that, when illuminated, have brilliant tops and gray bases. The edges of the cloud will sometimes appear ragged. Cumulus clouds form through convection in ascending thermals. [Latin: 'heap']

cumulus congestus *See* congestus.

cup anemometer (rotating-cups anemometer) The most commonly used ANEMOMETER, which utilizes the kinetic energy of the wind to measure wind speed. It consists of three, four, or sometimes six hemispherical cups mounted symmetrically on a vertical axis of rotation. The rotating cups either operate a counter with each rotation or drive an electric generator operating an electric meter calibrated in terms of the wind speed. Wind speeds suitable for this type of anemometer range between 5 and 100 knots. A propeller may be used instead of the cups (*see* propeller anemometer).

current *See* ocean current.

curvature A common feature of atmospheric flow, particularly in synoptic-scale circulation, when the wind blows more-or-less parallel to the isobars that curve around cyclones and anticyclones (highs), etc. The degree of curvature is expressed as the *radius of curvature*, measured from the center of circulation to the point in question; it is used in the calculation of the CYCLOSTROPHIC WIND and the GRADIENT WIND.

cut-off high A synoptic-scale area of high pressure that remains in the high latitudes after a highly wavy long-wave pattern (with large north–south and south–north excursions) changes to one in which the flow is mainly zonal (most often west to east). It often occurs as a result of BLOCKING. Cut-off highs are part of the same highly meridional wavy circulation pattern of which CUT-OFF LOWS are an integral component. They lie poleward of the cut-off lows, are similarly slow moving, and are associated with anticyclonic weather.

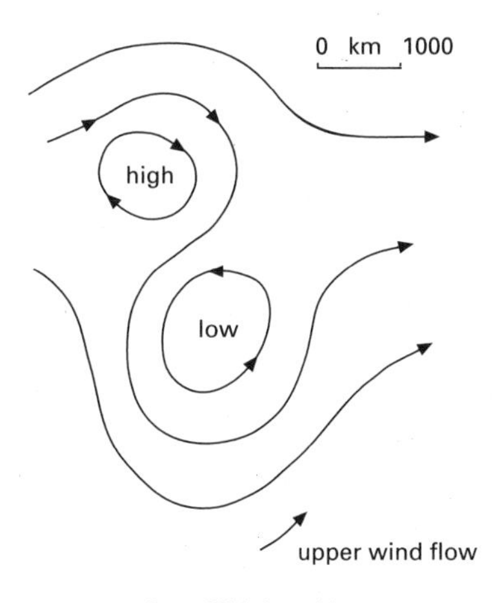

Cut-off high and low

cut-off low (closed low) A synoptic-scale feature, usually found in middle and subtropical latitudes, in which a normally slow-moving cyclonic circulation remains after an atmospheric long wave TROUGH has relaxed; i.e. the very wavy large-scale pattern of flow changes to one in which the flow is more west to east, parallel to lines of latitude. The result is that cut-off lows are left in lower latitudes, extending vertically from the surface into the upper TROPOSPHERE while the main west-east circulation is shifted toward the pole. The trapped cold air in the cut-off often sparks deep convection over the subtropical oceans.

cycle *see* hydrologic cycle.

cyclic storm A THUNDERSTORM that undergoes cycles or pulses of weakening and strengthening. Cyclic SUPERCELL thunderstorms may produce multiple tornadoes.

cyclogenesis The process of the generation of a cyclonic circulation or low-pressure system, or of the strengthening and intensification of cyclonic circulation around an existing low-pressure system. *Compare* cyclolysis.

cyclolysis The process associated with the disappearance of a cyclonic circulation or low-pressure system, or a weakening of the circulation around an existing system. *Compare* cyclogenesis.

cyclone 1. (low, low-pressure system, depression) A region of relatively low pressure, characterized by inward-spiraling flow that is counterclockwise in the northern hemisphere and clockwise in the southern hemispheres and leads to convergence. On a synoptic chart a cyclone appears as a set of closed isobars, circular or elliptical, around the central low pressure. Cyclones can extend up through the TROPOSPHERE, as for example in an EXTRATROPICAL CYCLONE, or be relatively shallow features, such as a HEAT LOW. They tend to be mobile features and are often associated with widespread cloud, precipitation, and wind. *The mid-latitude cyclone* (mid-latitude depression) is responsible for much of the weather that occurs in the middle-latitudes, especially during winter, and usually travels from west to east, crossing the US in 3–4 days. It is also important on the global scale in redistributing heat from subtropical to higher latitudes. It often begins as a small disturbance or wave (*see* frontal wave) on the POLAR FRONT; the wave develops into the warm sector of the cyclone, air begins to curve around in the cold sector, and cyclonic circulation develops and deepens. As the COLD FRONT travels faster than the WARM FRONT it begins to overtake it, forming an OCCLUDED FRONT. In the dissolving stage the warm sector is forced aloft and the surface fronts dissipate.

A cyclone in the tropics or subtropics may strengthen into a TROPICAL CYCLONE (a HURRICANE or TYPHOON).

2. The regional name in countries of the South Pacific and the Indian Ocean for a TROPICAL CYCLONE (a HURRICANE or TYPHOON).

cyclone wave *See* frontal wave.

cyclonic circulation (cyclonic rotation) The circulation or rotation of an air flow that is in the same sense as the Earth's rotation when viewed from above. In the northern hemisphere this is a counterclockwise inward spiral, while in the southern hemisphere it occurs in the clockwise sense. Such circulation is characteristic of CYCLONES and is associated with large-scale ascent and, if the air is moist enough, extensive cloud and precipitation.

cyclonic precipitation *See* frontal precipitation.

cyclostrophic wind A wind that blows around circular ISOBARS when they are very tightly curved. It is used to calculate the theoretical speed of the wind under conditions when the CENTRIPETAL ACCELERATION is much larger than the Coriolis acceleration (*see* Coriolis effect). TROPICAL CYCLONES are the kind of system for which this wind is the best approximation to their tightly curved circulation.

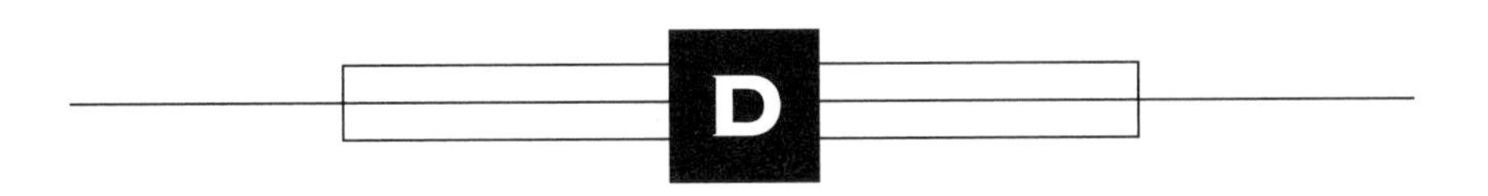

daily mean The mean value of the maximum and minimum temperatures recorded in one 24-hour period; i.e. the average temperature for a day.

DALR *See* dry adiabatic lapse rate.

Dalton's law A law that describes the behavior of a mixture of gases. The law states that the total pressure of a gas mixture is equal to the sum of the partial pressures that make up the mixture. Applied to the atmosphere: the total ATMOSPHERIC PRESSURE is the sum of all the partial pressures exerted by each individual gas in the atmosphere. Nitrogen, for example, with a concentration of 78%, contributes 780 hPa if the sea-level pressure is 1000 hPa. The law is named for the British chemist John Dalton (1766–1844).

damp haze A HAZE in which some of the hygroscopic dry particles become dampened when the relative humidity rises above about 75%, most often when the air cools in the evening and at night.

dart leader In a LIGHTNING discharge, the LEADER initiating a subsequent lightning channel after the main RETURN STROKE in a multistroke discharge. It travels more rapidly than the first STEPPED LEADER as it follows the remains of the original ionized channel. Dart leaders give lightning its flickering appearance, but not every lightning flash will produce one.

debris cloud In meteorology, the volume of particulate matter thrown up from the surface by a tornado to form a rotating cloud of debris; of soil and vegetation over agricultural land, for example.

deepening The decrease in value with time of the central and surrounding sea-level pressure within the circulation of a low-pressure system. The result is that mass is being exported from the total air column overlying the disturbance faster than it is being supplied. The process of deepening usually means that the winds are intensifying simultaneously. A large value for deepening would be sea-level pressure falling 10 hPa in three hours. *Compare* filling.

deforestation The process of removing vegetation from land areas, often for the purposes of using the land for agriculture or using the vegetation, often trees, for building materials, fuel, or commercial logging purposes. Deforestation is an important component in GLOBAL WARMING studies; forests are a major CARBON SINK, so their loss potentially adds to the likelihood of increased greenhouse gases; moreover, when burned or decomposed the vegetation releases the carbon, which combines with oxygen to form carbon dioxide. The removal of land-surface vegetation is also a potential hazard to local ecosystems, with the loss of animal habitats and removal of topsoil from land-water runoff, together with affects on the microclimate of the area. It is estimated that currently over ten million hectares of forest are lost ever year.

degree 1. A unit of measure on a temperature scale. *See* Celsius scale; Fahrenheit scale; Rankine scale.
2. A unit of measure of latitude and longitude: each degree is divided into 60 minutes.

degree-day A measure of the difference between the mean daily temperature and a specified reference temperature for a given day: degree-day = reference temperature – (maximum temperature + minimum temperature)/2. For a specified period, for example a month or a year, the number of degree-days is the sum of all degree-days for that period.

delta T The average rate of change of temperature with height (LAPSE RATE) within an atmospheric layer. Delta T is calculated by simply dividing the temperature difference between the top and bottom of the layer by the change in height. As delta Ts are indicators of possible THUNDERSTORM development, they are most frequently calculated for the layer between the 700 hPa and 500 hPa pressure levels, in order to provide a proxy indicator of the amount of instability evident in middle levels of the atmosphere (*see* absolute instability). Values greater than about 18 are indicative of sufficient instability for severe thunderstorm development.

dendrochronology (tree-ring analysis) The science of dating events or specimens by making use of the annual growth rings of trees. The annual width of the tree ring varies in relation to climate, such as the amount of precipitation and the temperature, among other factors. Dendrochronology has been used in the study of past climatic conditions and climatic trends, and also of past atmospheric pollution. Chronologies extending back to almost 10,000 years BP have been produced for bristlecone pine, individual specimens of which can live for over 4000 years. These chronologies have been used to recalibrate the CARBON-DATING process.

dendroclimatology The study of past CLIMATE using tree growth rings as an indicator of local climate conditions. The distance between tree growth rings is a function of the climate conditions experienced in that region. New wood grows between the old wood and the bark (cambium layer) so that when water is plentiful the tree devotes its energy to produce new large cells. During dryer periods, such as summer, the tree produces smaller cells giving the appearance of a ring.

density The mass of material per unit volume, usually expressed in kg/m^3. Like ATMOSPHERIC PRESSURE, the air density decreases exponentially with height, from around 1.2 kg/m^3 at the Earth's surface, to about 0.7 kg/m^3 at 5000 m (16,500 ft). The density of air also varies according to its temperature, with colder masses of air being denser than warmer ones. In water bodies, such as the oceans, the density of the fluid is affected by its temperature and salinity.

deposit gauge An instrument designed to collect and measure the products of atmospheric pollution.

depression 1. *See* cyclone.
2. *See* tropical depression.

depression tracks *See* storm tracks.

derecho In the US, a widespread convectively induced windstorm associated with a MESOSCALE CONVECTIVE SYSTEM (MCS) that can produce strong STRAIGHT-LINE WINDS across areas several hundred kilometers long and over 150 km (100 miles) wide. The word derecho was first used by Gustavus Hinrichs, an Iowa weather researcher, in 1888. In the warm season, the main derecho corridor is located in the southern Great Plains. During the cool season, derecho activity occurs in the southeast states and along the Atlantic seaboard. Temporally, derechos are primarily late evening or overnight events during the warm season, but are more evenly distributed throughout the day during the cool season. Marginal instability and strong synoptic-scale forcing favor derecho formation. [Spanish: 'straight']

descriptive meteorology The descriptive study of the atmosphere and weather phenomena, with little or no explanation of the processes or theory.

desert A region of the Earth's surface that has an arid climate (*see* dry climate), which can support little or no vegetation. Deserts occur in the high-pressure areas of the subtropics where air strongly subsides, warming adiabatically, and preventing condensation. Examples of such areas are the Sahara, Saudi Arabia, Iran, Iraq, California, and much of central Australia. Desert conditions are not restricted to the tropics and hot climates; cool deserts include the Gobi Desert in Mongolia. *See also* tropical and subtropical desert climate.

desertification The progressive destruction or degradation of existing vegetative cover to produce desert or desert-like conditions. This can occur through either lack of rainfall or through deforestation, overgrazing, or burning, or combinations of these factors. Desertification can lead to local CLIMATE CHANGE through an increase in albedo, a decrease in precipitation, and greater wind erosion of topsoil. It was responsible for the DUST-BOWL in the 1930s when drought and poor farming practices led to parts of the US Great Plains suffering severe soil erosion and dust storms. During the early 1970s, the Sahel region of Africa, bordering the southern Sahara, underwent desertification as a result of severe droughts, which, in combination with overstocking, caused famine, heavy losses of livestock, and crop failures. *Compare* desiccation.

desert wind A dust-laden airflow capable of entraining particles from the desert floor and transporting them aloft to other parts of the desert or sometimes vast distances beyond; examples include the HARMATTAN and SCIROCCO.

desiccation In climatology, the long-term reduction or disappearance of water from a region, such as occurs when precipitation levels decrease through climate change. *See also* desertification.

dew Moisture deposited in the form of water droplets on the ground surface or on objects close to ground level, such as blades of grass. Dew forms when nocturnal long-wave radiation from the Earth's surface cools the lowest layer of the atmosphere to below the DEW-POINT, allowing condensation to occur. Prerequisites for dew formation include calm conditions or very light wind speeds, less than 1 knot or 1 mph at a height of 2 m (6.5 ft), that allow the air to remain in contact with the ground for long enough to be cooled to dew-point; a high water-vapor content near the surface; and a suitable radiating surface. Calm and clear nocturnal weather conditions that follow relatively warm days during spring and autumn are the most favorable for dew formation. If the temperature drops below the freezing point, 0°C (32°F), following dew formation the water droplets solidify into ice forming *white dew*.

dew-point (T_d, dew-point temperature) The temperature to which air must be cooled, at constant pressure and moisture content, in order to saturate it. Continued cooling below the dew-point will cause condensation of water droplets if the atmospheric conditions are favorable. Dew-point provides a good indicator of the moisture content of the air. *See also* dew.

dew-point hygrometer (chilled-mirror hygrometer, condensation-type hygrometer) A HYGROMETER that measures the temperature at which dew forms (the dew-point). It works by artificially cooling a highly polished metal mirror surface at a constant pressure and constant vapor content until dew or frost forms on the surface. Cooling methods include the use of compressed carbon dioxide, dry ice, liquid air, or mechanical refrigeration.

dew-point temperature *See* dew-point.

diabatic (non-adiabatic) A thermodynamic process in which the temperature of an air parcel changes as a result of the exchange of heat (energy) with the surroundings. In a diabatic process:

$$\Delta Q \neq O.$$

Most thermodynamic processes near the Earth's surface are diabatic, owing to the continual mixing of air and turbulence. *Compare* adiabatic.

Diablo wind A wind that occurs below canyons in the East Bay hills (Diablo range) of northern California. Such a wind develops when high pressure is located inland over Nevada and lower pressure is situated along the central Californian coast. The Diablo wind is similar in character to the SANTA ANA wind experienced in southern California. In extreme cases, wind velocities can exceed 50 knots (58 mph).

diagnostic model In pollution studies, a method used to model wind fields that determines the transport and residence times of pollutants. It requires the solving of the continuity equation (conservation of mass) to determine wind fields, and a dense observational network to achieve reasonable results.

diamond dust A type of precipitation, occurring especially in high latitudes, composed of minute unbranched crystals of ice that appear to be suspended in the air or float slowly to the ground surface from high cloud or from an apparently cloudless sky and. Optical phenomena, such as SUN PILLARS and HALOES, may be seen in association with diamond dust; on cold nights when diamond dust occurs the reflections of street lights can produce pillars of light.

differential motion The commonplace observation that clouds at different levels normally move with different speeds and directions.

diffraction A bend in the direction of a wave when it passes through a small aperture or around an obstacle or barrier. Interference patterns following the obstruction result when the diffracted waves either reinforce each other or cancel each other out. In meteorology, the effects of diffraction are observed in optical phenomena, such as the BISHOP'S RING, CORONA, GLORY, FOGBOW, and IRIDESCENCE.

diffuse radiation (sky radiation) In meteorology, the component of incoming solar radiation that has been scattered by molecules of air, aerosols, particulate matter, clouds, etc., which falls on the surface in nearly equal amounts from nearly all parts of the sky during daylight.

diffuse reflection The reflection of electromagnetic radiation evenly in all directions.

diffusion The process by which properties of the atmosphere, such as HUMIDITY and HEAT, are mixed within the fluid of the air. Molecular diffusion is not significant compared to the role of turbulent diffusion that is carried out by EDDY motions in the air.

difluence (diffluence) The flow of a fluid, for example air, in which the STREAMLINES spread apart along the line of the motion. It is the opposite of CONFLUENCE. *See also* convergence; divergence.

digital image In remote sensing, an image produced by the conversion of the variations in radiation received from the remotely sensed scene into PIXELS (picture elements), which form the basic building blocks for the image. The radiance is expressed as a numerical value (the *digital number*) within a wide range of possible values. *Compare* analog image.

dimethylsulfide (DMS) An important sulfur-bearing gas produced from natural sources. Large amounts of the gas are generated by the metabolic processes experienced by certain oceanic algae (phytoplankton) and thence transferred into the TROPOSPHERE where the gas ultimately contributes significantly to the presence of sulfuric acid.

direct circulation A circulation system in which warm lighter air ascends and cold denser air descends, associated with the conversion of potential energy – related to the juxtaposition of air of different densities – into kinetic energy.

directional shear The rate of change of wind direction with height or along a horizontal plane. In the vertical plane, the surface wind is usually backed (*see* backing) in direction from the GRADIENT WIND in the free atmosphere. The degree of shear is a function of the roughness of the surface and the stability of the air. *See also* wind shear.

direct radiation (beam radiation) The short-wave radiation arriving directly on the surface of the Earth from the Sun; i.e. it has passed through the atmosphere without being absorbed, reflected, or scattered. Globally, it represents about 25% of the radiation received at the top of the atmosphere.

discontinuity In meteorology, a horizontal zone across which temperature, humidity, or wind velocity change very rapidly, for example a FRONT.

dishpan experiment A laboratory experiment to simulate convection and atmospheric movements in which a rotating dishpan (essentially a shallow cylinder) that has a small cylindrical center, is filled with water, between the central core cylinder and its circumference. The ring of water represents the atmosphere in a simple way, while the core cylinder is the pole and the outer rim the Equator. The experiment then looks at the role of different rotation rates of the cylinder and different Equator-to-pole temperature gradients on the way in which the fluid transports heat from the outer to the inner region.

dispersion diagram A diagram upon which the dispersion of points is plotted, for example rainfall against time. This enables the visualization of the behavior of rainfall for a station or region over time.

disturbance A general weather-producing system (with precipitation and strong winds for example) associated with a deterioration in conditions. Examples of such a feature on a mean-sea-level pressure chart would be a traveling FRONTAL CYCLONE or, on a much smaller scale, a THUNDERSTORM.

disturbance line A line of storms traveling west across West Africa usually associated with EASTERLY WAVES. They may be accompanied by high winds forming gust fronts – similar to microbursts from middle-latitude thunderstorms. Disturbance lines are most frequent during the AFRICAN MONSOON season. The storms are associated with uplift of very moist tropical air, possibly of Equatorial origin, and descent of drier Saharan air from above. Lines, which can be up to several hundred kilometers in length, may suddenly form and disappear within 2 hours or can sometimes be tracked for up to 24 hours.

diurnal variation (diurnal range) The range of a meteorological element, for example temperature, pressure, or relative humidity, between the maximum and minimum values observed during the course of a solar day.

divergence In general, the decrease over time in the flux of an atmospheric property in a unit volume. This might be, for example, the rate of gain of water vapor from a cubic meter box into which, with flow in one direction only, a lower concentration is entering through one vertical face than is leaving through the opposite one. The difference, or the divergence, must be related to the evaporation of water into the air within the box.

In the circulation of the atmosphere, it is the process in which there is an outflow or decrease in air mass from an area and consequent spreading of the field of flow (the opposite of CONVERGENCE); air streams increase in speed along their line of motion. Such flow in the lower TROPOSPHERE occurs in the spiraling outflow from ANTICYCLONES (highs), for example. This horizontal divergence is calculated by considering the horizontal rates of change of the components of the wind. Divergence is on average most marked in the lower and upper troposphere; it is at a minimum at around 600 hPa, the level of non-divergence. Divergence is compensated by vertical motion. *See also* difluence.

D-layer (D region) The lowest layer or region of the IONOSPHERE at a height of 50–90 km (30–55 miles) above the Earth's surface. The free electron density is at its lowest in this layer and the ionization disappears during the night.

DMS *See* dimethylsulfide.

Dobson spectrophotometer A ground-based instrument used for measuring ozone within the atmosphere. The instrument monitors the ozone content in the vertical column above it by comparing the ultraviolet light intensity at two wavelengths: one of which is strongly absorbed by ozone and the other that is weakly absorbed. The instrument was originally developed during the 1920s by the British meteorologist G M B Dobson (1889–1976), who was among the first scientists to study atmospheric ozone.

Dobson unit (DU) The unit of measure for atmospheric ozone. One unit is defined as equal to 0.01 mm of ozone at 1 atmosphere pressure and 0°C (32°F). It is named for the British meteorologist G M B Dobson (*see* Dobson spectrophotometer).

doctor A cooling refreshing wind that moderates the often unpleasant and sometimes unhealthy conditions of many tropical environments. It is most commonly applied to the HARMATTAN wind of West Africa; the regular summer SEA BREEZE along the coast of Western Australia is also often known as the doctor. The *Cape doctor* is a strong southeasterly wind that blows onto the South African coast.

doldrums The region of light and variable surface winds in the vicinity of the Intertropical Convergence Zone (ITCZ). In this region, in the age of sail, vessels would sometimes be becalmed.

dominant wind The wind that predominantly affects a particular location. The wind climate of a location may be controlled more by locally induced wind circulations as opposed to the PREVAILING WIND, which is generated as a result of the location of high- and low-pressure areas at the synoptic scale. For example, a town positioned in a deep valley is often affected by local valley winds that may blow in a totally different direction to the prevailing wind.

Doppler effect (Doppler shift) The effect in which the frequency of waves emitted by an object appears to change as the object moves relative to a stationary observer. This occurs, for example, with sound waves emitted by a whistle, say, on an approaching and then receding train; the apparent frequency of the waves increases as the train comes toward the observer, then decreases as it recedes. The same effect occurs in radiation backscattered to a RADAR by falling RAINDROPS or CLOUD DROPLETS. As a result of the Doppler effect raindrops moving away from the radar have a return frequency shifted down, while those moving toward the radar have a frequency shifted up. The effect is named for the Austrian physicist Christian Doppler (1803–53), who discovered it in 1842. *See also* Doppler radar.

Doppler radar A type of RADAR that takes advantage of the DOPPLER EFFECT by measuring the frequency of the signal scattered back to it from an object or source and detecting whether this is moving toward or away from the radar receiver. In meteorology it is used, for example, in the tracking of precipitation to enable deductions to be made regarding wind movements. Since raindrops are borne on the wind, the frequency shift observed at the radar will enable the mapping of the component of the wind flow toward or away from the radar antenna. Therefore, not only is precipitation mapped, but so too are the low- and middle-level zones of CONVERGENCE and DIVERGENCE from the deduced wind observations. Low-level convergence along a line may relate to a potential for THUNDERSTORM development, for example. In the US, the National Weather Service (NWS) operates a network of Doppler radar (NEXRAD), using Weather Surveillance Radar (WSR–88D), which provides a powerful system capable of measuring reflectivity echoes.

Doppler shift *See* Doppler effect.

downburst A strong localized DOWNDRAFT resulting in an outburst of very strong damaging winds at or near the ground level that can be experienced beneath THUNDERSTORMS or showers or sometimes in clear air. The wind surges out in the direction the storm is moving and can cause damage that is similar to that produced by a tornado; it may last for 5–30 minutes with gusts up to 60 m s^{-1} (130 mph). A large downburst extending over 4 km (2.5 miles) in the horizontal dimension is also sometimes known as a *macroburst*. A downburst less than 4 km across is called a MICROBURST.

downdraft An anomalously strong vertical current of descending air, usually within a storm, but sometimes in clear air. In the deep convective clouds of storms it is formed by the dual action of rain falling within the cloud, by it evaporating and therefore cooling the surrounding air and by the weight of the rain. When the downdraft meets the ground it fans out away from the parent cloud as a noticeably cool current of air with a gust front at its leading edge. Its temporal duration is usually short, with a spatial scale of 10–100 m (30–300 ft) in a storm. Downdrafts associated with clear air turbulence (CAT), however, are hazardous to aircraft and may lead to a plane falling thousands of feet in as little as one minute.

downslope flow A thermally induced downslope movement of cold air, which is often evident in mountainous terrain and usually occurs at night. Unlike such shallow downslope KATABATIC flows are the *downslope winds* (e.g. the warm FÖHN WIND), which are produced by the large-scale flow that extends far above the mountain tops.

down-valley wind A LOCAL WIND that blows down a valley in a mountainous environment. It is usually observed during the late nighttime and early morning period before sunrise when DOWNSLOPE FLOWS have ceased and a MOUNTAIN WIND fills the valley.

downwelling (sinking) In the oceans, the sinking of surface water, which is usually caused through density differences in sea water, as a result of cooling and/or increased salinity, or through convergence of water masses with different properties. The denser colder water tends to sink.

drainage wind *See* katabatic wind.

drizzle Light slowly falling PRECIPITATION in which the water droplets are tiny and resemble a fine spray. The diameters of drizzle droplets fall typically within the range of 0.2–0.5 mm (0.008–0.02 in). Given their small size, drizzle droplets are unable to fall to the ground if there is strong vertical air motion. A high relative humidity (with a DEW-POINT depression of less than 2°C) between the cloud and the Earth's surface is required to prevent the droplets evaporating before they reach the ground. Drizzle usually falls from low stratus clouds in which the base is lower than 300 m (1000 ft), and is often associated with the stratiform cloud of a WARM FRONT. *Heavy drizzle* mostly occurs from clouds formed through OROGRAPHIC uplift over upland areas. *See also* freezing drizzle.

dropsonde An instrument package that is released from an aircraft and borne by parachute to obtain soundings of the atmosphere below. The measurements recorded are temperature, atmospheric pressure, and humidity. Some versions carry GPS systems enabling additional wind measurements to be recorded (sometimes known as a *dropwindsonde*). Dropsondes are dropped into TROPICAL CYCLONES (HURRICANES), for example, to obtain information of the cyclone's intensity.

drought A period of dry weather of sufficient length for the lack of water to cause a hydrological imbalance in an area. Depending on the discipline, several definitions exist that attempt to quantify drought severity and magnitude. In the UK, an absolute drought was defined in 1919 as 15 consecutive days without measurable rain. In the US, drought is more commonly defined as an extended period of anomalous moisture deficit. This places more emphasis on the summer season with higher evapotranspiration rates from soil and vegetation in which drought is defined as a temporary negative deviation in environmental moisture status. Other definitions include, ABSOLUTE DROUGHT, PARTIAL DROUGHT, PHYSIOLOGICAL DROUGHT, and hydrological and agricultural droughts. The most versatile method of defining drought intensity is in terms of a deviation from the annual mean rainfall, water balance, or riverflow. Drought severity can then be measured by multiplying the drought intensity by the duration of the drought. *See also* Palmer Drought Severity Index.

drought cycle A pattern of periodic droughts caused by prolonged periods of insufficient rainfall.

dry adiabat A line drawn on an THERMODYNAMIC DIAGRAM that traces the path of a dry air parcel moving through the atmosphere adiabatically. A dry adiabat is also a line of constant POTENTIAL TEMPERATURE (θ), this being the temperature of an air parcel after it has been brought dry adiabatically to the 1000 hPa pressure level. Theta can be read directly off a thermodynamic diagram or calculated by means of a formula.

dry adiabatic lapse rate (DALR) The rate at which temperature changes in an unsaturated parcel of air that rises or sinks adiabatically (i.e. there is no exchange of heat with the surrounding air). As the air parcel rises, if condensation does not occur, the energy used by expansion causes the temperature of the parcel to drop at the dry adiabatic lapse rate, which is approximately 9.8°C km^{-1}. The dry adiabatic lapse rate is sometimes called a process lapse rate, as it refers to the thermal behavior of an air parcel as opposed to that of the surrounding environmental air.

See adiabatic lapse rate; lapse rate; saturated adiabatic lapse rate.

dry air **1.** In climatology, air of low RELATIVE HUMIDITY, below 60%.
2. In meteorology, air that contains no WATER VAPOR.

dry-bulb temperature The temperature recorded by a DRY-BULB THERMOMETER exposed to air and shaded from direct solar radiation.

dry-bulb thermometer A THERMOMETER with a dry bulb used to determine the ordinary temperature of the air. It is used in conjunction with a WET-BULB THERMOMETER in a PSYCHROMETER to measure relative humidity.

dry climate One of the major climate groups in the KÖPPEN CLIMATE CLASSIFICATION, designated as B. In this climate group, potential evaporation and transpiration exceeds precipitation on average throughout the year, and there is a constant deficiency in water. The group is subdivided into the *arid (desert) climates*, designated by BW, and the *semiarid (steppe) climate*, designated by BS. The dry climate group occupies over 26% of the total land area of the globe. *See also* desert.

dry deposition The settling out under gravity of pollutants, largely in the BOUNDARY LAYER. The pollutants are deposited on soil, vegetation, or water bodies at the Earth's surface and may account for the larger part of deposition in dry climates. In humid climates, the amounts removed by dry and wet deposition are approximately equal.

dry ice Solid carbon dioxide used in cloud-seeding operations. Dry ice particles are most often dropped from an aircraft into supercooled clouds, where they vaporize at a temperature of –78.5°C (–109°F). The cooling associated with this change of phase produces very large numbers of ice crystals that may then grow large enough to fall through the cloud – to finish ultimately at the surface as precipitation.

dry intrusion A feature that is typically a few hundred kilometers long and perhaps a hundred kilometers wide that marks the descent of very dry air from the lower STRATOSPHERE or upper TROPOSPHERE. It is important because it is now known that it plays a critical role in the rapid deepening of FRONTAL CYCLONES, mainly over the middle-latitude oceans. The downward-plunging dry air both spins up the pre-existing VORTICITY of the system and enhances the horizontal density gradients – both of which mean stronger winds. The dry intrusion becomes a component part of the cloud and water vapor pattern presented by a rapid deepener (*see* explosive cyclogenesis). The feature was first recognized during the latter part of the 20th century, on water vapor images from weather satellites.

dry lightning *See* heat lightning.

dry line (dew-point front) A boundary that separates a moist air mass from a dry air mass. It refers especially to the generally north-south orientated boundary that often extends across the central and southern High Plains of the US in spring and summer, separating warm and moist air moving north out of the Gulf of Mexico from the usually warmer and drier air from the west. The dry line often moves to the east in the afternoon and moves back westward at night. It is usually a narrow zone across which the DEW-POINT temperature can fall by as much as 9°C per kilometer, and in which the moister air rides up and over the drier air. These conditions, coupled with low-level CONVERGENCE, can lead to the development of severe storms that may become tornadic along the dry line or in the moist air to the east of it.

dry-line storm A storm that develops along a DRY LINE, which indicates a boundary between warm dry air and warm moist air, marking a region of instability. Dry-line storm is sometimes used synonymously with a LOW-PRECIPITATION STORM that characteristically develops along dry lines.

dry season A period exceeding a month that occurs at about the same time every year when no or very little precipitation is

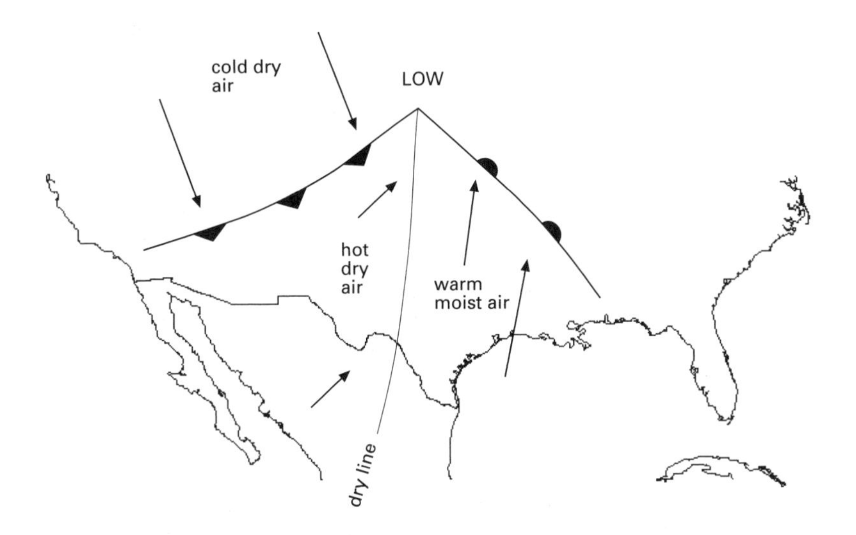

Dry line

recorded. It is usually applied to tropical climates that experience a wet-and-dry annual rainfall regime (*see* tropical wet-dry climate), classified as Aw in the KÖPPEN CLIMATE CLASSIFICATION scheme. In the tropics, winter is the dry season and is associated with the withdrawal of the zenithal Sun into the opposite hemisphere. Simultaneously, the regional subtropical high-pressure cell develops under the influence of a dominant Hadley Cell meridional circulation. The length of drought, in general, increases toward the respective tropic, until it stretches through the entire year and ceases to be a seasonal feature.

dry slot An elongated filament-like intrusion of very dry stratospheric air that plunges down into the TROPOSPHERE following COLD FRONTS or even overrunning such features. It can also refer to the region of tropospheric subsidence, involving a drying out of the air, that very often occurs after the passage of a cold front.

Stratospheric dry slots are known to occasionally be associated with EXPLOSIVE CYCLOGENESIS in middle latitudes.

dry snow A type of powdery SNOW with a density less than 0.1 kg m^{-3}. It contains very small ice crystals that do not bond together and is common in continental interiors away from maritime influences.

DU *See* Dobson unit.

duplicatus A cloud variety consisting of superposed sheets or patches of cloud at slightly different levels; it can occur with some low, middle, or high clouds including STRATOCUMULUS, ALTOCUMULUS, ALTOSTRATUS, CIRROSTRATUS, and CIRRUS. *See also* cloud classification.

dust Microscopic dry particulate matter held in suspension within the atmosphere. The dust may originate, for example, from volcanic eruptions, dust storms raising quantities of dust into the atmosphere from the surface, and smoke particles. The smallest particles can act as AITKEN NUCLEI. Dust can produce SCATTERING and haze.

dust-bowl A region affected by drought and subsequent dust storms. The expression was first used, in 1935, to describe the region in the Great Plains of the south-central US in which a combination of drought and poor farming practices led to crop failures, loss of topsoil, and severe dust storms. *See also* desertification.

dust devil (dust whirl, sand pillar) A localized whirl of dust in a desert area, in which the particles are swept around a central vortex, reaching heights of up to 900 m (3000 ft) above the Earth's surface. The dust devil is caused through intense local heating of the ground surface by the Sun, which results in a strong convection current being formed. Dust devils typically travel across the desert floor at speeds of 4–13 knots (5–15 mph), although velocities in excess of 30 knots (35 mph) have been observed. In general, dust devils do not pose a serious hazard, even to aviation, because they are only a few meters in diameter and gradually die away. Terminology varies with location: *dancing devil* (southwestern US); *sand auger* (Death Valley, California); *devil* (India); *desert devil* (South Africa).

dust plume An elongated plume of polluted air that extends downwind of a major city. Due to the warmth of the city relative to its surroundings, warm air rises convectively over the city center, creating a *dust dome* over the urban area in which aerosol concentrations may be more than one thousand times higher than in a rural environment. If the regional wind speed is greater than 8 knots (9 mph), this dust dome extends downwind in the form of a dust plume. As a result, the city's pollutants are spread over the surrounding countryside: for example, Chicago's dust plume is sometimes visible 240 km (150 miles) downwind of the city.

dust storm An event in which solid particles of dust and debris are suspended by strong winds, with an associated dramatic reduction in visibility. The US National Weather Service defines a dust storm as having wind speeds equal to or greater than 26 knots (30 mph) and visibility of 0.8 km (0.5 miles) or less. Dust storms are usually restricted to arid middle-latitude regions in which there are sudden onsets of strong winds; poor agricultural practices can exacerbate their occurrence. During the 1930s frequent dust storms were a feature of the Dust Bowl of the central southern US, which resulted from the removal of topsoil by agriculture and lack of rainfall leaving barren dusty areas. The Caucasus (southern Russia) and central Australia are particularly prone to dust storms. *See also* sandstorm.

dust veil index (DVI, Lamb's dust veil index) A numerical index, devised in 1963 by Hubert H. Lamb, a British climatologist, that ranks volcanic dust veils from volcanic explosions in terms of the mass of material initially injected together with the duration and maximum spread of the veil. Single volcanic explosions such as Krakatau and Mt Agung, rank as 1000 (the reference dust veil index). The index is an indication of the effects of volcanic dust on climate.

dust whirl *See* dust devil.

DVI *See* dust veil index.

dynamic climatology The study of CLIMATE and its relation to dynamical explanations associated with climate patterns or deviations from long-term climate trends. This branch of climatology has developed since the 1960s and uses models of atmospheric processes to simulate climate and climate change.

dynamic meteorology (dynamical meteorology) One of the main branches of METEOROLOGY in which the laws of physics and mathematics are applied to studies of processes and motion within the atmosphere. Computer models are used as tools in the study of the complex processes, for example, in modeling the general global circulation. *See also* synoptic meteorology.

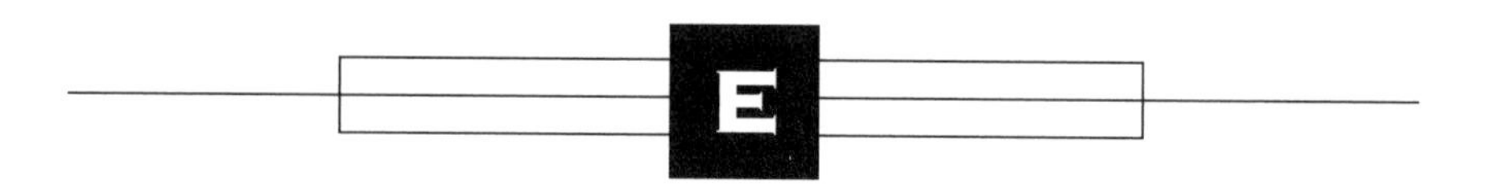

E

Earth observation satellite A SATELLITE focused on mapping the detail of the Earth's surface, including natural resources, rather than clouds and water vapor, for instance. Earth observation satellites tend to be in polar orbits and have very much better RESOLUTION (down to a few tens of meters for the smallest PIXEL dimension) than the meteorological satellites. *See also* European Remote Sensing Satellite.

Earth rotation *See* rotation of the Earth.

earth science Any one of the sciences concerned with the study of the Earth, including CLIMATOLOGY, METEOROLOGY, geology, geomorphology, geophysics, geochemistry, hydrology, and oceanography.

Earth Summit *See* United Nations Framework Convention on Climate Change.

East Australian Current A warm current in the Pacific Ocean that flows southeast along the east coast of Australia. It is an example of an eastern BOUNDARY CURRENT.

easterlies The belts of generally persistent winds that blow from the east toward the west; this includes the easterly TRADE WINDS and the POLAR EASTERLIES.

easterly wave (tropical wave) A synoptic-scale wave found on either side of the Equator between about 5° and 15° of latitude. The waves move westward within a deep easterly flow (flowing from the east) at about 3–4 km (2–2.5 miles) above the surface, and may show an inverted 'V' structure when viewed either by satellite or on isobaric charts. They travel west at an average speed of 5–10 m s^{-1} and take 35–40 days to circumnavigate the Earth. In Africa, easterly waves are very important weather-producing systems, bringing fairly widespread rain over West Africa, where they are associated with DISTURBANCE LINES, in the northern summer. They form along the very strong temperature gradient between the hot Saharan air to the north and the cooler Gulf of Guinea air to the south, and occur typically every four-to-five days. Usually they travel from Cameroon right across the region and out across the Atlantic Ocean. Some develop ultimately into HURRICANES that track across the Atlantic Ocean and may hit the Caribbean, Central America, and sometimes the US.

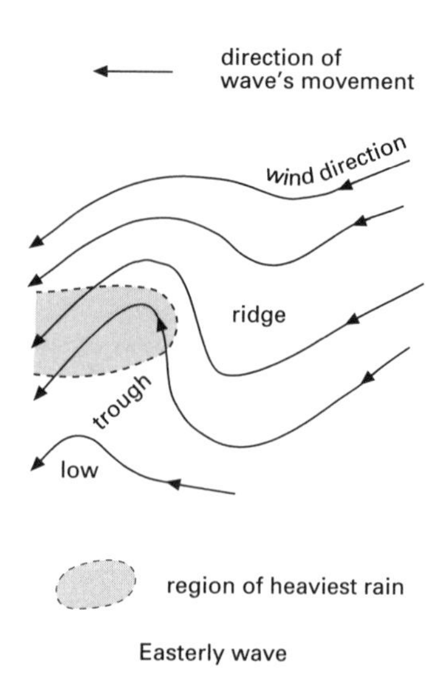

Easterly wave

Other easterly waves of note are those that run right across the tropical North Pacific, sometimes evolving into TYPHOONS.

eastern boundary current *See* boundary current.

EBM *See* energy balance model.

eclipse The partial (*partial eclipse*) or total (*total eclipse*) obscuration of one heavenly body by another. In a *solar eclipse*, the Moon obscures the light from the Sun, producing a shadow on the Earth. In a *lunar eclipse*, the Earth is aligned between the Sun and the Moon, and the shadow of the Earth falls on the Moon.

ECMWF *See* European Centre for Medium-Range Weather Forecasts.

ecoclimate The climate in relation to the flora and fauna of an area.

ecological climatology (ecoclimatology) The branch of BIOCLIMATOLOGY that studies the relationship of living organisms to climate, such as the geographical distribution of plants and animals in relation to climate.

eddy In the atmosphere, a distinct mass within a turbulent fluid that retains its identity and behaves differently for a short period within the general larger volume flow. An eddy thus ranges in size from microscale TURBULENCE (1 cm for example) to many hundreds of kilometers in the form of FRONTAL CYCLONES and ANTICYCLONES. The smallest scale eddies are critical in the process of, for example, heat and water vapor transfer from the Earth's surface into the air, while frontal cyclones transport heat toward the poles.

eddy diffusion (turbulent diffusion) The mixing of a property of the atmosphere by EDDY motions.

eddy viscosity A means by which turbulent motion within the atmosphere overcomes the resistance to deformation of the fluid; i.e. how viscous the fluid is. This means that, for example, the effective mixing of momentum by the air's turbulent flow depends directly on the decrease of the wind speed with height and a measure of how easily mass is transported upward by the fluid motion.

edge enhancement **1.** In remote sensing, an image-processing technique to increase contrast between adjacent tones and emphasize edge features or lines on an image. **2.** An automatic technique used to find edge features on an image. It is based on comparing an area of PIXELS to find gradients or sudden changes in value from one to its neighbor, sometimes along a line of contrasting pairs of pixels. An example might be the sharp change across a coast. This change would be distinct from the more even values over land and over sea, either side of the coastal edge.

effective precipitation (EP) That part of the total PRECIPITATION (P) input remaining after EVAPORATION (E) losses and thus available for the growth of vegetation. It is expressed by the formula:

$$EP = P - E.$$

effective temperature **1.** The temperature of a planet, such as the Earth, that it would have if it behaved like a black body and absorbed all incoming radiation and re-radiated this back to space.
2. In the US, a thermal comfort index that attempts to measure the comfort or discomfort of temperature. The effective temperature relates to the sensation of warmth or cold experienced by the human body as being dependent on the temperature, humidity, movement of air, and amount of clothing worn. *See also* temperature humidity index.

Ekman depth *See* Ekman effect.

Ekman effect **1.** In the oceans, the change in direction of current and the lower velocity that occurs in water layers with increased depth. Wind stress sets in motion the water movement in the surface layer of the ocean. The surface water is directed theoretically at an angle of 45° (in reality usu-

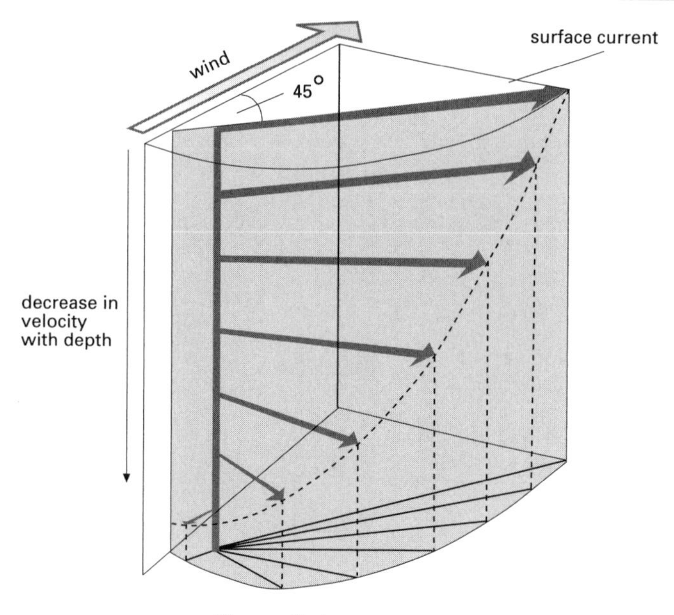

Ekman effect

ally 20–40°) to the wind, to the right in the northern hemisphere, as a result of the CORIOLIS EFFECT. With depth the flow is increasingly deflected to the right and velocity decreases away from the effects of the wind. In the southern hemisphere the converse occurs where the flow is deflected to the left while the velocity decreases with depth. The Ekman effect is negligible at the Equator. The *Ekman depth* (Ekman layer) is the depth of water within which the current change occurs; it is around 100 m (330 ft) depending on latitude and the wind speed. The net transport direction within the Ekman layer is at an angle of 90° to the wind; this is the *Ekman transport*. The effect was developed in 1902 as a mathematical model – the EKMAN SPIRAL.

2. In the atmosphere, the way in which the winds within the friction layer in the lower TROPOSPHERE gradually change speed and direction moving from the surface up to the free atmosphere, where friction does not play any role in the air's motion. The successive end points of the wind vectors form a spiral shape up through the friction layer (*see* Ekman spiral). The wind speed gradually increases with height as the influence of friction weakens. In the northern hemisphere, the wind gradually turns to the right of its surface direction (in the southern hemisphere the effect is to the left), spiraling in that sense up to where it is in balance as the GEOSTROPHIC WIND. The angle that the wind makes across the ISOBARS near the surface varies with its roughness, so that it is around 25° to 30° for continental surfaces and 10° to 20° over the sea.

Ekman layer In the atmosphere, the layer of transition between the surface boundary layer, where shearing stress is constant, and the free atmosphere. The direction of the wind alters as it approaches the Earth's surface due to the increasing influence of friction. At the top of this layer is the GEOSTROPHIC WIND, which is located approximately 500–1000 m (1600–3300 ft) above the Earth's surface in the middle latitudes of the northern hemisphere. Below this level develops the EKMAN SPIRAL, which describes the backing (changes in the counterclockwise direction) of the wind from the direction of the geostrophic wind. Friction acts to decrease wind speed and thus reduces the effect of the Coriolis force (*see* Coriolis effect); this causes the wind to blow across isobars in the direction of the pressure gradient (i.e. toward low pressure). The angle between the geostrophic wind direction and the surface wind increases with an increase in frictional drag; for example, over land the angle of difference is much greater than over sea.

Ekman spiral The theoretical mathematical model that was originally formulated in 1902 by the Swedish physicist, Vagn Walfrid Ekman (1874–1954), to explain the currents resulting from wind blowing over the ocean (*see* Ekman effect), which was first observed in the motion of ice floes by the Norwegian explorer, Fridtjof Nansen (1861–1930). The model has been subsequently applied to the variations in wind direction and speed between the surface and level of geostrophic flow in the atmosphere. The spiral refers to the shape defined by the successive end points of the vectors of the oceanic current flow with increased depth, or of the wind vectors with increased altitude.

Ekman transport *See* Ekman effect.

E-layer (E region, Heaviside layer, Heaviside–Kennelly Layer) The layer or region of the IONOSPHERE, at heights of 90–140 km (55–90 miles) above the Earth's surface, that marks the lower region of a distinct maxima of electron concentration and that is capable of reflecting radio waves back to Earth. The upper level is marked by the F-LAYER.

electromagnetic radiation (EMR) Energy propagated through space or a medium in the form of waves associated with electric and magnetic fields oscillating at the same frequency. The transverse wave motion acts at right angles to the direction of propagation of the energy, the speed of which is about 3×10^8 meters per second (known as the speed of light). The *electromagnetic spectrum* is the range of electromagnetic radiation having wavelengths measured in kilometers down to micrometers. The three sections that are of importance in radiation climatology and the Earth's atmosphere are the two short-wave bands of ULTRAVIOLET (UVA, UVB, and UVC) and the visible band for SOLAR RADIATION, and the infrared (IR) or LONGWAVE RADIATION predominantly emitted from the Earth's surface.

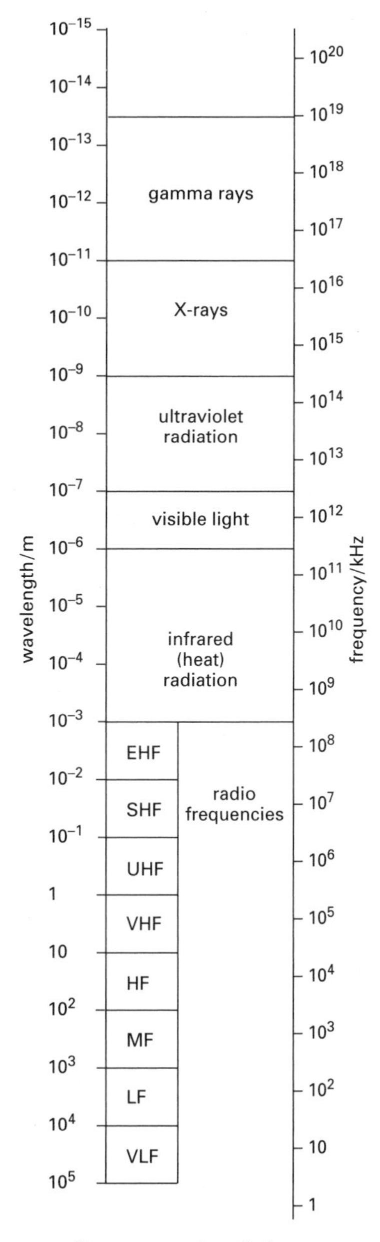

Electromagnetic radiation

electromagnetic spectrum *See* electromagnetic radiation.

element In climatology, one of the main atmospheric conditions, i.e. precipitation, humidity, temperature, atmospheric pressure, and wind.

elevated convection CONVECTION that occurs in an elevated layer that is not in contact with the Earth's surface. Elevated convection most frequently occurs when the air near the ground is relatively cool and stable, but that aloft is UNSTABLE. In such situations, stability indices based on surface data, such as the LIFTED INDEX, usually underestimate the degree of instability. Elevated convection is less likely to lead to severe weather conditions than surface-based convection.

El Niño The unusually warm surface water conditions that occur in the central and eastern Pacific at periodic intervals.

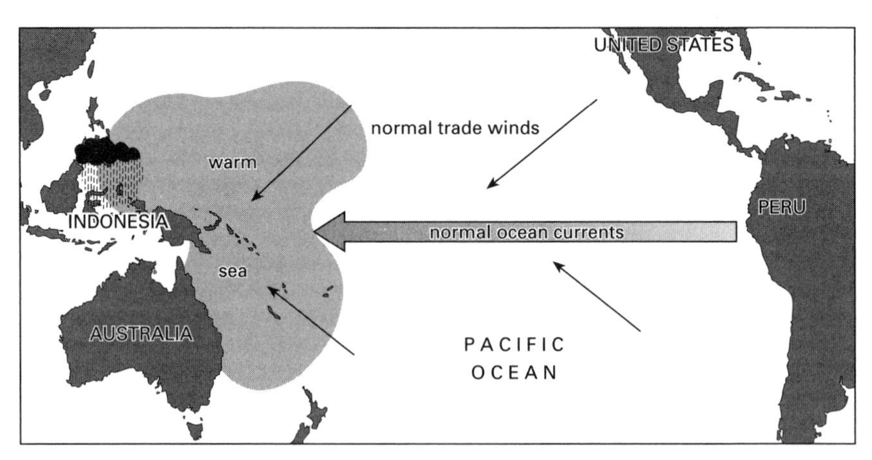

Under normal conditions the trade winds push water from east to west and warm water accumulates around Indonesia

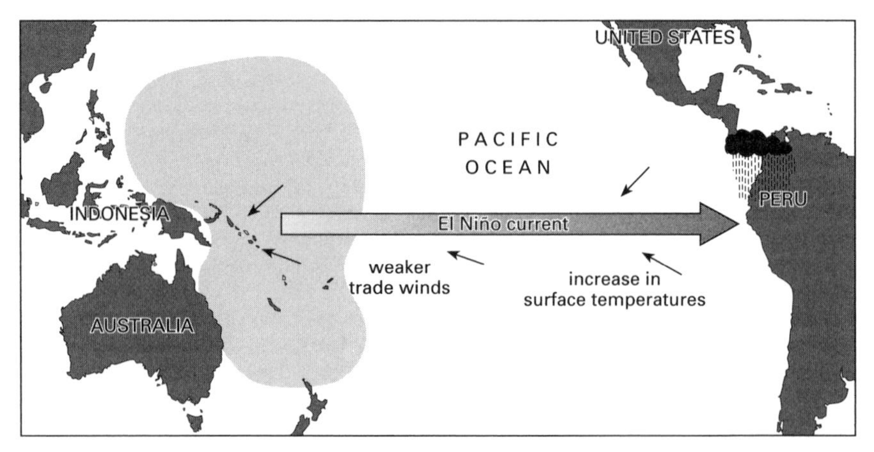

In El Niño years, the trade winds weaken and the warm water flows back toward the coast of Latin America

El Niño

Originally, the name was applied to the warm ocean current that flows south along the coasts of Ecuador and Peru in South America, occurring seasonally around Christmas or near the beginning of the year. Generally, this current persists until the end of March but on a timescale of 2 to 7 years it becomes unusually warm and persists for up to 18 months. It is to this major event that the name now applies. During an El Niño event the normally cold water to the west of South America and in the central Pacific becomes much warmer, while the waters in the western Pacific are cooler.

El Niño now often applies more broadly to the whole of the disruption to the ocean–atmosphere system that takes place, i.e. the EL NIÑO SOUTHERN OSCILLATION (ENSO) phenomenon, with the repercussions it has for climate and weather around the world. The Southern Oscillation represents the atmospheric reaction to the changes in sea surface temperature and is manifest by a seesawing of pressure over the tropical Indo-Pacific region. *See also* La Niña. [Spanish: 'Christ child' or 'little boy', because of its original association with Christmas]

El Niño Southern Oscillation (ENSO) The large-scale atmosphere–ocean interaction event that occurs irregularly across the tropical Pacific Ocean every 2 to 7 years, linking the SOUTHERN OSCILLATION and EL NIÑO. The Southern Oscillation is one facet of the phenomenon that in the atmosphere is marked typically by the gradual eastward migration of a significantly warm ocean water pool from its normal location in the western Equatorial Pacific. This event is stimulated by the relaxation of the Pacific TRADE WINDS that may even reverse into westerlies. This leads to the genesis of, and eastward migration of, what is known as a *Kelvin wave* – a large patch of slowly moving very warm water that is a very shallow dome or wave shape some 15 cm (6 in) high but a few hundred kilometers either side of the Equator. The anomalously warm water has dire local consequences as it migrates to affect regions that are normally dry. It sparks off very active THUNDERSTORMS that can culminate, after their journey across the Pacific, in disastrously torrential downpours in Peru. In addition, the unusually located deep CONVECTIVE CLOUDS transport large amounts of heat into the upper TROPOSPHERE into regions in which the JET STREAMS are consequently strengthened and sometimes shifted. This can and does have an influence on anomalous weather patterns in regions of the world that are very far from the tropical Pacific.

ELR *See* environmental lapse rate.

emagram *See* skew-*T* log-*P* diagram.

emissions trading A regulatory scheme in which credits or permits are allocated to parties emitting pollutants (e.g. a firm emitting sulfur dioxide) by a regulatory organization. Those parties that reduce their emissions are able to sell or trade their credits to those parties that are unable to reduce their emissions. The concept was proposed in the KYOTO PROTOCOL to enable parties to trade the assigned amounts of their GREENHOUSE GAS emissions allowances in order to fulfill their emissions commitments.

emissivity The ratio of the actual power per unit area emitted by a surface or body to that of a BLACK BODY (perfect radiator) at any given temperature. The Earth's surface, oceans, and clouds have an emissivity value close to 1 (black body), while high cirrus clouds have a value of 0.2.

EMR *See* electromagnetic radiation.

energy balance *See* energy budget; global energy balance; radiation budget.

energy balance model (energy budget model, EBM) A simple MODEL used to analyze the solar radiation incident on the Earth and to predict variation of surface temperature with latitude. For this purpose, atmospheric motions are omitted, although surface air temperature and surface albedo are used. The model can be used to consider the Earth as a whole to give the global average temperature. Energy balance models are used, for example, in studies of climate change.

energy budget The surface energy inputs and outputs that form a part of the energy balance of the Earth–atmosphere system. Most of the exchanges of energy take place on the Earth's surface, which is therefore called the 'active surface'. In its simplest form the energy budget may be expressed as:

$$Q^* + H + LE + G = 0,$$

where Q^* is net radiation, H is sensible heat, LE is latent heat (L is the latent heat of vaporization of water and E is the amount of water evaporated), and G is the heat flow into the ground. *See also* global energy balance; radiation budget.

enhanced greenhouse effect *See* greenhouse effect.

enhanced V A V-shaped region of colder cloud tops seen on infrared satellite images of a thunderstorm ANVIL. This region signifies an area of strong updraft and therefore potentially severe weather; the evidence on the satellite data provides a useful tool that can be used to detect and monitor thunderstorms.

ensemble method (ensemble forecasting) In weather forecasting, a relatively recent innovation in NUMERICAL WEATHER PREDICTION in which the initial state of the atmosphere, upon which the model prediction is run, is not taken as fixed. This means that, at the EUROPEAN CENTRE FOR MEDIUM-RANGE WEATHER FORECASTS (ECMWF) for example, 51 different forecasts are run simultaneously, each one beginning with subtly different initial conditions – it is this large collection that constitutes the 'ensemble' in the method. The reason for the method is to attempt to incorporate the role of 'chaos' in the quality of the weather forecast, because a small perturbation that cannot possibly be observed may play a significant role in the evolution of the weather globally or regionally.

The ensemble method produces 51 forecasts for each daily operational run, then automatically groups them into predictions (for day three, for example), that all point broadly in the same direction regarding the future weather conditions. On some occasions a large grouping will indicate the same trends, no matter what the wide range of changed initial conditions were. At other times there will be a number of clusters, each of which suggest quite different forecast outcomes. The method is employed to give forecasters an indication about the confidence they can have in how the weather will evolve over a period of days. This degree of confidence will vary from day-to-day and geographically on any particular day.

ENSO *See* El Niño Southern Oscillation.

entrainment The process in which air from the environment surrounding a developing cloud is dragged into the ascending CONVECTION current within the cloud. This has the effect of reducing the current's buoyancy; sometimes the cloud system may dissipate completely if very dry air is introduced, leading to evaporation of the cloud droplets. The area in which this takes place is the *entrainment zone.*

entrance region That part of an upper tropospheric JET STREAM in which the horizontal flow of air accelerates markedly into the elongated core of a jet. The right entrance of a jet stream is commonly associated with cyclonic features in the lower troposphere, while the left entrance is linked to anticyclonic features.

entropy (*S*) A measure of the energy within a closed thermodynamic system that is unavailable to do work. It is a measure of the disorder of the system – the greater the disorder, the higher the entropy. Thus, the entropy of a gas is higher than that of a liquid because its constituent atoms and molecules are in a less ordered state.

environmental hazard A process, caused by human activity or NATURAL HAZARD, that has a detrimental effect on the environment and/or human life. Examples of human-induced environmental hazards include the Exxon Valdez oil spillage of March 1989, when 11 million gallons of oil poured into the Prince William Sound, Alaska, affecting 750 km (470 miles) of coastline. Many wildlife and fish species still have not recovered over 12 years after the disaster.

environmental lapse rate (ELR) The actual rate at which temperature changes with height throughout the TROPOSPHERE as recorded, for example, by a radiosonde. A positive LAPSE RATE is usually observed in the atmosphere indicating that temperature decreases with height. On average this is approximately 6.5°C km^{-1} but the rate varies according to local factors, for example height, season, time of day, surface characteristics, and is a function of the prevailing weather conditions. The environmental lapse rate is lower nearer to the ground surface, during the winter season, and over continental areas. Lapse rates are recorded, for example, during radiosonde ascents and plotted on thermodynamic diagrams.

environmental temperature sounding The determination of the changes in temperature with height. A vertical profile of temperature is usually obtained using balloon-borne instruments or measured using remote-sensing equipment. *See also* balloon sounding.

Environment Canada A major federal organization that includes the METEOROLOGICAL SERVICE OF CANADA, the country's source for meteorological information. In addition it undertakes a legally enforceable environmental protection role, including problems related to pollution.

eolian (aeolian) Pertaining to the wind; for example, materials eroded, transported, and deposited on the Earth's surface by the wind are referred to as *eolian deposits*. Eolian processes are most common in arid environments and on exposed beaches.

EP *See* effective precipitation.

epilimnion The upper warmer layer that lies above a THERMOCLINE and below the surface of a body of water, such as a lake or sea.

equable In climatology, denoting a climate type in which there is little variation.

equation of motion In meteorology, a set of equations based on Newton's second law of motion that relates the total force acting on a unit mass of air to the acceleration induced. The accelerations are of the westerly, southerly, and upward components of atmospheric motion. The acceleration with regard to the two horizontal components is related to the horizontal PRESSURE GRADIENT FORCE, the Coriolis force (*see* Coriolis effect), and the frictional force.

Equator The great circle around the Earth, along the 0° parallel of latitude. *See also* thermal equator.

equatorial climate The climate of the low latitudes between about 10°N and S. The climate is dominated by the Intertropical Convergence Zone (ITCZ). Temperatures are consistently high with little range. Rainfall and humidity are also high.

Equatorial Countercurrent An ocean current that flows east between the westward-flowing North and South EQUATORIAL CURRENTS. Westward flow of the Equatorial Currents results in a pressure gradient produced by the higher sea surface, to the west of the tropical ocean basins. To balance this gradient, the low-velocity Equatorial Countercurrent transports excess water eastward.

Equatorial Current Broad ocean currents in the equatorial regions, which are driven by the trade winds and generally directed westward, flowing through the major oceans: the *North Equatorial Current* and the *South Equatorial Current*. Separating the two currents is the narrow EQUATORIAL COUNTERCURRENT, which flows eastward at the warmest part of the Equator. At the Equator, an EQUATORIAL UNDERCURRENT is also produced but at a depth of about 100 m (330 ft). In the Indian Ocean the North Equatorial Current only flows during the northeast monsoon (November to March); it is in summer replaced by the Southwest Monsoon Current, which flows eastward, when the southwest monsoon is occurring.

Equatorial orbit A satellite orbit that occurs in the Earth's Equatorial plane.

Equatorial trough An approximately east–west orientated and elongated pressure minimum that extends around the Earth at very low latitudes and is associated with low-level CONVERGENCE and extensive cloudiness. It is located in the vicinity of the Equator, lying between the SUBTROPICAL ANTICYCLONES of the northern and southern hemispheres and is indistinguishable from the INTERTROPICAL CONVERGENCE ZONE (ITCZ) when the latter is close to the Equator. It exhibits some north–south seasonal migration so that over tropical continents it moves a significant distance away from the Equator. The Equatorial trough may be detected up to a height of 5.8 km (3.6 miles) but is most pronounced near the surface. It coincides with the DOLDRUMS over the oceans.

Equatorial Undercurrent (Cromwell Current) The undercurrent that flows at a depth of approximately 100 m (330 ft) below the surface and transports water eastward and toward the Equator.

Equatorial westerlies The westerly winds that occur when the NORTHEAST and SOUTHEAST TRADE WINDS have to flow across the Equator into the INTERTROPICAL CONVERGENCE ZONE. If this occurs more deeply into the opposite hemisphere by more than about 5° of latitude, they are deflected into Equatorial westerlies. During a strong El Niño Southern Oscillation event, there can be extensive westerlies along the Equator across part of the Pacific Ocean.

equatorial zone *See* tropics.

equinoctial gale A strong wind that occurs at or around the vernal (March 21) or autumnal (September 22) equinoxes. The view that strong winds occur with a greater frequency than the long-term average at the equinoxes is frequently expressed, but this opinion is not generally supported by analyses of past data in such countries as the US and the UK; it therefore still lacks a plausible physical explanation.

equinox **1.** Either of the two points at which the ecliptic intersects the celestial equator.
2. Either of two dates at which the Sun appears to cross the celestial equator. The *vernal (or spring) equinox* occurs on 20 or 21 March, and the *autumnal equinox* on 22 or 23 September in the northern hemisphere; this is reversed in the southern hemisphere. Day and night are of equal duration at the equinoxes.

equivalent potential temperature (theta-e) The EQUIVALENT TEMPERATURE transformed dry adiabatically to the 1000 hPa pressure level. It can be calculated using a THERMODYNAMIC DIAGRAM by following the DRY ADIABAT from T_e at a given pressure level down to the 1000 hPa level, this being approximately equal to mean-sea-level pressure.

equivalent temperature The temperature (T_e) obtained when an air parcel is expanded adiabatically, at constant pressure, until the parcel's water vapor content has been condensed out and the latent heat of condensation used to raise the air temperature.

ERS *See* European Remote Sensing Satellite.

ESA *See* European Space Agency.

etesian wind (meltemi) A moderate to strong northwesterly to northeasterly wind that blows in the eastern Mediterranean, especially in the Aegean Sea area, between about mid May and mid October, occurring most frequently in July and August. It is caused by the steep pressure gradient that exists between an area of high pressure over the Western Mediterranean and an elongated trough of low pressure that extends westward from NW India. The etesian wind typically attains its highest velocity of around 30 knots (35 mph) during the late afternoon, when convection over the hot land temporarily increases the pressure gradient; gradually declines, and dies away during the evening and night. While the etesian wind can be refreshing, its strength and dryness mean that windbreaks (e.g. a row of cypress trees) have to be planted on the northern side of fields to prevent damage to agricultural production.

EUMETSAT The European weather satellite organization, based in Darmstadt, Germany. It is sponsored by the weather services of 17 member states and, in addition, has agreements with three cooperating states within Europe. Its prime objective is to establish, maintain, and exploit European systems of operational weather satellites. The first Meteosat satellite was launched by the EUROPEAN SPACE AGENCY (ESA), in 1977; EUMETSAT took over the responsibility for the Meteosat system from ESA in 1987. The Meteosat Second Generation (MSG) satellite system, coming into operation in the first decade of the 21st century, is to provide more frequent and comprehensive data, with the launch of the first satellite (by ESA) in 2002. In addition, the development of the Polar System will introduce the Metop polar-orbiting system of satellites.

European Centre for Medium-Range Weather Forecasts (ECMWF) An international medium-range weather forecasting organization, located in Reading, UK. It is sponsored by its 17 European member states and has cooperation agreements with four other European states. Established by a Convention in 1973, it has produced operational MEDIUM-RANGE WEATHER FORECASTS since 1979. Its principal objectives are to develop numerical methods for medium-range weather forecasting; to prepare medium-range weather forecasts for distribution to the meteorological services of the member states; to carry out scientific and technical research to improve the forecasts; and the collection and storage of appropriate meteorological data.

European Remote Sensing Satellite (ERS) One of the two satellites operated by the European Space Agency (ESA) in near-polar orbit to perform remote sensing of the details of the Earth's surface, especially oceans, ice-caps, and coastal regions, for environmental monitoring purposes. *ERS-1* was launched in 1991, followed by *ERS-2* in 1995. Advanced microwave techniques are used to collect measurements and images; the data collected includes measurements of wind speed and direction, surface temperatures, cloud cover, atmospheric water vapor levels, and ozone levels. Its scatterometer data have been used by operational weather forecast centers to deduce sea-level winds over vast data-void areas.

European Space Agency (ESA) An organization that operates Europe's space program. It was formed in 1975 and has 15 member states, with its headquarters in Paris, France. The organization is composed of four establishments and has, overall, the role of developing and promoting European activities in space. The *European Space Operations Center* (ESOC) is located in Darmstadt, Germany, while the *Research and Technology Center* (ESTEC) is in Noordwijk, The Netherlands. ESA launched the first Meteosat satellite in 1977 and is developing, with EUMETSAT, the Meteosat Second Generation (MSG) satellite system.

eustasy A global variation in sea level brought about by a change in the amount of water in the oceans. The factors involved include the melting or growth of glaciers, ocean temperature fluctuations at depth resulting in density changes and an expansion or contraction of the world's oceans, and alterations in basin shape caused by tectonic movements.

evaporation The process by which a liquid (in meteorology and climatology this is invariably WATER) is transformed to a gas (WATER VAPOR); it is the opposite process to CONDENSATION. In meteorology, evaporation may also encompass SUBLIMATION, by which solid water (ice) is converted to water vapor with no intermediate liquid water stage. For climatological purposes evaporation refers to the loss of water from the Earth's surface to the atmosphere, usually expressed in millimeters per day, month, or year. Evaporation is a major component in the HYDROLOGIC CYCLE.

evaporation fog (mixing fog) A fog formed by the mixing of two unsaturated volumes of air at different temperatures. The mixing process leads to saturation and condensation of fog droplets under pertinent conditions, for example, when warm rain falls into a cooler layer of air near the surface, as occurs in warm fronts.

evaporation pan A pan used to hold water during observations to determine the rate of evaporation at a given location. Evaporation pans are usually set above the ground on a small plinth, while *evaporation tanks* are buried in the ground. *See also* evaporimeter.

evaporimeter (atmometer) An instrument used for measuring the rate of evaporation of water into the atmosphere. There are two main types. The first type measures the evaporation rate from a free water surface. In the *open pan evaporimeter* a tank is sunk into the ground so the water surface is at ground level and evaporation is measured using a micrometer gauge. Allowances for precipitation and drainage are also taken into consideration. The second type of evaporimeter, of which there are several designs, measures the evaporation rate from a continuously wet porous surface. The *Piché evaporimeter* uses an inverted graduated cylinder of water, the mouth of which is covered with a filter paper disk. As water evaporates off the filter paper, the level of the water reduces and the amount of evaporation can be measured directly. *See also* lysimeter.

evapotranspiration The combined processes of EVAPORATION from the Earth's surface and TRANSPIRATION from plant vegetation that transfer water in the form of WATER VAPOR to the atmosphere. It is a fraction of POTENTIAL EVAPOTRANSPIRATION.

evapotranspirometer An instrument used for measuring the rate of EVAPOTRANSPIRATION. It consists of a tank containing soil, in which vegetation is planted, and designed so that all water added to the tank and water left following evapotranspiration can be measured.

exit region That region of a JET STREAM in which the air is leaving the elongated narrow core of maximum winds and slowing down markedly. It is commonly associated with cyclonic features in the lower troposphere below the left-exit and anticyclonic features below the right-exit region.

exosphere The outermost region of the Earth's atmosphere, extending from approximately 700 km (430 miles) above the Earth's surface. Gas densities are very low, mean free paths are large, and collision frequencies are low enough for particles to achieve sufficient velocity from a collision to escape from the atmosphere.

explosive cyclogenesis A very quickly deepening FRONTAL CYCLONE, usually over middle-latitude oceans, in which the rate of deepening (or the fall of central pressure) is at least 24 hPa in one day. This rate of deepening defines the system as a *bomb*. Such disturbances are associated principally with very damaging winds.

exposure The exposure of a site to SOLAR RADIATION; this depends upon its latitude, elevation, aspect, slope, and lack of shielding by obstacles (for example, buildings and vegetation).

extended-range weather forecasting *See* medium-range weather forecasting.

extraterrestrial radiation Shortwave radiation, almost exclusively from the Sun, reaching the outer limits of the atmosphere (the top of atmosphere or TOA). This flow of radiation, the SOLAR CONSTANT, has a value that varies with the radiant output of the Sun but is around 1367 W m^{-2}.

extratropical cyclone A traveling CYCLONE (low-pressure system) in middle and higher latitudes, outside the tropics. It normally refers to a frontal system but is also applicable to a non-frontal depression.

extremes of climate The highest or lowest values for a meteorological parameter, or an exceptionally large range between the highest and lowest values for a meteorological parameter, over a specific time period, such as a month, season, or year.

eye (eye of the storm) The generally circular central region of an intense TROPICAL CYCLONE, usually about 15–20 km (9–12 miles) in diameter, with a small horizontal pressure gradient. This region is characterized by light winds and no rain, and a clear sky or broken clouds. As the eye passes over a location the violent winds associated with the storm reverse their direction. *See also* hurricane; storm surge.

eye of the wind A nautical term used to describe the direction from which the wind is blowing.

eye wall cloud The narrow zone of cloud within the band of convection that surrounds the eye of a TROPICAL CYCLONE (HURRICANE or TYPHOON). It is essentially an upright cyclinder that is some 12–15 km (7–9 miles) deep, consisting of a ring of very vigorous CUMULONIMBUS clouds. It is the narrow zone across which the hurricane's horizontal pressure gradient is at its steepest; as a result the strongest winds occur underneath the eye wall cloud, as do the highest rainfall totals and severest thunderstorms.

Fahrenheit scale A standard temperature scale widely used in the US; it is based on the freezing point of water at 32°F and the boiling point of water at 212°F with 180 equal divisions between these two fixed points. The scale was developed originally in 1714 by Gabriel Daniel Fahrenheit (1683– 1736), a German physicist, basing the zero point on the temperature of an equal ice–salt mixture. Although the Fahrenheit scale was widely used it has been largely replaced internationally, and for scientific purposes in the US, by the CELSIUS SCALE and by thermodynamic temperatures measured in KELVINS. The conversion formula from Celsius to Fahrenheit is:

$°F = (°C \times 1.8) + 32.$

fair-weather cumulus LOW CLOUDS with flat white or gray bases and 'bubbling' white tops that are not very developed vertically. Fair-weather cumulus clouds are generally widely spaced with blue sky apparent between them. They are quite common on pleasant summer afternoons when surface heating on a sunny warm day sparks off shallow CUMULUS clouds that drift innocuously across the sky. Fair-weather cumulus may be used to describe both HUMILIS and MEDIOCRIS species of cumulus.

fall *See* autumn.

fallstreak *See* virga.

false cirrus The CIRRUS cloud that was originally the anvil topping a CUMULONIMBUS cloud but has become detached as the original cloud dissipates. Such false cirrus occurs often in the afternoon and evening of a thundery day, vestiges of the storms that have died out.

false color In remote sensing, the representation on an image of data collected in wavelengths that are not normally visible. The non-visible electromagnetic spectrum is shown as one or more of the red, green, or blue components; for example, data in the infrared wavelength is frequently displayed as visible red. The presence of a particular range of sea-surface temperatures in a remotely sensed scene may be represented by using different colors not natural to such a scene. The warmest sea may, for example, be false-colored in red to emphasize its location.

fanning In pollution studies, emissions from a chimney stack under predominantly stable atmospheric conditions, usually in the presence of a weak to moderate temperature inversion, where horizontal dispersal of pollutants is several times greater than vertical dispersal.

fata morgana An optical phenomenon, a complex type of superior MIRAGE, in which distant objects may appear as distorted images resembling walls, castles, or turrets, as a result of temperature inversions within the lowest layers of the atmosphere. The phenomenon, which occurs sometimes in the Strait of Messina, Italy, was there named for Morgan le Fay (from Arthurian legend), who was believed to use sorcery to produce such illusions. [Italian: Morgan the fairy]

fathom A nautical measurement of depth equivalent to 6 ft (1.8 m).

feeder band 1. (outer convective band) In tropical meteorology, one of several lines or bands of thunderstorms that spiral around the center of a TROPICAL STORM or TROPICAL CYCLONE.
2. (inflow band) In a thunderstorm, a band of low cloud, one of several lines of such clouds ahead of, and moving into, the core region of the storm. The feeder bands move with the low-level flow and may be an indication of the strength of the warm moist air flowing toward the main updraft.

Ferrel cell (middle-latitude cell) The large-scale wind circulation cell that exists in the vertical north–south plane in middle latitudes. It is characterized by flow toward the poles in the lower TROPOSPHERE, often as southwesterly winds in the northern hemisphere (northwesterlies in the southern hemisphere) that have their core in the SUBTROPICAL ANTICYCLONES. The air in the Ferrel cell then ascends within FRONTAL ZONES. In the upper troposphere there is a return component that flows back toward the subtropics, where it meets the upper outflow from the HADLEY CELL in a zone of CONVERGENCE. It descends below this to replenish the air that flows out from the surface high. The cell is named for the US meteorologist William Ferrel (1817–91), who originally described a middle-latitude circulation cell.

fetch The length of the surface area of sea or ocean over which a wind blows in one direction unimpeded by land; this is an important factor that affects the height of the waves.

fibratus A species of CIRRUS and CIRROSTRATUS cloud that appears as a thin veil, detached straight filaments, or irregularly curved filaments, none of which end in hooks or tufts. *See also* cloud classification.

fiducial point An accepted fixed point of reference for an instrument or scale. For example, the point to which an instrument must be reset before it can be used.

filling The phenomenon of increasing barometric pressure within a low-pressure region. Thus, a FRONTAL CYCLONE (frontal depression) can 'fill' but an ANTICYCLONE (high) cannot. A rapid rate of filling would be 10 hPa in three hours; it is more typically 1 or 2 hPa every three hours.

fire A hazard, caused either by natural disaster or human activity, in which fire spreads over wide areas destroying vegetation, wildlife, and property, releasing quantities of smoke and pollutants into the atmosphere. Fire can have positive in addition to negative impacts. It has been used by humans to modify the landscape, clear forests for agriculture, improve the quality of grazing land, and encourage the rapid regeneration of plant species. A wildfire in a forest or bush area may be caused, for example, by human carelessness or lightning strikes, often preceded by drought. It can be worsened by strong winds that help to spread the fire over large areas. Regions especially susceptible to wildfire are those that have a marked dry season with hot temperatures.

The incidence of EL NIÑO SOUTHERN OSCILLATION events has also been related to increased fire risk in those parts of the western Pacific that have reduced convection and precipitation when it occurs. During 1997 extensive forest fires in Indonesia, many of which were started to clear forest but subsequently grew out of control, had an impact on many parts of SE Asia, including Malaysia, Singapore, and Thailand, as a result of smoke spreading over the region creating severe pollution and affecting weather. In late 1997 and early 1998 Australia suffered widespread bush fires. These events were partly influenced by the 1997–98 ENSO. *See also* crown fire; National Fire Danger Rating System.

firn (névé) Granular snow that has survived at least one summer melt season. It is intermediate in the conversion of snow into pure glacial ice by the process of *firnification*, its density being in the range of 0.4 to 0.89 kg m^{-3}.

firn line (firn limit, annual snowline) A line on a glacier that marks the upper limit of the region in which winter snowfall melts during the summer ablation season. It is often clearly marked on many glaciers, separating hard blue ice below from snow above.

firn wind *See* glacier wind.

flanking line A component of a tornadic supercell that occurs on the rear or trailing side of such a system. It is formed of increasingly tall CUMULUS and cumulus congestus cloud (*see* congestus) that feed into the massive SUPERCELL. The flanking line is defined by the rear-flank downdraft that is part of the supercell's circulation. The WALL CLOUD is usually found underneath the area in which the flanking line intersects the tallest growing cloud in the system. *See also* tornado.

flash *See* lightning.

flash flood A sudden and destructive FLOOD occurring with little or no prior warning caused usually by excessive rainfall and/or rapid snow and ice melt. Other factors that can increase the likelihood of a flash flood include changes in land use within a watershed, which can accelerate runoff rates, and changes in basin shapes downstream. For example, basins that become narrower downstream reduce the capacity of the channel. Flash floods are most common in mountainous regions but can also occur over flat terrain. The sudden failure of a dam can also produce a flash flood. In the US the National Weather Service issues a *Flash Flood Watch* when it is possible that rainfall will cause flash flooding in a particular area; a *Flash Flood Warning* is issued if a flash flood is occurring or is imminent in the area.

F-layer (F region) A layer or region of the IONOSPHERE in which there are high concentrations of ions and electrons; it extends from about 150 km (90 miles) above the Earth's surface to about 1500 km (900 miles). Two F-layers are identified. The *F_1-layer* tends to have stable ion and electron concentrations; it is found at around 160 km (100 miles) but its height varies systematically depending on the latitude, the time of day, the season; it also varies with sunspot activity. The *F_2-layer* (also known as the *Appleton layer*), found at around 300 km (200 miles), is less predictable and is influenced by the solar and lunar tides and the Earth's magnetic field.

floccus A species of cloud in which the constituent CUMULIFORM units appear like small tufts that have an overall ragged look, often with VIRGA. It is applied to CIRRUS, CIRROCUMULUS, and CIRROSTRATUS cloud. *See also* cloud classification.

Flohn climate classification A genetic CLIMATE CLASSIFICATION proposed in 1950 by H. Flohn. The classification is based on the global wind belts and precipitation characteristics and recognized seven major categories of regional climates. *See also* Köppen climate classification; Strahler climate classification; Thornthwaite climate classification.

flood An overflowing of a body of water from its natural confines produced by excessive water input into a river, or a temporary incursion of ocean or sea system onto areas not normally submerged. Floods can be caused, for example, by very intense precipitation on to saturated ground, the melting of snow packs, the collapse of dams (natural and man-made), high tides, TSUNAMIS, and STORM SURGES.

flood forecasting (flood prediction) The process of predicting river or coastal flooding, defined as inundation of land areas that are not normally submerged. Fluvial flood prediction is part of the routine operations undertaken by hydrologists. They require meteorological input in the form of observed and predicted PRECIPITATION patterns, including their speed and direction of movement, and the intensity of the rain or snowfall. Predicted precipitation patterns may be obtained from NUMERICAL WEATHER PREDICTION. The actual patterns are mapped by precipitation RADAR networks, and the digital data are fed automatically into hydrological flood prediction models. In addition, the forecast models require quantification, for example, of the soil moisture levels within a particular catchment, and the evaporation losses during the prediction period. They must also be set up to incorporate the melting of snowpack including its extent and depth, its location, and how rapidly it occurs. Flash flooding (*see* flash flood) is a very dangerous phenomenon that occurs on a relatively small-scale; it is difficult to predict since it is most commonly related to vigorous THUNDERSTORMS that can break out virtually anywhere – over a hilly or mountainous area, for example – on a hot spring or summer's day. The forecast can point to the high risk of such floods, but cannot easily pinpoint the location or the time at which they will occur. The storms that produce them can often develop in as little as tens of minutes.

The forecasting of coastal flooding relies on numerical weather prediction. Low-lying coasts can be flooded by swollen rivers that issue into the sea across them or by STORM SURGES that move onshore in association with HURRICANES (TROPICAL CYCLONES) or TROPICAL STORMS. The location, height, and timing of such surges are well handled by routine weather forecast models, and in the US are most likely to affect the coastline from southern Texas as far as the mid-Atlantic states. Storm surges can also occur with deep FRONTAL CYCLONES that track with a particular speed and route across relatively enclosed ocean basins. This is the case around the shores of the North Sea, for example, where low-lying areas of the eastern UK, northern Germany, and The Netherlands are susceptible.

flood peak estimation The process of predicting the timing and height of the maximum levels in stream discharge at various points down the main channel of a flooding river. These factors will depend partly on how the channel's shape changes along its course, and how large the contribution from groundwater storage is as the flood peak moves downstream.

flood prediction *See* flood forecasting.

Florida Current A northward-flowing section of the GULF STREAM that extends from the Straits of Florida to Cape Hatteras, North Carolina. It is characterized by low salinity and temperatures above 6.5°C (44°F).

fluorocarbon A synthetic compound of carbon and fluorine that also often contains other elements, such as hydrogen, chlorine, or bromine. The fluorocarbons include the CHLOROFLUOROCARBONS (CFCs), HYDROCHLOROFLUOROCARBONS (HCFCs), HYDROFLUOROCARBONS (HFCs), and PERFLUOROCARBONS (PFCs).

foehn pause *See* föhn pause.

foehn wall *See* föhn wall.

foehn wind *See* föhn wind.

fog A dense mass of minute water droplets suspended in the atmosphere that obscures the lower layers of the atmosphere and reduces horizontal visibility to less than 1000 m (0.621 mile). It is defined as *thick fog* if visibility is 100 m (330 ft) or less. Fog is normally caused by the condensation of water vapor that results when air is cooled to below its DEW-POINT and is generally associated with calm weather conditions. Formation is aided when there is a concentration of suspended smoke and/or dust particles in the lower atmosphere; such particles act as condensation nuclei and may cause fog to form before the saturation point is reached. There are several types of fog, most being defined according to the mechanism of cooling. *See* advection fog; evaporation fog; frontal fog; hill fog; ice fog; radiation fog; sea fog; steam fog; upslope fog. *See also* smog.

fogbow (white rainbow, mist bow, cloud bow) A faintly colored or whitish arch, the inner edge of which is bluish and the outer reddish, seen opposite the Sun in fog or mist. It is similar in formation to the rainbow and is due to refraction and reflection, together with some limited diffraction, of light from the Sun or Moon by small water droplets, the diameters of which are less than 100 μm (0.004 in).

fog-drip (fog precipitation) Precipitation in the form of moisture from a FOG bank that has been intercepted by vegetation. It is common along the coast of arid regions where humidity is constantly high and there is a cold ocean current offshore (e.g. S California and Peru). Other forms of precipitation in these regions are scarce and fog-drip may be sufficient to sustain vegetation in such arid environments.

föhn pause (foehn pause) A temporary cessation of a FÖHN event at the ground due to the formation or intrusion of a cold air layer, which lifts the föhn off the Earth's surface.

föhn wall (foehn wall) A stationary line or bank of clouds that is an indicator of the existence of a FÖHN WIND over a mountain range. The name was first applied to the cloud observed in the European Alps. The air ascending the south side of the Alps as a southerly wind will lead to condensation upslope as far as, and overlapping, the line of the Alpine watershed. This elongated cloud that runs along the crest line of the mountains can be seen from the downslope side, appearing like a wall of cloud that does not move as the wind flows through it. *See also* chinook arch; helm wind.

föhn wind (foehn wind, fon wind) A warm and dry wind that blows down the LEEWARD side of a mountain range. The word föhn was originally used in the European Alps for a southerly wind, which can result in a sudden increase in temperature (sometimes by 8° to 11°C) in the north–south orientated valleys of the northern Alps; this can lead to rapid snow melt and avalanches. In the Reuss Valley of Switzerland, a föhn wind typically blows on 48 days per year, occurring most frequently during the spring when a depression moves to the north of the Alps. Consequently, warm air is drawn in from the south and forced to ascend the southern slopes of the Alps at the saturated adiabatic lapse rate (~0.5°C/100 m), resulting in condensation and precipitation, with an associated release of latent heat. The mass of cloud enveloping the mountain is called the FÖHN WALL. This airflow is then warmed through compression at the dry adiabatic lapse rate (1.0°C/100 m) on the LEEWARD side of the range. Temperatures at a given altitude are thus higher on the leeward side of the range when compared to their WINDWARD-side equivalent. This same *föhn effect* is given different names in other parts of the world: BERG WIND (South Africa); CHINOOK (Rocky Mountains, North America); NOR'WESTER (e.g. New Zealand); SAMOON (Iran); SANTA ANA (California, US); and ZONDA (Argentina).

forced convection CONVECTION that is a product of mechanical turbulence, which occurs due to the development of eddies as an airflow crosses an uneven surface. This is distinct from free (thermal) convection originating from the heating of the surface.

forcing (forcing mechanism) A mechanism or process that affects the balance of a system and forces a response or feedback. For example, a forcing mechanism, such as a volcanic eruption or anthropogenic-induced enhanced GREENHOUSE EFFECTS, may alter the energy balance between the Earth's incoming solar radiation and outgoing infrared radiation.

forecast A statement (in verbal or map form) of the anticipated state of the weather at a particular place or over a region, up to the global scale, over a specific period. The word was introduced by Robert FitzRoy (1805–65), first director of the Meteorological Department of the Board of Trade in the UK. Forecasts are defined according to the periods they cover. *See* long-range weather forecasting; medium-range weather forecasting; nowcasting; short-range weather forecasting.

forest fire *See* fire.

forked lightning Visible LIGHTNING in which individual branches of lighting channels can be seen, often as the lightning passes from a cloud to ground (*see* cloud-to-ground discharge). The forked tortuous path of a lightning discharge may also be evident in inter-cloud events (*see* cloud-to-cloud discharge).

Fortin barometer An instrument used for measuring atmospheric pressure; the most common type of MERCURY BAROMETER. It consists of a long glass tube (sealed at one end) that is evacuated, filled with mercury, and inverted with the open end in a reservoir or cistern of mercury. The height of the mercury is a measure of the atmospheric pressure acting upon it. The scale within the cistern has to be reset to zero before use of the barometer.

fossil fuel A fuel derived from organic matter that has been deposited over long periods of geologic time; it includes coal, oil, natural gas, oil shales, and tar sands. All fossil fuels on combustion produce carbon dioxide (CO_2), which inevitably contributes to the enhanced GREENHOUSE EFFECT.

fractocumulus cloud A cloud type or species that appears as ragged irregular shreds of cloud at a low level. They are shallow CUMULUS clouds that do not have a smooth appearance and occur when the cloud is beginning to evaporate and dissipate into the air around it, or when strong winds are fragmenting it.

fractostratus cloud A cloud type or species that consists of ragged fragments of low gray or dark-gray cloud that forms in strong winds below NIMBOSTRATUS rain clouds, when it is also known as SCUD. It may be located close to the ground surface. It can also form with the break up of a sheet of STRATOCUMULUS.

fractus A cloud type or species consisting of ragged irregularly shaped patches or shreds of CUMULUS or STRATUS. *See* fractocumulus cloud; fractostratus cloud. *See also* cloud classification. [Latin: 'broken']

Framework Convention on Climate Change (FCCC) *See* United Nations Framework Convention on Climate Change.

frazil ice Tiny platelets or needles of ice that form in rapidly flowing or turbulent rivers and lakes in which the temperature is below freezing point. Frazil particles adhere to each other in the supercooled water, rising to form spongy accumulations or floc on the surface. The movement of water prevents the ice crystals from forming a continuous sheet. Frazil ice is common in Canadian rivers where the term was first applied.

free atmosphere The part of the atmosphere lying above the PLANETARY BOUNDARY LAYER in which the influence of surface friction and heat fluxes are negligible.

freezing and thawing indices Indices used to measure the severity of climate. They are generally defined as the cumulative number of degree-days during which air temperatures are below and above 0°C (32°F). The *total annual freezing index* is used, for example, to predict permafrost (perennially frozen ground) distribution, to project maximum depth of frost penetration, and to estimate thicknesses of ice on sea, lakes, and rivers. The *total annual thawing index* is also used in the prediction of permafrost, in addition to estimating the depth of thaw of frozen ground.

freezing drizzle Precipitation comprising supercooled water droplets with diameters of less than 0.5 mm (0.02 in) that descends through relatively warm air in the lower layers of the atmosphere in which a temperature inversion is established but, on impact with the ground surface or objects with a temperature at or below 0°C (32°F), freezes to form GLAZE. *Compare* freezing rain.

freezing level The height above mean sea level at which the air temperature is 0°C (32°F), or the pressure altitude above a specific location of the 0°C (32°F) isotherm. This is not necessarily the level at which liquid water droplets becomes solid ice; as a result of SUPERCOOLING liquid water may remain unfrozen to temperatures as low as –40°C (–40°F).

freezing nucleus A tiny atmospheric particle on which, at temperatures below 0°C (32°F), the freezing of supercooled water droplets occurs generating an accumulation of ice. The nuclei, which have a similar crystalline structure to ice, may be, for example, very fine wind-blown soil particles, volcanic dust, certain bacteria, or actual ice crystals. The temperature at which freezing occurs is variable depending on the nuclei but is typically between –25°C (–13°F) and –35°C (–31°F); without the presence of freezing nuclei the supercooled water droplets in a cloud would not freeze above –40°C (–40°F). The presence of freezing nuclei is important if the BERGERON–FINDEISEN PROCESS of precipitation generation is to operate. *See also* cloud condensation nucleus.

freezing point The temperature at which a liquid changes phase to become a solid. For pure water at standard atmospheric pressure this is at 0°C (32°F) or 273.15 K. However, in reality a liquid may not freeze at this temperature if the pressure varies from the standard atmospheric pressure, if impurities are present, or through SUPERCOOLING.

freezing rain Precipitation comprising supercooled water droplets with diameters in excess of 0.5 mm (0.02 in) that descends through relatively warm air in the lower layers of the atmosphere where a temperature inversion is established but, on impact with the ground surface or exposed objects where the temperature is at or below 0°C (32°F), freezes to form GLAZE. *Compare* freezing drizzle.

fret A regional name in the UK for an ADVECTION FOG that drifts inland from the North Sea across northeast England. It occurs with the cooling of a moist inshore flow and is associated with particularly miserable weather with suppressed maxima and often persistent drizzle. *See also* haar.

friction The mechanical force of resistance associated with the relative motion of the boundaries of two adjacent substances (solids or fluids). In the atmosphere, friction is important in the flow of air across the Earth's surface and has the effect of decreasing the wind speed relative to that ideally associated with a given pressure gradient (GEOSTROPHIC WIND). It also leads to the deflection of the air across isobars so that there is frictional inflow into a low and frictional outflow from a high within the PLANETARY BOUNDARY LAYER (friction layer). This layer is typically about 1 km (0.6 mile) deep, but can vary considerably in time and space. The influence of friction is partly related to the EKMAN SPIRAL.

frigid zone (polar zone) One of the three historic climatic zones, based on temperature and sunshine characteristics, into which the world was divided by the classical Greeks; it consisted of the high-latitude areas bounded by the Arctic or Antarctic Circle. Temperatures in the zone are cold all year round as a result of extended periods of winter darkness and the low summer Sun. *See* temperate zone; torrid zone. *See also* climate classification.

front The moving leading edge of an extensive region of warm or cold air in the extratropics, associated with other weather changes, such as barometric pressure, humidity, and wind speed and direction. Fronts are shallow sloping zones of transition that separate two different AIR MASSES – MARITIME POLAR from MARITIME TROPICAL, for example. They are also zones of deep CLOUD and often widespread PRECIPITATION.

Fronts were defined originally soon after World War I by workers led by Vilhelm Bjerknes (1862–1951) in Norway at the BERGEN SCHOOL. The word 'front' was used as an analogy to the front line in the recent war. *See* ana-front; cold front; katafront; occluded front; stationary front; warm front. *See also* Antarctic front; Arctic front; Atlantic polar front; polar front.

frontal cyclone (frontal depression) A low-pressure system or CYCLONE in extratropical regions that has associated warm, cold, and occluded fronts. Frontal cyclones are characterized by cyclonic inflow, extensive layer cloud, and widespread precipitation.

frontal fog Fog formed in association with frontal precipitation that falls from the warmer air above the FRONT into the cooler unsaturated air below it. When this happens very near to the surface, the chilling of the cooler air by the evaporation of raindrops falling through it can lead to saturation and, sometimes, fog. It occurs most commonly in the shallow layer of cold air just ahead of the WARM FRONT in which warm air lifts over the cold front.

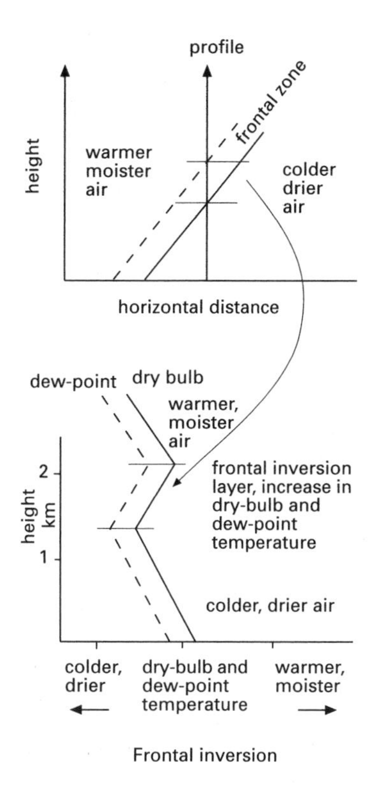

Frontal inversion

frontal inversion A layer within the TROPOSPHERE within which both DRY-BULB TEMPERATURE and DEW-POINT TEMPERATURE increase with height. These increases are associated with a change, moving upward, from cooler drier air in the polar air below a FRONT to the relatively warmer and moister tropical air above it. It is a stable layer that can act as a lid to convective cloud. The inversion is usually some few hundreds of meters deep.

frontal precipitation (frontal rainfall, cyclonic precipitation) PRECIPITATION that is generated as a result of large-scale uplift and cooling of warm moist air along the frontal zones of a low-pressure system, usually in mid- and high-latitude regions. Its characteristics are dependent on the type of front: rainfall from a WARM FRONT is typically more prolonged but rather drizzly; heavier bursts of rain are usually associated with a COLD FRONT.

frontal system *See* frontal cyclone.

frontal wave An undulation on a FRONT that travels along the POLAR FRONT as a wave-like perturbation, rather like a wave running along a length of string when 'flicked' at one end. The young frontal wave consists of a WARM FRONT and COLD FRONT but gradually starts to occlude (OCCLUDED FRONT) as time progresses.

frontal zone The zone of transition between two adjacent AIR MASSES. *See* front.

frontogenesis The process by which a FRONT is formed, often by the juxtaposition of two different AIR MASSES that flow toward one another in such a way that the horizontal temperature gradient is strengthened. *Compare* frontolysis.

frontolysis The process by which a FRONT is suppressed or dissipates and disappears, usually in association with marked DIVERGENCE and SUBSIDENCE. *Compare* frontogenesis.

frost **1.** A state in which the air temperature at Stevenson screen level (1.25 m; 4 ft) falls to at or below the freezing point of water, i.e. 0°C (32°F).
2. Deposits of ice crystals that form on exposed surfaces and objects, such as grass

FUJITA TORNADO INTENSITY SCALE

strength	*wind speed* mph	km h^{-1}	*damage experienced*
F0	40–72	65–116	light damage
F1	73–112	117–180	moderate
F2	113–157	181–255	severe
F3	158–206	256–334	considerable
F4	207–260	335–420	devastating
F5	261–318	421–515	incredible

and trees, as a result of direct sublimation of water vapor from the air when the frost-point temperature is reached. *See also* air frost; glaze; ground frost; hoar frost; rime.

frost hollow *See* frost pocket.

frost pocket (frost hollow) A generally low-lying location in which frosts, or more severe frosts, tend to occur more frequently than in the surrounding areas. These locations are commonly associated with a hollow or narrow valley in which cold air accumulates at night due to katabatic flows of cold air down the slopes and is either trapped or flows out slowly. *See also* katabatic wind.

frost-point (frost-point temperature) The temperature below 0°C (32°F) at which moisture in the air condenses as a layer of frost on any exposed surface. It is the temperature to which a sample of air (containing water vapor) must be cooled, at constant pressure and humidity, to achieve saturation with respect to an ice surface. It is measured with a frost-point HYGROMETER. *See also* dew-point.

frost-point temperature *See* frost-point.

frost smoke *See* Arctic sea smoke.

frozen ground *See* permafrost.

F scale *See* Fujita Tornado Intensity Scale.

Fujita–Pearson Scale *See* Fujita Tornado Intensity Scale.

Fujita Tornado Intensity Scale (F scale, Fujita–Pearson Scale) A reference scale running from F0 to F5 to classify the strength of tornadoes, based on maximum wind speed and damage caused. It was devised in 1971 by the Japanese–American meteorologist Tetsuya Fujita (1920–98) and Allen Pearson, based on wind speeds measured in miles per hour.

The wind speed within a tornado is highly variable and difficult to measure, consequently a tornado's intensity often has to be inferred from the damage it causes. See Table.

full-physics numerical model A three-dimensional computer model used in meteorology that simulates the physical characteristics of the meteorological fields, especially mesoscale motion present within the BOUNDARY LAYER. The model is an adaptation of the GENERAL CIRCULATION MODEL (GCM) and is used in conjunction with air-quality models.

fumigation In pollution studies, the emission of pollutants from a chimney stack under turbulent atmospheric conditions dominated by gusty winds. Relatively high concentrations of pollutants may be periodically brought down to the surface, interspersed by periods of low concentration.

funnel cloud The visible rotating, generally long and thin, tube-like cloud associated with tornadic circulation that extends down from the cloud base but does not come into contact with the ground surface. Funnel clouds can occur in different shapes and sizes, including a twisted contorted rope-like shape, a vertical wide, symmetric feature, and like an elephant's trunk hanging from the base of a towering CUMULONIMBUS cloud. The condensation of water droplets in these clouds is linked to the very low pressures within their circulation. *See also* tornado.

funneling The process by which an airflow is constrained by the presence of a valley, leading to convergence, higher wind speeds, and uplift. A similar effect can be seen in the air between an advancing front and a mountain barrier.

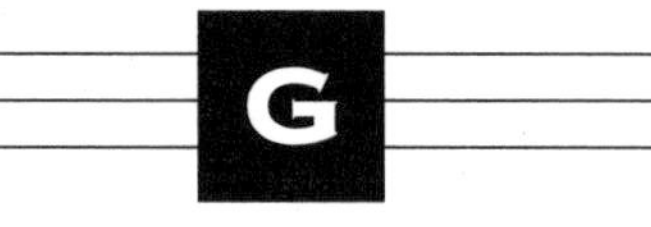

Gaia hypothesis A theory, derived from a concept proposed by the British scientist James Ephraim Lovelock (1919–) and the American biologist Lynn Margulis in the 1970s, that the entire Earth is a self-regulating ecosystem with feedback mechanisms between the abiotic (nonliving) and biotic (living) components. Thus the climate and biogeochemical cycles are regulated by the living organisms on Earth. The theory was named for the Earth goddess of ancient Greece.

gale **1.** A strong wind, the velocity of which when recorded at a height of 10 m (33 ft) and averaged over a 10-minute period, is classified on the BEAUFORT SCALE as force 7 (28 knots; near gale) or more. The World Meteorological Association defines the various categories of a gale as: *near gale* (28–33 knots), *gale* (34–40 knots), *strong gale* (41–47 knots), and *storm* (48–55 knots).
2. A commonly used name for any strong wind.

gap wind (wind gap) A wind that blows along a valley or col in a ridge. In the US, for instance, a gap wind is experienced in the town of Harrisburg, Pennsylvania.

GARP *See* Global Atmospheric Research Program.

garua The MIST or DRIZZLE that often envelops the Peruvian Costa like a thick blanket of cloud. Given that extreme aridity characterizes this coastline, the garua provides nearly all of the region's rainfall, which in many locations (e.g. Peru's capital Lima) is frequently less than 25 mm (10 in) per annum.

Gaussian plume model A continuous dispersion computer model of a pollution plume from a chimney stack under near neutral atmospheric conditions in which pollutants are dispersed equally both vertically and horizontally. Cross-sections show the typical 'bell-shaped' characteristics of a Gaussian or normal distribution. *See also* Gaussian puff model.

Gaussian puff model A computer model that seeks to simulate pollution dispersal from a stack under near neutral atmospheric conditions in which DRY DEPOSITION occurs at the surface; this is in contrast to the GAUSSIAN PLUME MODEL. The number of 'puffs' impacting downstream may be described by a Gaussian distribution under CONING conditions.

GAW *See* Global Atmospheric Watch.

GCM *See* general circulation model.

genera *See* cloud classification.

general circulation (global circulation) The very large-scale averaged atmospheric circulation pattern that results from the thermal and pressure gradients that occur on the same large scale. These are forced by the inequalities of incoming solar and outgoing terrestrial radiation between the Equator and poles. The general circulation evolves significantly over the seasons so that it is more intense in the middle and higher latitudes during the winter than the summer, for example, in response to the stronger imposed tropics-to-pole thermal contrast in the cold season. One critical role of the general circulation is to mix air from the lower and higher latitudes in order to offset the radiation imbalance.

The THREE-CELL MODEL characterizes the principal features of the general circulation in the vertical plane. There are significant seasonal shifts of the cellular pattern, particularly of the HADLEY CELL, which migrates toward higher latitudes in the summer in unison with the middle-latitude FERREL CELL, and toward lower latitudes in the winter. The middle latitudes are characterized by traveling FRONTAL CYCLONES (frontal depressions) and ANTICYCLONES (highs), notably across the oceans, while the wintertime continents are typified by extensive areas of high pressure. The global circulation also involves the subcontinental scale changes that are associated with the seasonal evolution of the MONSOON. The best known occurs over southern Asia (*see* Asian monsoon), although a separate significant example also affects extensive areas of West Africa (*see* African monsoon). In the middle and upper TROPOSPHERE, the large-scale mean circulation features include the LONG WAVES in middle latitudes, the POLAR FRONT, the SUBTROPICAL JET STREAM, and the TROPICAL EASTERLY JET STREAMS, the last of which is seasonal. The long waves exhibit an INDEX CYCLE that is related to fluctuations in the pattern of weather at the surface in middle latitudes in particular – these mark changes of the general circulation over periods of several weeks. The global oceans play a major role in the general circulation; the warm currents transport warm water to higher latitudes to ameliorate sea temperatures within the tropics. At the same time, this warm water is imported into higher latitudes to elevate otherwise cooler sea temperatures. Cold currents export cooler water into lower latitudes and suppress temperatures there below a higher possibility.

general circulation model (GCM) A global computer MODEL that represents the Earth's climate in three dimensions (four including time), with the most sophisticated models simulating physical, thermodynamic, and dynamic activity. Most general circulation models use equations to describe processes over land, ocean, sea ice, and atmosphere while *coupled ocean–atmosphere models* resolve the fluxes between the ocean and atmosphere. General circulation models can be used to

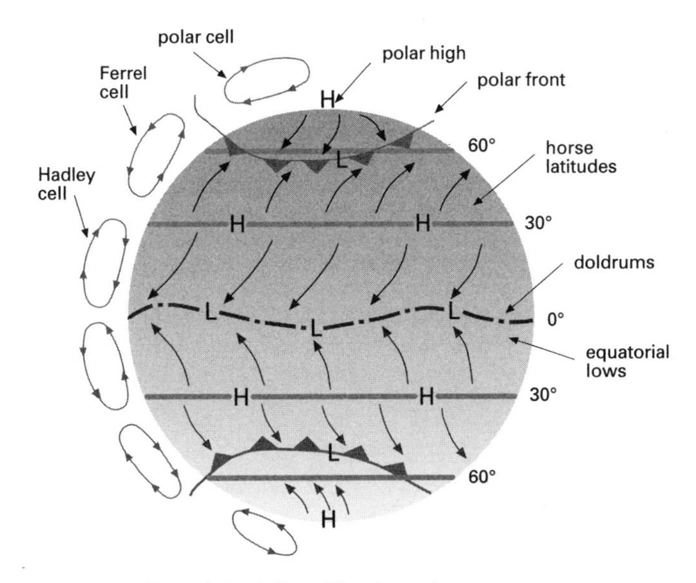

General circulation of the atmosphere

analyze potential effects of climate change using changes in greenhouse gases or other boundary conditions, or they may be used to simulate past climatic conditions for paleoclimatology studies. *See also* Atmospheric Model Intercomparison Project; climate model.

generator cell The shallow patch that produces or generates ice crystals in the tropospheric region in which CIRRUS cloud form. The cells are the sources for such crystals that fall through a layer that experiences WIND SHEAR so that they tend to be stretched out as they fall into a zone in which the wind speed increases. The generator cells are the upper fibrous ends to the hooks of cirrus that are part of the elongated filaments of the same cloud.

genitus A word added to a cloud type used in CLOUD CLASSIFICATION to signify that a new kind of cloud genus has been formed from the original cloud genus. For example, stratocumulus cumulogenitus is STRATOCUMULUS formed by the spreading of CUMULUS.

geographical information system (GIS) A computer system that includes the hardware, software, and data for capturing, storing, updating, manipulating, analyzing, and displaying geographical data. Within climatology and meteorology GIS has been used, for example, to produce maps on a global scale showing the incidence of natural hazards, including tsunamis and storm surges, tropical storms and cyclones, extratropical storms, and winter storms.

geopotential height The altitude of a given point or layer in the atmosphere in relation to the POTENTIAL ENERGY of a unit mass at that altitude above sea level.

Geostationary Meteorological Satellite (GMS) A series of satellites operated by the Japanese Meteorological Agency. It is one of the operational meteorological GEOSTATIONARY SATELLITES (currently five or six) positioned in orbit around the Earth, and is located above the Equator at 140°E.

Geostationary Operational Evironmental Satellite *See* GOES.

geostationary orbit A satellite orbit in which the satellite keeps pace with the Earth's rotation about its polar axis. This occurs in the Equatorial plane; the orbit must be eastbound, in the same direction as the Earth's spin, and have a period that is the same as the Earth's, i.e. one day. In order to achieve this, the satellite has to orbit at the geostationary elevation of about 35,800 km (22,230 miles) above the surface. It is moving very rapidly in space, but just keeping pace with the Earth so that it appears to hover above the same SUB-SATELLITE POINT. *See also* geostationary satellite.

geostationary satellite A satellite positioned at about 35,800 km (22,230 miles) above the Earth's surface in a GEOSTATIONARY ORBIT. From this high altitude, geostationary satellites can view the whole Earth disk and make frequent scans (about every 30–60 minutes). The spatial resolution of images from the geostationary satellites is about 2.5 km (1.5 miles) per pixel. The main *meteorological geostationary satellites* are positioned around the Equator to give full coverage of the Earth. Usually five or six geostationary satellites are in operation at any one time and these include the two GOES-series satellites, operated by the US NATIONAL OCEANIC AND ATMOSPHERIC ADMINISTRATION (NOAA; GOES-EAST over the US and South America at 75°W, and GOES-WEST over the Pacific at 135°W); METEOSAT operated by EUMETSAT (over Europe/Africa at 0°); GMS operated by the Japanese Space Agency (over Japan/Australia at 140°E), INSAT, operated by the Indian space agency (over the Indian Ocean), and GOMS operated by Russia (76°E). China also has a geostationary satellite, Fengyun–2 (over China and the Indian Ocean).

geostrophic balance *See* Coriolis effect.

geostrophic current In the oceans, a lateral rotary current that results from the balance between the horizontal pressure gradient force and the Coriolis force (*see* Coriolis effect). The current that results flows parallel to the pressure gradient. The major surface currents, for example, the Gulf Stream and West Wind Drift, are approximate geostrophic currents.

geostrophic wind (geostrophic flow) A theoretical steady non-accelerating wind that blows parallel to the isobars and results from the balance between the real horizontal PRESSURE GRADIENT FORCE and the fictitious but important CORIOLIS EFFECT. It is valid strictly for frictionless straight parallel flow. A geostrophic wind scale is sometimes printed on weather maps to aid the quick estimation of the wind speed from a mean-sea-level pressure analysis. The geostrophic wind (V_g) is defined as:

$$V_g = (2\omega \sin\theta)^{-1} . \partial p / \partial n,$$

where ω = angular velocity of the Earth (7.27×10^{-5} radians per second), θ is the latitude, and $\partial p/\partial n$ is the horizontal pressure gradient in Pa/m (1 mb = 100 Pa).

geosynchronous orbit An orbit in which a satellite orbits the Earth once in 24 hours and follows the same path.

GEWEX *See* Global Energy and Water Cycle Experiment.

ghibli (gibli) A hot and dry SCIROCCO-type wind that blows from south to north across the North African country of Libya. It is often experienced in the country's capital, Tripoli.

GIS *See* geographical information system.

GISP 2 *See* Greenland Ice Sheet Project 2.

glacial period *See* ice age.

glaciation 1. In meteorology, the forming of ice crystals from supercooled water droplets. When this occurs in the upper layer of a cumulonimbus cloud, the cloud is said to have a glaciated upper layer.
2. In climatology, the covering of land by glacial ice. *See also* ice age.
3. The process of glacier growth and modification of the land surface by glaciers.

glacier wind (glacier breeze, firn wind) A cold KATABATIC-type wind that blows down a valley glacier in the daytime, particularly during the summer. Velocities are greatest within 2 m (6 ft) of the surface. The glacier wind is a result of the cooling of air in contact with the ice; the higher air density over the glacier causing this air to sink. It is also observed on a much larger scale as *gravity winds* descending from the high-elevation ice sheets of Antarctica and Greenland.

glaze (glazed frost, clear ice) A coating of clear smooth ice, sometimes of considerable thickness, that forms as a result of rain or drizzle hitting a ground surface where the temperature, and that of the lower atmosphere, is below 0°C (32°F), the freezing-point of water. In North America this phenomenon is also referred to as *silver thaw*, so called because of the period of rapid warming that often follows a severe frost. The term BLACK ICE is used for glaze on a road or sidewalk. A period of heavy rainfall that gives rise to a deposit of glaze is frequently referred to as an ICE STORM in North America. *See also* freezing drizzle; freezing rain.

glazed frost *See* glaze.

Global Atmospheric Research Program (GARP) Formerly, a very large comprehensive program of research, sponsored by the World Meteorological Organization (WMO) that aimed to stimulate an improved understanding of mainly large-scale features of the atmospheric circulation. In the summer of 1974 it conducted the *GARP Atlantic Tropical Experiment* (GATE), an international experiment to further understanding of the tropical atmosphere and its role in the GENERAL CIRCULATION of the atmosphere.

The research covered a vast area from tropical Atlantic Africa to South America.

Global Atmospheric Watch (GAW) A WORLD METEOROLOGICAL ORGANIZATION (WMO) program that uses over 300 observation stations worldwide to provide data, assessments, and other information on the atmosphere's physical characteristics. The program was established in 1989 and provides data on greenhouse gases, ozone, radiation, optical depth, precipitation chemistry, chemical and physical properties of aerosols, and other meteorological parameters.

global circulation *See* general circulation.

Global Data-Processing System (GDPS) A system operated by WORLD WEATHER WATCH (WWW) to process meteorological data obtained from the GLOBAL OBSERVATION SYSTEM (GOS), and to make such data available, through the GLOBAL TELECOMMUNICATIONS SYSTEM (GTS), to the member countries of the WORLD METEOROLOGICAL ORGANIZATION (WMO). The GDPS provides meteorological analyses and a range of forecast products from nowcasts, through short-, medium-, long-, and extended range forecasts, to climate forecasts. The three-tier structure of the system includes global (*see* World Meteorological Center), regional, and national meteorological centers.

Global Energy and Water Cycle Experiment (GEWEX) A program that observes and models the hydrologic cycle and energy fluxes in the atmosphere, land surface, and upper oceans. It was established in 1988 by the WORLD CLIMATE RESEARCH PROGRAM (WCRP) and is primarily concerned with the effects of climate change.

global energy balance The annual energy balance of the Earth–atmosphere system; the result of all the linkages between the constituent energy components (radiation, sensible heat, latent heat, etc.) at all scales, including all feedback mechanisms. The annual energy inputs and outputs must approximate to zero over time if there is to be no net heating or cooling in the system. Averaged over a year for the whole of the Earth, the incoming and outgoing radiation is in balance. Averaged over latitude, incoming radiation is at a surplus between the latitudes of 40°N and 40°S (i.e. more energy is stored than re-radiated), while between the 40° latitudes and the poles there is a deficit, in which outgoing terrestrial radiation is larger than the incoming solar radiation. The redistribution of this energy through the atmospheric and oceanic circulation systems from low latitudes to high latitudes maintains the global energy balance. *See also* general circulation.

Global Observation System (GOS) A system operated by WORLD WEATHER WATCH (WWW) to provide high-quality and standardized meteorological observations from the facilities of the member countries of the WORLD METEOROLOGICAL ORGANIZATION (WMO). Observations are provided, for example, by stations on land and at sea, from aircraft, and from meteorological satellites, *See also* Global Data-Processing System; Global Telecommunications System; World Meteorological Center.

Global Ozone Monitoring Experiment (GOME) An instrument carried on the EUROPEAN REMOTE SENSING SATELLITE, ERS–2 (which was launched in 1995), to measure a range of atmospheric trace constituents, especially the levels and distribution of ozone globally. It consists of a nadir-viewing spectrometer that measures the solar radiation scattered in the ultraviolet and visible spectral range in the Earth's atmosphere, to derive measurements of such constituents as ozone, nitrogen dioxide, water vapor, bromine oxide, and other trace gases.

global positioning system (GPS) A satellite-based coordinate positioning tool and navigation system that can rapidly and accurately determine the latitude, longitude, and altitude of a point on or above the Earth's surface. It is based on a constellation of 24 satellites orbiting the Earth at a very high altitude and uses a form of triangulation based on the known positions and distances of three satellites relative to the surface of the Earth. First developed by the US Department of Defense to provide the military with a state-of-the-art positioning system, GPS receivers are now small enough and economical enough to be used by the general public. In meteorology and climatology, GPS receivers are increasingly used, for example, in RADIOSONDES, and have experimentally been used in the measurement of integrated (total column) precipitable water vapor.

Global Sea Level Observing System (GLOSS) An international program, coordinated by the Intergovernmental Oceanographic Commission (IOC), to establish sea level networks to provide high-quality data to climate, oceanographic, and coastal sea-level research. A network of nearly 300 sea level stations has been established around the world for long-term climate change and sea-level monitoring.

Global Telecommunications System (GTS) The international system for the dissemination of meteorological information, which is part of the WORLD WEATHER WATCH (WWW) of the WORLD METEOROLOGICAL ORGANIZATION (WMO). The system links together the three WORLD METEOROLOGICAL CENTERS (Melbourne, Moscow, and Washington), 15 Regional Telecommunications Hubs, and numerous national meteorological centers in a network of integrated meteorological telecommunications centers. The use of satellites is central to the system of data collection and data distribution at global, regional, and national level. *See also* Global Data-Processing System; Global Observation System.

global warming A warming of the Earth's surface temperature, which is responsible for changes in global climate patterns. While the term may be used broadly to encompass past warming events, such as interglacials, it is more commonly used to refer to recent global warming due to the enhanced GREENHOUSE EFFECT and anthropogenic influences. The quantities of many GREENHOUSE GASES within the atmosphere are rising, especially of CARBON DIOXIDE. During the past 200 years it has been estimated that carbon dioxide levels in the atmosphere have risen by 25–30% as a result primarily of changes in land use (e.g. deforestation) and the burning of fossil fuels (e.g. coal, oil, and natural gas). In addition, levels of methane, another greenhouse gas, doubled in the 100 years up to the year 2000. Since the late 19th century the globally averaged temperature of air at the Earth's surface has risen by 0.3–0.6°C (0.5–1°F). Complex computer models have been developed to try to predict the changes to climate that may occur as a result of increased greenhouse gas emissions and anthropogenic influences. One prediction is that if greenhouse gas emissions continue on a business-as-usual basis, the estimated average global temperature rise will be between 1° and 3.5°C (2° and 6°F) by the end of the 21st century. Some of the consequences envisaged by such warming include a rise of global sea levels of about 0.50 m (1.6 ft) during the 21st century, boundary shifts in the world's vegetation zones, extension of desertification, and a reduction in extent of polar ice and glaciers.

global warming potential (GWP) An index that measures the relative potential of various gases to contribute to greenhouse warming, and avoids the necessity to directly calculate the changes in atmospheric concentrations. The reference gas is carbon dioxide, and the global warming potential of other gases is given as a ratio of the radiative forcing that would result from the emission of that gas over a certain time period. It is calculated as the ratio of the radiative forcing that would result from the emission of 1 kilogram of a greenhouse gas to that from emission of 1 kilogram of carbon dioxide over a fixed time period (usually 100 years).

glory An optical phenomenon consisting of one or more colored concentric rings, which are red on the outer edge and violet on the inner edge; they surround the shadow of the observer's head, appearing on cloud, fog, or mist. The rings are seen at the antisolar point against the white background. The phenomenon is frequently seen by pilots around the shadow, which appears on clouds below, of the aircraft. Glories occur as a result of diffraction and reflection.

GLOSS *See* Global Sea Level Observing System.

GMS *See* Geostationary Meteorological Satellite.

GMT *See* Greenwich Mean Time.

GOES (Geostationary Operational Environmental Satellite) A series of operational meteorological GEOSTATIONARY SATELLITES positioned in orbit around the Earth and operated by the NATIONAL OCEANIC AND ATMOSPHERIC ADMINISTRATION (NOAA). Two are in position at any one time above the Equator: *GOES-EAST* at 75°W over the US and South America, and *GOES-WEST* at 135°W over the eastern Pacific Ocean. The first satellite in the series to be operated by NOAA (GOES–1) was launched in 1975.

GOME *See* Global Ozone Monitoring Experiment.

GPS *See* global positioning system.

gradient In meteorology, the rate of change of an atmospheric property (e.g. temperature, humidity, or pressure), usually in a horizontal or vertical direction. A FRONT is a zone across which there is a relatively steep horizontal gradient of temperature and humidity, for example. In contrast, within an AIR MASS such gradients are weak. *See also* pressure gradient.

gradient wind The theoretical wind that, unlike the GEOSTROPHIC WIND, is pertinent to curved motion around a high or a low. It represents the balance between three forces acting upon the movement of air in a horizontal plane: the PRESSURE GRADIENT FORCE, the Coriolis force (*see* Coriolis effect), and the CENTRIFUGAL FORCE, the latter reflected by the curvature of the isobars around a pressure system. The gradient wind refers to frictionless motion.

The gradient wind speed (V m s^{-1}) is defined as:

$$V^2/r = (2\omega\sin\theta)(V_g - V),$$

where r is the radius of curvature of the air's path (m) and V_g is the geostrophic wind speed. The V^2/r term is positive (negative) when the motion is around a low (high) pressure area. Insertion of the requisite numbers will lead to a quadratic equation in V.

grass minimum temperature A measure of the lowest temperature recorded overnight from a standard MINIMUM THERMOMETER, the bulb of which is in contact with the tips of blades of short grass, 25–50 mm (1–2 in) above the ground.

grass minimum thermometer A MINIMUM THERMOMETER in which the bulb is in contact with short grass, in an open area. In this case the minimum thermometer records the night-time minimum grass temperature.

graupel (snow pellet, soft hail) Small white opaque ice grains 2–5 mm (0.08–0.2 in) and approximately spherical or conical in shape, that resemble tiny snowballs. Graupel form by accretion when supercooled water droplets in the atmosphere come into contact with falling ICE CRYSTALS or SNOWFLAKES; the droplets freeze on contact, covering the crystals or snowflakes. The term graupel is also applied to pellets of partially melted and refrozen snow, each with a soft core. Snow pellets often fall in showers of snow; unlike SNOW GRAINS, they often bounce and sometimes break up when they hit the ground. *Small hail* refers to grains having a similar diameter but glazed with clear ice, which is more likely to be associated with rain than snow. The word graupel is derived from German, in which it means soft hail. *See also* hail.

gravity wave A type of atmospheric wave in which the pertinent forces are those of gravity and buoyancy. They are manifested as wave motion on the interface between adjacent layers within the atmosphere, for example, and can be set off by the air flowing over a developing CUMULUS cloud – in much the same way as a hill or mountain can stimulate downstream undulations in the flow.

green flash A glimmer of bright green light that occasionally is seen to fleetingly color the edge of the Sun as it disappears from view on the horizon at sunset or just as it appears at sunrise. It is most likely to be seen at locations with a low distant horizon, for example, over the ocean. The green flash is caused by differential refraction of light waves. One view is that the green flash is illusory as a result of looking at the Sun.

greenhouse effect The warming effect that results from short wave (infrared, visible, and ultraviolet) solar radiation being largely able to pass unhindered to the surface of the Earth, where it is re-radiated as longer wave (infrared) radiation (as happens in a greenhouse). This outgoing long-wave radiation is partially absorbed by GREENHOUSE GASES (especially water vapor and carbon dioxide) in the atmosphere. The failure of most of the radiant energy to escape means that it is 'recycled' and retained as heat in the lowest part of the atmosphere, an important component in maintaining the Earth's surface temperature. Without greenhouse warming the Earth's average surface temperature would be around –18°C (0°F) and unable to support life.

The natural effect of the recycling of radiant energy is, however, being enhanced by increases in the concentration of greenhouse gases, particularly of carbon dioxide (CO_2), owing to human activity, notably since the beginning of the Industrial Revolution (around 1700 AD). The burning of FOSSIL FUELS (oil, coal, and natural gas) and the clearing of land and burning of vegetation, in particular, contribute to the rise of carbon dioxide concentrations. It is predicted that the *enhanced greenhouse effect* resulting from the increased concentrations of greenhouse gases will generate increased GLOBAL WARMING and contribute to global climate change. Computer models have been used to attempt to predict the amount of warming that may result from further increases in greenhouse gas levels. One complication is the uncertainty as to how global warming will affect levels of water vapor, which have so far remained relatively constant, in the atmosphere.

greenhouse gas A gas that is mainly transparent to solar radiation but absorbs strongly in the infrared part of the electro-

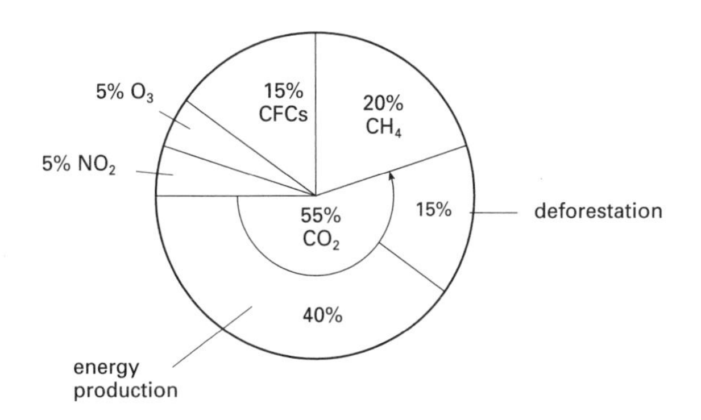

Greenhouse effect: the relative effects of various greenhouse gases taking into account their abundance and effectiveness as absorbers.

magnetic spectrum and re-emits such long-wave radiation in all directions. The absorption of this infrared radiation leads to warming of the air. A number of gases contribute to the natural GREENHOUSE EFFECT, the most important of which is water vapor. The other greenhouse gases include CARBON DIOXIDE (CO_2), METHANE (CH_4), NITROUS OXIDE (N_2O), CHLOROFLUOROCARBONS (CFCs), PERFLUOROCARBONS (PFCs), HYDROFLUOROCARBONS (HFCs), and OZONE. Increases in the concentrations of carbon dioxide, in particular, and methane, together with the presence of the anthropogenic greenhouse gases is contributing to an intensification of the greenhouse effect. Under the KYOTO PROTOCOL, restrictions were to be applied to emissions of six of the greenhouse gases. *See also* global warming.

greenhouse period A period during the geologic history of Earth during which there were no glaciers, and sea levels were high. *Compare* icehouse period.

Greenland Ice Core Project (GRIP) A European research project organized through the European Science Foundation aimed at understanding past climatic conditions, based on ice cores obtained from the Greenland ice sheet. Its duration was from January 1989 to December 1995 and its main aim was to retrieve and analyze a 3000-m (9800-ft) long ice core drilled through the Greenland ice sheet at the summit, 28 km (17 miles) east of the GREENLAND ICE SHEET PROJECT 2 site.

Greenland Ice Sheet Project 2 (GISP 2) A US research project, conducted from 1989 to 1994, to extract ice cores from the Greenland ice sheet for studies of atmospheric and past climatic conditions. Located 28 km (17 miles) west from the GREENLAND ICE CORE PROJECT site, near the Greenland summit, GISP 2 recovered a 3045-m (9990-ft) long core, with the potential to yield over 110,000 years of global change history.

Greenwich Mean Time (GMT) The timescale based on the apparent motion of the mean Sun with respect to the meridian of 0° longitude at Greenwich, England. The standard times for different areas of the world are calculated from this; 15° longitude representing 1 hour in time. In 1928 the term UNIVERSAL TIME (UT) was adopted for Greenwich Mean Time when used for scientific purposes. The timescale has been replaced by COORDINATED UNIVERSAL TIME (UTC) to which it is approximately equivalent.

GRIP *See* Greenland Ice Core Project.

ground clutter RADAR echoes received at the radar antenna that have been reflected from the Earth's surface or buildings within some 30 km (20 miles) of the antenna. They are fixed features, and are often removed automatically from a radar display in order not to confuse the operational image of, for example, precipitation.

ground fog *See* radiation fog.

ground frost A condition in which the temperature of the ground surface measured at short-grass level is at or below 0°C (32°F). Ground frost is frequently used in weather forecasts to indicate that a frost may occur at ground level although the air temperature remains above 0°C (32°F); in such circumstances, an AIR FROST does not occur at the same time.

ground heat flux The vertical movement of energy across the surface of the Earth. The flux is positive (downward) during the day and negative (Earth to atmosphere) at night in the absence of warm or cold ADVECTION. It is an important component (G) of the ENERGY BUDGET.

groundhog day In North American weather folklore, the day (February 2) on which, if a groundhog emerging from its burrow sees its shadow, winter will last another six weeks; if not there will be an early spring. The rationale behind this is that if the day is clear and sunny it is probably cold and anticyclonic, conditions that are likely to persist, whereas no shadow implies a cloudy milder day.

ground ice 1. (anchor ice) Minute needles of frazil ice that adhere as spongy accumulations attached or anchored to the bed of a body of moving water, which itself remains unfrozen.
2. A body of clear ice found in the ground surface of the permafrost zone in which temperatures have remained continuously below 0°C (32°F); for more than two years.

ground mist A MIST formed by radiative cooling of damp air at night under clear calm conditions in low-lying areas and valley floors.

ground stroke *See* cloud-to-ground discharge.

growing season The period of the year during which air temperatures remain sufficiently high to permit plant growth. In the US, it is defined as the period between the last killing frost in spring and the first killing frost in the autumn, i.e. between the last day in spring with a minimum temperature of 0°C (32°F) or lower, and the first day in the autumn with a minimum temperature of 0°C (32°F). In the UK a temperature of 6°C (43°F) is preferred to 0°C (32°F) due to the different temperature regime on the eastern seaboard of the North Atlantic Ocean. The average length of the growing season may be obtained from the curve of annual variation in the mean temperature and may be expressed in growing degree units.

GTS *See* Global Telecommunications System.

Guinea Current A warm current in the Atlantic Ocean, an extension of the EQUATORIAL COUNTERCURRENT, that flows east along the Guinea coast of west Africa, into the Gulf of Guinea.

Gulf Stream A warm ocean current that flows in the North Atlantic along the coast of east North America between Cape Hatteras, North Carolina, and the Grand Banks of Newfoundland, Canada. The name is frequently extended in common usage to describe the system of connected warm Atlantic Ocean currents, including the FLORIDA CURRENT, the Gulf Stream, and the NORTH ATLANTIC DRIFT. The Gulf Stream system is fed by the westward-flowing North EQUATORIAL CURRENT and flows north from the Gulf of Mexico, through the Straits of Florida, and along the east coast of the US, where it is approximately 80 km (50 miles) wide and has an average speed of 3.5 knots (4 mph). It leaves the coast at around 40°N and becomes at around 45°W the North Atlantic Drift, which crosses the Atlantic Ocean toward northwestern Europe.

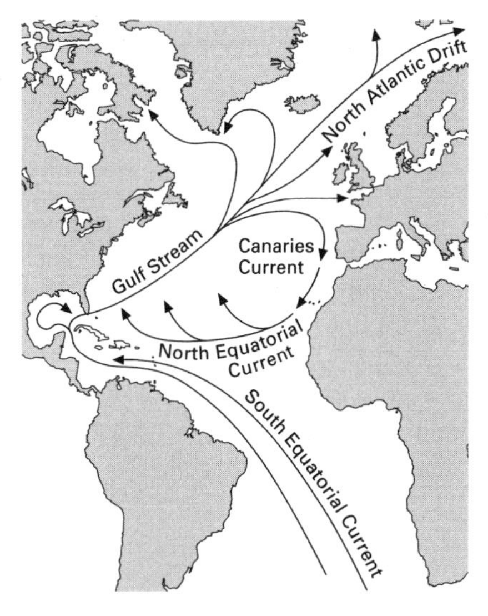

Gulf Stream

gust A rapid but short-lived increase in wind speed experienced, in general, close to the ground. It is caused primarily by mechanical disturbance of an airflow and *gustiness* increases when the surface is aerodynamically rough, for example, as occurs in towns and cities.

It can also be generated by wind shear (e.g. in CLEAR-AIR TURBULENCE) and by rapid changes in temperature with height.

gust front The leading edge of a sudden and often dramatic increase in wind speed (and direction) linked to, for example, the outflow from a THUNDERSTORM in the form of a MICROBURST that has reached the surface and fanned out sideways. A SQUALL is more prolonged than the increased wind associated with a gust front. A HABOOB is an example of a sand and dust-laden current that follows a gust front in desert regions.

GWP *See* global warming potential.

gyre In oceanography, the large circular flows of water within the ocean basins that generally rotate clockwise in the northern hemisphere and counterclockwise in the southern hemisphere. There are five major gyres in the Earth's oceans: two each in the Pacific and Atlantic Oceans (North and South) and one in the Indian Ocean.

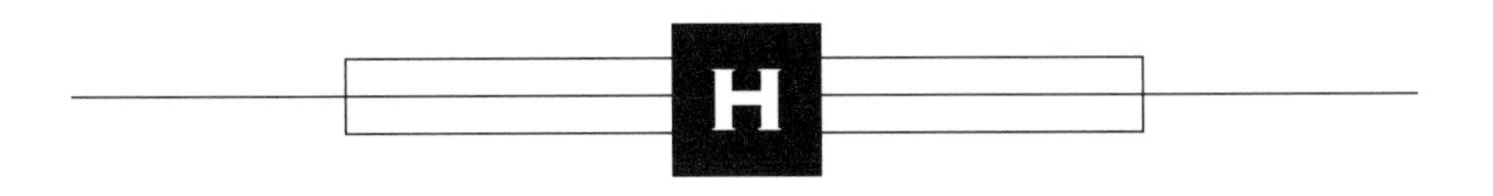

haar A regional term in Scotland and northeast England for ADVECTION FOG that has drifted inland from the North Sea across southeast and eastern Scotland and northeast England. It occurs most often in the spring and early to mid summer, and is generally associated with miserable weather. *See also* fret.

haboob A type of dust storm that occurs in northern Sudan, sending clouds of sand and dust into the atmosphere and resulting in poor visibility. It is most common in the afternoon and evening between May and September, but can occur at any time of the year. The dust storm is accompanied by a noticeable drop in temperature together with a sudden increase in wind speed and change in direction. The haboob is mostly a dry squall, and is thought to result from downdrafts in large cumulonimbus clouds.

Hadley cell (tropical cell) A major feature of the GENERAL CIRCULATION of the atmosphere, confined mainly to the tropics. There are two Hadley cells that occur as TROPOSPHERE-deep features in the north–south vertical plane. Their low-level signature is the NORTHEAST and SOUTHEAST TRADE WINDS that are best marked over the tropical oceans; it is these two currents that converge into the Intertropical Convergence Zone (ITCZ). This elongated zone of CONVERGENCE is associated with very deep CUMULONIMBUS clouds within which there is a net upward transport of very warm moist air. Aloft, above the low-level convergence, the air flows away northward and southward from an east–west elongated zone of DIVERGENCE in the upper TROPOSPHERE. The air in the two upper currents gradually cools and sinks to the lower troposphere at around 30° latitude, on average, in either hemisphere. The deep subsidence in this region is associated with the SUBTROPICAL ANTICYCLONES, from which one low-level current flows Equatorward, as the trades, to complete the cell. Hadley cells migrate seasonally, so are not often symmetrically positioned about the Equator. They are furthest poleward during the respective hemisphere's summer.

A simple thermal cell was originally proposed in 1735 by George Hadley (1685–1765) to explain the movement of tropical warm air toward the poles and cold polar air toward the Equator, and also the origin of the trade winds.

Hadley Centre for Climate Prediction and Research A major national center in the UK, funded by the Department of the Environment, that provides the UK with up-to-date assessments of changes in climate. Opened in 1990, it is part of the MET OFFICE, the national meteorological service of the UK. Its role is to understand the processes responsible for, and to simulate, the present-day climate through the development of state-of-the-art computer models. In addition, it is a major international center for the simulation of past and future global climate over a century or so, for monitoring actual climatic fluctuations, and to explain such features.

hail A form of PRECIPITATION comprised of hard pellets of ice (*hailstones*) that have a concentric structure. A hailstone normally has a diameter of between 5 and 50 mm (0.2–2 in) but some have been recorded weighing in excess of 1 kg (2.2 lb). The largest hailstone recorded in the US fell in Coffeyville, Kansas, on September 3, 1970; it had a diameter of some 14 cm (5.6 in) and weighed 757 g (27 oz). Hail causes considerable damage to crops and property.

Associated with the rapid ascent of moist air, a hailstone usually develops around an ice nucleus in a cumulonimbus cloud in which strong updrafts and downdrafts are characteristic; it is sometimes associated with thunderstorms. The ice nucleus acquires fresh layers of ice by collision and coalescence with supercooled water droplets as it is borne upward, creating a concentric structure – clear layers when the humidity is low and opaque layers when the humidity is higher alternating within the hailstone. Several such layers may form before the hailstone acquires the terminal velocity to overcome the pronounced updrafts in the cloud and fall to the ground. *See also* graupel.

hair hygrometer A mechanical HYGROMETER that measures changes in RELATIVE HUMIDITY by the changes in the length of a strand of human hair, which expands in length as relative humidity increases.

halo One of a number of optical phenomena that appear as colored or whitish rings and arcs around the Sun or Moon when seen through an ice-crystal cloud, typically cirrostratus, or through ice crystals suspended within the atmosphere, such as diamond dust. When colored, the rings or arcs are red on the innermost edge and violet or blue on the outside edge (the opposite to the color sequence in a CORONA). Haloes occur as a result of the REFRACTION of light rays by the ice crystals. Points on the halo are an angular distance of 22° (the most common halo) or 46° from the Sun or Moon.

halocarbon A chemical compound consisting of carbon, possibly hydrogen, and any of the halogens, i.e. chlorine, fluorine, bromine, or iodine. *See also* chlorofluorocarbon; halon; hydrochlorofluorocarbon.

halon (bromofluorocarbon) A group of chemical compounds: CHLOROFLUOROCARBONS (CFCs) that have been compounded with bromine (to be used as fire retardants, for example). They are associated with changes in the concentration of stratospheric OZONE and many were phased out under the MONTREAL PROTOCOL.

harmattan A dry and often strong northeasterly wind that blows from the Sahara desert to the Gulf of Guinea, affecting countries such as Mali, Senegal, Guinea, and Sierra Leone in West Africa. It is hot, dusty, and unpleasant when experienced inland in the Sahel zone but the dryness of the harmattan can bring welcome relief from the high humidity experienced along the coastlands of the Gulf of Guinea. In light of its beneficial health effects, the harmattan is referred to locally as the DOCTOR. Usually, but not exclusively, affecting locations north of the Equator, the wind's southern limit in the northern winter is typically 5°N; this limit is 17°N in the northern summer.

Hawaiian high *See* North Pacific high.

haze A suspension of very small dry particulate matter in the atmosphere such as pollutants and/or natural aerosols (dust, salt particles, and smoke, for example), that produces a milky sky. It is used officially as a term if the concentration of particles is sufficient to give the sky an 'opalescent' look. The particle sizes are such (less than 1 μm) that they contribute to dramatic colors at sunrise and sunset by their differential scattering of sunlight traveling through the hazy air.

HCFC *See* hydrochlorofluorocarbon.

heat The energy that is transferred from one body or region to another as a result of a difference in temperature. It is measured in joules or calories.

heat balance A component part of the energy balance. In a limited sense, the heat balance may provide an estimate of the energy balance of the Earth's surface by converting other components into units of heat. This was a favored method of analysis in the mid-twentieth century. *See also* global energy balance.

heat capacity The amount of heat required to change the temperature of a substance by 1°C. Substances with a high heat capacity, such as water, require a large amount of heat to produce a small rise in temperature. Thus changes in ocean surface temperatures are slow, compared with those in the atmosphere. The *specific heat capacity* is the amount of heat required to raise the temperature of unit mass of a substance by 1°C, usually expressed in joules per kilogram.

heat index (HI, apparent temperature) An index, devised by the US National Weather Service, to provide a warning of conditions in which a combination of high temperatures and high relative humidity, as in a heat wave, can be dangerous to human health. It gives, in °F, a measure of what the recorded temperature in combination with the relative humidity feels like in reality. The *heat index chart* can be used to find the HI; for example, if the air temperature is 96°F (35.5°C) and the relative humidity is 55%, then the HI is 112°F.

heat island A dome of warm, frequently polluted, air above an urban area in which the temperature is higher than that in the surrounding rural areas. Urban areas, irrespective of their size, may be warmer than their surroundings since they are centers of energy storage (buildings) and energy activity (such as domestic heating, industrial processing, and transportation). A study in Atlanta, during a one-month period in July 1996, found that temperatures averaged 8–10°F higher in the urban area than in the surrounding rural areas. This can create localized weather effects, such as convection, producing thunderstorms. The heat island effect is most pronounced around sunset during spells of calm weather. This occurs when the greater vegetation surface area in rural locations allows greater radiative cooling to take place.

heat lightning Lightning observed from such a considerable distance away that any associated thunder is inaudible. Heat lightning is frequently associated with the misconception that the warmth of a cloudless summer evening is, of itself, sufficient to cause the diffuse lightning in the absence of a thundercloud.

heat low (thermal low) A synoptic-scale region of low pressure or CYCLONE formed by strong heating of land masses, for example, in spring and summer in subtropical regions. The marked warming of the surface leads to a vertical expansion of the overlying air, and an associated fall of surface pressure. There are strong seasonal heat lows over the western Sahara and southwestern US, and also occasional developments over the Iberian peninsula.

heat wave A prolonged period of abnormally hot and humid weather. These events can pose serious health issues as the young and elderly, in particular, are vulnerable to the impact of excessive heat.

Heaviside–Kennelly layer *See* E-layer.

hectopascal (hPa) A unit of ATMOSPHERIC PRESSURE that is a multiple (100) of the PASCAL, the SI unit of pressure. One hectopascal (hPa) is equal to 1 MILLIBAR (mb). While the millibar (mb) is the c.g.s. unit still often used in everyday situations (e.g. by meteorological observers and in television weather forecasts), it has been replaced by the hectopascal in most scientific work and in publications, such as scientific journals.

Heinrich event An event characterized by the marked discharge of large numbers of icebergs from the eastern North American ice sheet during glacial periods. They occur at the end of cooling periods, about every 7000–10,000 years, and are generally indicative of a shift toward warmer conditions. Evidence for these events from marine core samples in the Atlantic was first observed by the German scientist Helmut Heinrich in 1988.

Heinrich layer Sediments occurring in marine deposits in the North Atlantic that contain a high proportion of ice-rafted debris to foraminifera deposits, and thus provide evidence of surges in the North American ice sheet, and increased calving of icebergs, pointing to frequent shifts in the Earth's climate.

helicity In a moving fluid, such as air, the component of spin of a particle on the direction of propagation. It is used in the forecasting of the formation of MESOCYCLONES.

helm wind A strong, cold, and gusty east to northeast wind in northern England that blows down the western slopes of the Cross Fell Range (maximum elevation 893 m (2930 ft)) into the Vale of Eden; this being the valley that separates the Pennine Mountain Range from the Lake District. It occurs most often in late winter and spring, and when it blows, a stationary *helm cloud* (a BANNER CLOUD) envelops the mountain top. Downwind in the Vale of Eden and parallel to the helm cloud, a thin roll of cloud (*helm bar*) forms as a result of eddies and turbulence in the lee of the high escarpment (an example of LENTICULARIS).

heterogeneous nucleation **1.** The formation of a pure water droplet in the atmosphere by condensation from a supersaturated vapor with the aid of atmospheric aerosols (*see* cloud condensation nucleus).
2. The formation of an ice crystal in the atmosphere from liquid water that collects and freezes around a foreign particle, for example dust, clay, or an atmospheric aerosol. It occurs mostly at temperatures less than –10°C (14°F). *See* freezing nucleus.

Compare homogeneous nucleation.

heterosphere The region of the atmosphere found above the TURBOPAUSE that contrasts markedly with the underlying HOMOSPHERE. The structure of the heterosphere depends on temperature, which varies in response to solar activity. The lower part is characterized by large numbers of free oxygen atoms produced by the photodissociation of diatomic oxygen. The heavier gases by molecular weight tend to settle in the lower levels of the heterosphere; the higher levels are dominated by the lighter gases, such as helium and hydrogen.

HFC *See* hydrofluorocarbon.

HI *See* heat index.

high *See* anticyclone.

high clouds In CLOUD CLASSIFICATION, those clouds that have a base height above the surface of roughly 5000 m (16,000 ft) or more (in middle latitudes). This criterion varies seasonally (higher in the summer) and latitudinally (higher at low latitudes). High clouds are composed predominantly of ice crystals and/or supercooled water droplets and are usually thin and white but reflect the colors of red, orange, and yellow at sunrise and sunset. They all have the prefix CIRRO-. *See also* cirrocumulus; cirrostratus; cirrus.

high-index circulation A type of atmospheric circulation that occurs when the air flow on a large scale is principally westerly, i.e. when it has a large value of the ZONAL INDEX.

highland climate The climate of the highland regions of the world, such as the Andes of South America, Cascades, Sierra Nevada, and Rocky Mountains of North America, the Himalayas, and central Borneo and New Guinea. The climates of the highlands vary considerably but in general, temperature, humidity, and pressure all decrease with height, and receipt of solar radiation is increased by the greater transparency of the atmosphere, due to lower dust content with height.

high-latitude climate The group of climates of the high-latitude regions of the world identified in the STRAHLER CLIMATE CLASSIFICATION: the boreal forest climate, tundra climate, and ice sheet climate, which are dominated by the polar and arctic air masses.

high-precipitation storm (HP storm) A SUPERCELL thunderstorm that produces very heavy precipitation often with prolonged DOWNBURST events, major FLASH FLOODS, and damaging hail episodes. The heaviest precipitation falls from the trailing side of the MESOCYCLONE within the supercell. *Compare* low-precipitation storm.

high-resolution picture transmission (HRPT) A method of encoding and transmitting high-resolution imagery, which is broadcast continuously, from the POLAR-ORBITING SATELLITES. *Compare* automatic picture transmission.

high-resolution satellite imagery The best quality images provided by satellite sensors. For meteorological satellites, the high-resolution images typically have PIXELS of about 1 km square at the SUBSATELLITE POINT. There is no great need operationally to have a resolution better than this for mapping cloud and water-vapor features.

high Sun 1. The period of around one hour either side of the maximum elevation of the Sun in the sky in summer in middle-latitude climates.
2. In general usage, the high elevation of the Sun in low latitudes.

hill fog Fog that occurs in hilly upland areas when the cloud base of STRATUS cloud is lower than the high ground; the tops of hills are occasionally shrouded in stratiform cloud for many hours – or sometimes for longer. *Compare* upslope fog.

hill wave *See* lee wave.

hoar frost (white frost) A deposit of patterned ice crystals with a feathery and needle-like appearance that forms on surfaces near the ground, especially on vegetation, when the temperature of the surface is below 0°C (32°F) and the air is saturated (i.e. at the FROST-POINT). Hoar frost is composed partly of ice crystals that are deposited directly from water vapor on to the surfaces by sublimation. Radiation cooling of the surface on calm and clear nights ensures that any condensation that occurs freezes when the temperature drops below 0°C (32°F). *See also* frost; dew.

hodograph A diagram using polar coordinates upon which wind vectors are plotted. The center of the chart represents the position of the station toward which all winds blow. Once the vectors are plotted, the points are joined. This diagram represents the differences in wind speed and direction between different layers in the atmosphere.

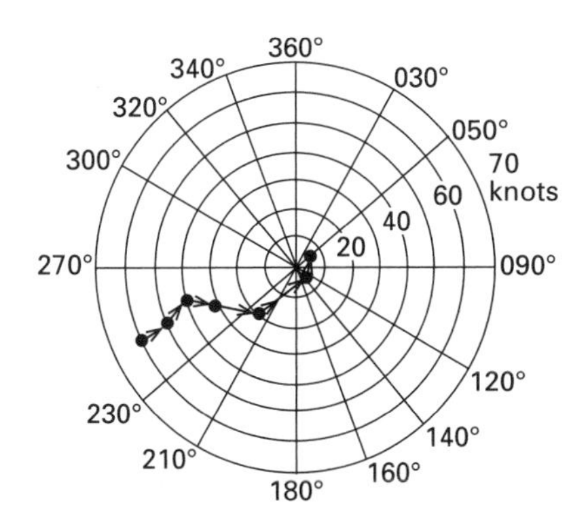

Hodograph

homogeneous nucleation 1. The formation of a pure water droplet by condensation from a supersaturated vapor without the presence of atmospheric aerosols to act as nuclei. Water droplets rarely form in natural clouds by homogeneous nucleation.
2. The formation of an ice crystal from liquid water molecules without the presence of atmospheric aerosols to act as nuclei. Homogeneous nucleation is most effective in high-level clouds at temperatures below –40°C (–40°F).
Compare heterogeneous nucleation.

homosphere The well-mixed region in the atmosphere lying below the TURBOPAUSE. The composition of this layer, which extends from the Earth's surface to around 85 km (50 miles), tends to be constant with an even mix of gases including nitrogen, oxygen, and argon (water vapor, and small concentration gases, such as carbon dioxide and ozone, are exceptions). This region contrasts sharply with the overlying HETEROSPHERE.

hook echo A RADAR echo pattern in the form of a distinct small-scale feature resembling a hook that frequently appears on radar images in tornadic situations. The hook shape is related to a narrow band of precipitation that is being drawn in to part of a tightly curved vortex in the flow associated with a SUPERCELL. Such hook echoes are sometimes, but not invariably, indicators of the presence of a TORNADO; they are always associated with severe weather and turbulence.

horizontal visibility *See* visibility.

horse latitudes The latitudes of the SUBTROPICAL ANTICYCLONES, at around 30° to 40° across the oceans. The name is said to originate from the practice of throwing horses overboard if the trading sailing vessels of the time were becalmed for considerable periods. This occasionally occurred in the light winds near the centers of the highs.

hot tower Large cumulonimbus clouds found at the INTERTROPICAL CONVERGENCE ZONE that draw latent and sensible heat primarily from underlying warm ocean surfaces up into the troposphere. Collectively, these hot towers represent the ascending limb of the HADLEY CELL.

hPa *See* hectopascal.

HRPT *See* high-resolution picture transmission.

Humboldt Current *See* Peru Current.

humid continental climate The climate of the areas between about 30° and 60°N in central and eastern North America and Asia, within the POLAR FRONT zone, in which there is conflict between the polar and tropical air masses. Summers are warm and winters cold with large temperature ranges between the seasons. Precipitation is abundant, generally cyclonic in the winter, with a summer maximum from convectional rainfall. Tornadoes occur most often in the early summer, especially in the US. The weather associated with this climate is characteristically changeable. It is denoted by Dfa, Dfb, Dwa, and Dwb in the KÖPPEN CLIMATE CLASSIFICATION.

humidity The concentration of WATER VAPOR in the atmosphere, expressed in meteorology in a variety of ways. The main operational method of measuring it is by sensing the WET-BULB TEMPERATURE, normally measured within a weather screen. Comparing this temperature with the simultaneously measured DRY-BULB TEMPERATURE is the basis for calculating the air's RELATIVE HUMIDITY or the nearness to saturation of the air. A value of 100 means the air is saturated, while one of 30 to 40 is typical for the air in a house. Meteorologists find an absolute measure of the amount of water vapor very useful. Its concentration in grams per kilogram, for example, is essential for calculations relating to how much precipitation may occur. *See also* absolute humidity; specific humidity.

humidity mixing ratio *see* mixing ratio.

humid subtropical climate The climate of the eastern parts of the continents between 20° and 30° latitude N and S, in North America (extending from the Carolinas to E Texas), South America (parts of Brazil, Uruguay, and Argentina), Australia, and parts of China, Taiwan, and southern Japan. Moist air (mT) flows from the west sides of the subtropical anticyclones located over the oceans. Precipitation is generally uniform throughout the year; temperatures in the warmest months are around 27°C and in the coldest month 5°–12°C. In winter frontal cyclones develop along the polar front; TORNADOES occur in North America along the zone between the polar and tropical maritime air. The climate is denoted by Cfa and Cwa in the KÖPPEN CLIMATE CLASSIFICATION.

humilis A cloud type or species of low CUMULUS in which the clouds are relatively small and do not grow much vertically. They are normally scattered, have sharp outlines with white or light-gray flat bases all at the same level, and flattened tops – the vertical growth is usually limited by the presence of a SUBSIDENCE INVERSION. Cumulus humilis is often also known as FAIR-WEATHER CUMULUS. *See also* cloud classification.

hurricane The name (of Spanish/Portuguese origin) for a TROPICAL CYCLONE with wind speeds in excess of 64 knots (73 mph), force 12 on the BEAUFORT SCALE, that is particularly associated with storms occurring in the North Atlantic and Gulf of Mexico, as well as the eastern seaboard North Pacific Ocean. In the North Atlantic these storms usually develop off the west coast of North Africa and move west/northwest toward the Caribbean and southern US. Hurricanes develop over warm water of at least 27°C (81°F) permitting appreciable evaporation of water vapor and subsequent latent heat release on cooling, which generates their motion. Their rotation results from the CORIOLIS EFFECT, therefore they cannot form on the Equator where the Coriolis acceleration is

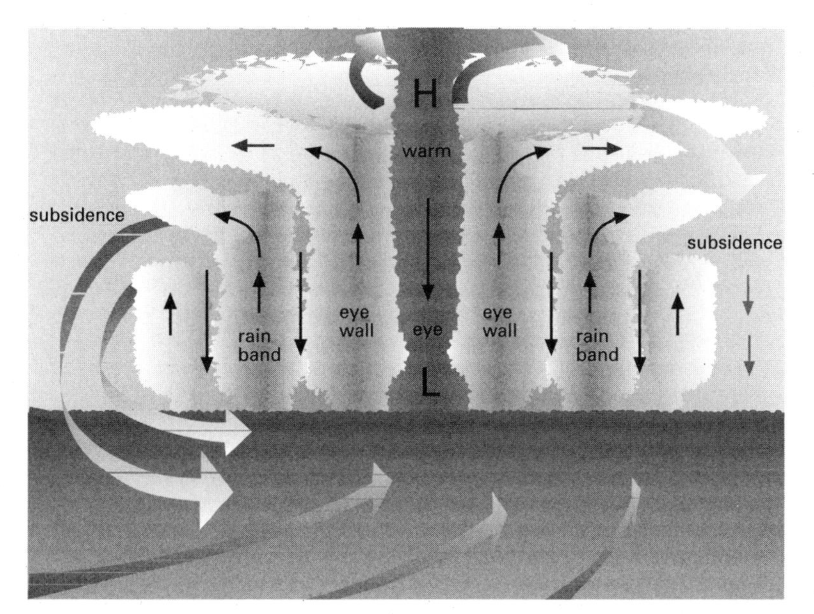

Vertical cross-section of a hurricane (tropical cyclone)

zero. Their development is in the INTERTROPICAL CONVERGENCE ZONE where the trade winds meet. This zone changes with the season from about 15°N between July and October to 10°–15°S between January and March. These periods represent the principal cyclone season in the respective hemispheres.

Hurricanes range from 80 km (50 miles) to 800 km (500 miles) across and the winds form an almost circular vortex with a slight inward deflection near the ocean surface. Sustained wind speeds in excess of 87 knots (100 mph) are common in full-scale storms and intense storms of this type are persistent over the sea but dissipate when deprived of the oceanic moisture source when they reach large land or even small islands. Central pressures below 880 hPa have been recorded at the center of hurricanes although a 960 hPa central pressure with peripheral pressure of about 1020 hPa is more common. These values are no more extreme than a severe mid-latitude CYCLONE but an intense tropical cyclone is much smaller in diameter with a much steeper pressure gradient. *See also* typhoon.

hurricane modification An attempt to alter the intensity of hurricanes. The most common technique is that of seeding used in the US in the 1960s, 1970s, and 1980s in attempts to modify the structure of hurricanes. This technique utilizes silver iodide, which causes supercooled water to freeze removing some of the latent heat energy used by the hurricane as fuel; as a result the hurricane is weakened. In recent years hurricane modification has been viewed with caution as hurricanes play an important role in balancing the Earth's energy budget. Altering hurricanes could therefore have wide-reaching implications.

hurricane strength scale *See* Saffir–Simpson Scale.

HWRP *See* Hydrology and Water Resources Program.

hydraulic jump An abrupt wave that occurs in the flow of a fluid. It occasionally occurs in the atmosphere when, for example, the middle and upper tropospheric flow approaches a mountain chain aligned at right angles to the wind's direction. There is then a relatively marked jump in the fluid as it passes up and over the elongated ridge. The fairly steep ascent on the upwind side, followed by more-or-less horizontal flow in the upper troposphere downwind, is often marked by a very extensive sheet of CIRROSTRATUS that streams away from the mountain ridge.

hydrocarbon A range of organic compounds formed of hydrogen and carbon, for example, fossil fuels. Some hydrocarbons are serious air pollutants.

hydrochlorofluorocarbon (HCFC) A compound containing hydrogen, chlorine, fluorine, and carbon. The HCFCs were introduced as temporary replacements for the CHLOROFLUOROCARBONS (CFCs), which were to be phased out under the MONTREAL PROTOCOL. The HCFCs are themselves also ozone-depleting substances, but not at the levels of the CFCs, and they have shorter atmospheric lifetimes. Their production is limited by the Montreal Protocol with phasing out of their use by developed countries in 2003.

hydrofluorocarbon (HFC) A compound containing carbon, hydrogen, and fluorine. The hydrofluorocarbons were introduced as replacement compounds for chlorofluorocarbons (CFCs), which are being phased out under the MONTREAL PROTOCOL, and are used, for example, as refrigerants in refrigeration and air-conditioning units and in aerosols. They do not destroy ozone but are powerful GREENHOUSE GASES and were mentioned in the KYOTO PROTOCOL as one of six key greenhouse gases.

hydrologic cycle (hydrological cycle, water cycle) The continuous events through which water passes in the Earth-atmosphere system, encompassing EVAPORATION from water bodies and land and TRANSPIRATION from land-based plants, cloud formation through CONDENSATION,

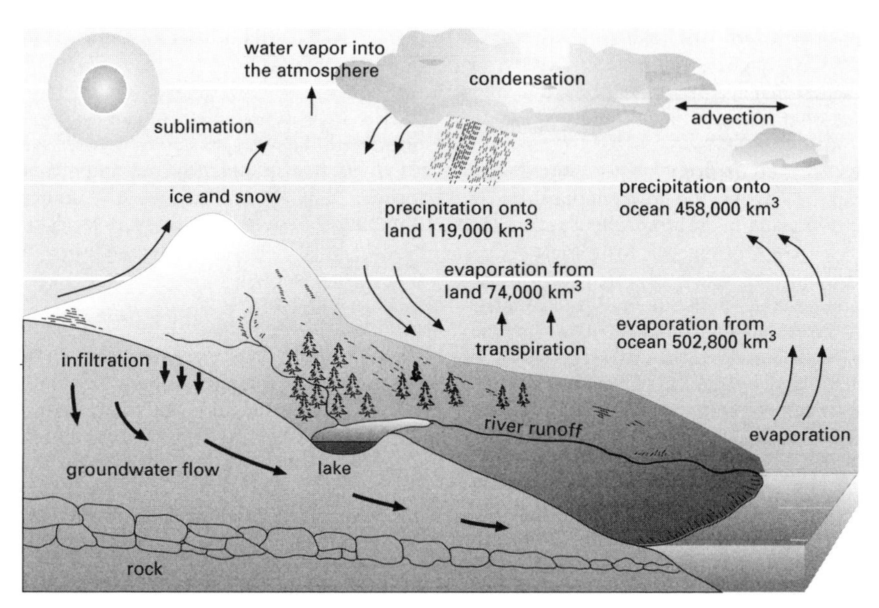

Hydrologic Cycle: the annual turnover of water on Earth

PRECIPITATION to the Earth's surface, accumulation in glaciers or soil, runoff from land surfaces to rivers, lakes, and seas, and re-evaporation. Over 99% of the Earth's water is stored in the oceans, glaciers, and icecaps. The atmosphere stores a minute 0.001% and the remainder is stored on land in lakes, rivers, and ground water.

While the total quantity of water in the global hydrologic cycle, estimated to be some 1.4×10^{18} cubic meters, remains constant, the hydrologic cycle of a region will be largely determined by its climate. It will also be influenced by the local topography, geology, and vegetation. The frequency and nature of the precipitation, the variations in local temperature and humidity, and the extent of its exposure to solar radiation are clearly of great importance in the interaction between the weather and a local hydrologic cycle. Human activities of various kinds are playing an increasing role in determining the flows in local hydrologic cycles.

hydrologic network A network of hydrologic monitoring sites incorporating stations measuring, for example, precipitation and other meteorological elements, groundwater flow, and streamflow. The data collected may be used in flood forecasting.

hydrology The scientific study of the properties of water on and within the ground, especially with relation to the effects of PRECIPITATION and EVAPORATION upon the occurrence and properties of water on or below the land surface, including that held in rivers and lakes. Hydrology encompasses studies of runoff, ground water, soil moisture, the hydrological balance, and snow and ice accumulation.

Hydrology and Water Resources Program (HWRP) A program of the WORLD METEOROLOGICAL ORGANIZATION (WMO). Its aim is to promote the measurement of basic hydrological elements from networks of hydrological and meteorological stations and the collecting, publishing, and storage of data, and to further close cooperation between meteorological and hydrological services. It has an overall objective of applying hydrology to the sustainable development and use of water and related resources; to the mitigation of water-related disasters; and to effective environmental management at national and international levels.

hydrometeorology The interdisciplinary study of the interrelationships between water in the atmosphere and on land as it moves through the hydrologic cycle. It includes the study of the water budget, atmospheric moisture and precipitation, infiltration and groundwater flow, streamflow characteristics, fluvial geomorphology. Hydrometeorology has applications, for example, in the design of flood-control measures and planning of dams and reservoirs.

hydrometeors All the forms of water associated with CONDENSATION or SUBLIMATION of atmospheric WATER VAPOR. Examples of hydrometeors include rain and drizzle, which fall as liquid precipitation; hail, graupel, sleet, and snow, which fall as frozen precipitation; virga, which evaporates before it reaches the ground; fog and cloud, which are suspended in the air; drifting snow; and dew and hoar frost.

hydrosphere The Earth's water in any of its states: liquid, solid, or gaseous. The hydrosphere includes land, sea, and atmospheric water sources. In general, the amount of water in the hydrosphere is constant. About 71% of the Earth's surface is covered by water, of which about 97% of this is contained in the oceans and seas, while the remaining 3% can be found on land in the form of polar snow and ice, ground water, rivers, and in the flora and fauna.

hydrostatic equation An equation that quantifies the rate of decrease of ATMOSPHERIC PRESSURE with height in the atmosphere. In a *hydrostatic atmosphere*, the downward force due to gravity balances the upward PRESSURE GRADIENT FORCE, so that the net resolution of forces acting in a vertical plane is zero. There is thus no vertical acceleration in a hydrostatic atmosphere. If p is the pressure in PASCALS, z is the height (m), ρ is the air density (kg/m^3), and g is the acceleration of free fall (9.8 m s^{-2}), then:

$$\partial p/\partial z = -\rho g.$$

The hydrostatic equation can be combined with the ideal gas equation to derive the *hypsometric equation*, an important formula that allows the vertical distance (THICKNESS) between two pressure levels to be calculated if the mean temperature of that layer is known:

$$Z_2 - Z_1 = [(R_d T_v)/g][\log_e(p_1/p_2)]$$

where Z_2 and Z_1 are the height of the top and bottom of the atmospheric layer respectively, R_d is the gas constant for dry air (287 J K^{-1}kg^{-1}), T_v is the mean virtual temperature of that layer (K), and p_1 (p_2) is the pressure at the bottom (top) of the layer. For example, to find the 1000–500 mb thickness when the mean virtual temperature of this layer is 282 K:

$$Z_{500} - Z_{1000} = (287 \times 282/9.8)\log_e(1000/500).$$

hyetograph **1.** A map or chart displaying the temporal or areal distribution of precipitation.
2. A type of self-recording RAIN GAUGE.

hygrogram *See* hygrograph.

hygrograph A HYGROMETER that continuously measures and records atmospheric humidity. The strip chart produced is known as a *hygrogram. See also* hair hygrometer.

hygrometer An instrument used for measuring the humidity of air. The most common type of hygrometer is the dry- and wet-bulb PSYCHROMETER. Other types include the HAIR HYGROMETER (the simplest form of the instrument), the DEW-POINT HYGROMETER, and those that use the moisture-absorbing properties of certain chemicals.

hygrometric table A table used to obtain vapor pressure, relative humidity, and dew-point from wet- and dry-bulb temperatures.

hygroscopic nucleus The very small

particle that forms the core onto which water vapor condenses during the process of cloud-droplet and ice-crystal formation. They have by definition an affinity for water (hygroscopic). *See* cloud condensation nucleus.

hygrothermograph (thermohygrograph) An instrument consisting of a combined HYGROGRAPH and THERMOGRAPH, continuously measuring atmospheric humidity and temperature and recording these values on a single chart.

hypsithermal *See* altithermal.

hythergraph A climatic graph showing the relationship between selected meteorological elements; for example, temperature and rainfall or temperature and humidity.

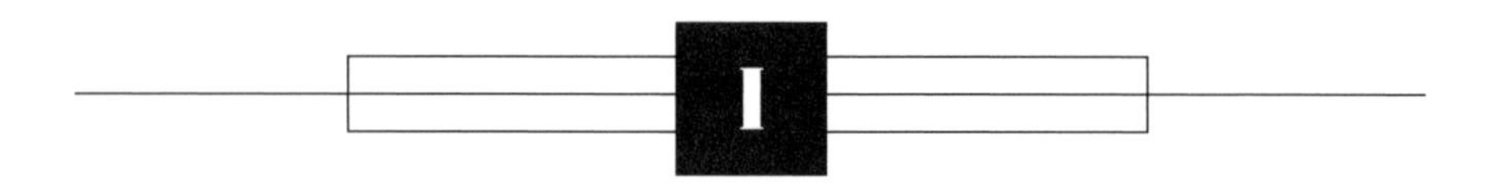

ICAO standard atmosphere *See* standard atmosphere.

ice Water in its solid crystalline state. It is formed from the freezing of liquid water, which, for pure water at 1013.24 hPa pressure, occurs at 0°C (32°F), or by direct sublimation from the gaseous state of water (i.e. water vapor). Within the atmosphere, ice occurs in a number of forms, including DIAMOND DUST, GRAUPEL, HAIL, ICE CRYSTALS, ICE FOG, ICE PELLETS, ICE PRISMS, and SNOW. It is deposited onto surfaces as GLAZE, HOAR FROST, ICING, and RIME. Within and above water bodies it forms as FRAZIL, grease ice, and PANCAKE ICE. On land and/or water, accumulations of ice form extensive ice sheets, ice caps, ice shelves, glaciers, ice bergs, and pack ice.

ice age (glacial period, glaciation) A period of extensive glaciation during the Earth's geologic history, during which large areas were covered by vast ice sheets and glaciers. The most recent ice age was that of the late Cenozoic (often known as the Ice Age), which is estimated to have ended around 10,000 years ago, although it is believed that the present time may be an interglacial. At least seven ice ages have occurred throughout geologic history, the first occurring 2500 million years ago during the Archean, and three more in the Proterozoic (between 900 and 600 million years ago). Others occurred toward the end of the Ordovician, and in the Carboniferous-Permian periods. The ice ages have generally lasted for around 100,000 years with oscillations of ice advance (glacial conditions), and of ice retreat (interglacial conditions).

A combination of processes is believed to be responsible for ice ages. These include the location and extent of continents and oceans, and mountain building (resulting from continental drift and plate tectonics), the changes in incoming solar radiation resulting from variations in the Earth's orbit (*see* Milankovitch cycles), and variations in atmospheric concentrations of carbon dioxide and other greenhouse gases.

ice blink A bright white or yellowish-white glare seen on the underside of low cloud that is produced by light being reflected onto the cloud from an ice-covered surface, such as pack ice. On a water surface, the appearance of ice blink can indicate the presence of ice that is beyond the range of vision when seen in contrast to the darker appearance of clouds above water (water sky).

ice-cap climate *See* ice-sheet climate.

ice core A cylindrical core removed from a section of ice for the purpose of studying past climate conditions. Ice cores from glaciers or sea ice may be examined for stratification, to determine seasonal changes, or chemically analyzed to estimate changes in carbon dioxide and trace gases over a period.

ice crystal A minute spicule of ice that forms from water in the atmosphere on an ICE NUCLEUS at temperatures below the freezing point of 0°C (32°F). Ice crystals have a variety of shapes, depending on the temperature range at their formation, that includes needles, hexagonal prisms, and stars, and they increase in number by splintering. Their growth occurs by the diffusion of water vapor onto them – and they collide with other ice crystals to form snowflakes.

ice fog A type of fog consisting of a suspension of ice crystals in the atmosphere that forms when mild maritime air blows across an ice or snow-covered surfaces. If conditions are extremely cold, condensation will occur as ice crystals rather than water droplets to produce a fog. These occur mostly in Arctic regions, occasionally in coastal stretches of Alaska, for example. Ice fog can also form artificially in these areas, for example from the burning of hydrocarbon fuels and from the exhaust gases of aircraft and motor vehicles.

icehouse period A time during the Earth's geologic history during which glaciers were at their most extensive, and sea levels were low. *Compare* greenhouse period.

Icelandic low The feature on averaged mean-sea-level pressure maps that expresses where the traveling CYCLONES (lows) of the North Atlantic reach their deepest (minimum) value on average. It does not mean a low is always centered over or near Iceland, but the region does lie on the major storm track of the North Atlantic. It is at its most intense during the northern-hemisphere winter and early spring when it is generally centered over Iceland and S Greenland. During summer it is weaker and may separate into two: west of Iceland and over the Davis Strait between Greenland and Baffin Island. *See also* Aleutian low.

ice nucleus A minute solid particle suspended in the atmosphere that provides the surface onto which an ICE CRYSTAL can form. They are less numerous than CLOUD CONDENSATION NUCLEI and are mostly particles that possess a geometry/lattice structure similar to that of ice crystals. Some clay minerals like kaolinite have such a structure and become active nuclei at temperatures of around –15°C (5°F).

ice pellets A form of PRECIPITATION consisting of small spherical or irregularly shaped pellets of ice with diameters of less than 5 mm (0.2 in). They are formed from frozen RAINDROPS or from SNOWFLAKES that have almost totally melted and then refrozen.

ice prisms Tiny unbranched ice crystals that form at temperatures below –30°C (–22°F) in the atmosphere. They may have the form of needles, platelets, or hexagonal columns.

ice rind A stage in the growth of SEA ICE during which ice crystals combine to form a brittle skin.

ice-sheet climate (ice-cap climate) A climate in which mean monthly temperatures throughout the year are below 0°C (32°F), occurring north and south of about 65° latitude over the ice caps of Greenland and Antarctica and over the frozen Arctic Ocean, which are the source regions for the ANTARCTIC AIR MASS and ARCTIC AIR MASS. Temperatures are extremely low, especially in the Antarctic region, where –89.2°C (–128.6°F) was recorded at the Russian meteorological station at Vostock in 1983 (the coldest ever recorded temperature; *see* cold pole), and precipitation is light, occurring as snow, or ice crystals or pellets. The climate is designated EF in the KÖPPEN CLIMATE CLASSIFICATION.

ice storm Conditions during which ice forms on exposed surfaces, commonly including power and telephone cables and their supports. Supercooled water droplets (i.e. water droplets that are liquid below the usual freezing temperature of bulk water) freeze on contact with surfaces. Although relatively rare in Europe, there are ice storms in eastern Canada and the northeastern states of the US more frequently.

icicle A suspended spike of clear ice formed by the freezing of dripping water.

icing The development of a coating of ice on an exposed object, such as the windshield of a motor vehicle or the wings of an aircraft. Icing occurs when supercooled water droplets in a cloud or fog come into contact with the surface to form RIME or when droplets freeze on impact, leading to GLAZE.

ideal gas laws The laws obeyed by ideal (or perfect) gases that relate their volume changes to variations in temperature and pressure; BOYLE'S LAW and CHARLES' LAW are the principal ideal gas laws. They can be combined into the single equation, $PV = RT$ for a single mole of an ideal gas, where P is its pressure, V its volume, and T its THERMODYNAMIC TEMPERATURE. R is the *universal gas constant*, which has the value 8.315 $J\ K^{-1}mol^{-1}$. These gas laws are not perfectly obeyed by real gases; they strictly apply only to ideal gases.

IFOV *See* instantaneous field of view.

IGAC *See* International Global Atmospheric Chemistry Program.

imaging A technique using remote sensing to record and interpret images of features, such as the surface of the Earth, from the sensing of various ranges of electromagnetic radiation emitted by the source. For example, an image of the distribution of clouds over a large region at one instant, may be mapped from a satellite by sensing the thermal infrared radiation being emitted into space by the clouds and cloud-free surfaces in view. The resulting map is an image or a representation of the natural scene.

inches of mercury (in Hg) A unit of measure of surface ATMOSPHERIC PRESSURE used in the US. Inches refers to the height of the column of mercury in the tube of a mercury barometer, a response to the force of atmospheric pressure. Average air pressure at mean sea level is 29.92 inches. Atmospheric pressure aloft is measured in HECTOPASCALS (hPa) or MILLIBARS (mb); 1 hPa (or mb) = 0.2953 (in Hg).

incus A supplementary cloud feature (*see* cloud classification) that refers specifically to the spreading anvil CIRRUS that streams out of and away from the upper reaches of CUMULONIMBUS clouds. [Latin: 'anvil']

index cycle The way in which the large-scale circulation in the free TROPOSPHERE tends to evolve over some weeks from a high ZONAL INDEX, when the circulation is principally west to east, to one of a low zonal index, when it is much more parallel to the lines of longitude.

Indian Ocean The third largest of the Earth's major oceans. It is located between Africa, India, and Australia, and covers an area of approximately 68,500 sq km (26,500 sq miles). A unique reversal occurs in the currents of the northern Indian Ocean; in summer during the southwest MONSOON the Southwest Monsoon Current flows eastward, while in winter the northeast monsoon occurs with winds and currents flowing from the northeast to the southwest. A counterclockwise gyre flows in the south Indian Ocean.

Indian summer A warm period of anticyclonic weather with clear skies and cool nights frequently occurring during October or November. The term was first used in the US in the late 18th century and in the UK at the beginning of the 19th century. Traditionally, it marks the end of the last vestiges of summer weather and usually follows the first frost. No climatic reason has been identified as to why such a period of warmth should occur in most years.

indifferent equilibrium *See* neutral stability.

industrial smog *See* smog.

inert gas A chemically inactive gas; inert gases include the gases argon, neon, helium, and xenon, which are present in the small quantities in the atmosphere.

inferior mirage *See* mirage.

inflow band *See* feeder band.

inflow jets Localized near-surface jets of air that flow into the base of TORNADOES.

infrared radiation ELECTROMAGNETIC RADIATION in the approximate wavelength from 0.7 µm (700 nm) to 1000 cm. In the electromagnetic spectrum it is between the red end of the visible light and microwaves; near-infrared (0.7–4.00 µm) is in that part of the spectrum closest to visible light whereas far-infrared is closest to microwave radiation. Although about 50% of solar radiation falls within the infrared band, the term is often loosely used as an alternative for longwave radiation (terrestrial radiation).

infrared remote sensing The technique that images the Earth and its atmosphere in infrared wavebands. Meteorological satellites, for example, produce cloud images by sensing the planet in the THERMAL INFRARED, and water vapor images by sensing in the WATER VAPOR waveband. Both of these lie in the infrared part of the electromagnetic spectrum.

infrared satellite imagery Images of the Earth and its cloud cover within specific infrared BANDS, obtained by satellites. Two such bands commonly used in weather imaging are the THERMAL INFRARED window and the WATER VAPOR. These sense the intensity of infrared radiation that is being emitted by the Earth's surface and its clouds, and the water vapor concentration in the middle TROPOSPHERE, respectively. Infrared imaging has the advantage over visible satellite imagery in that, because the surface, clouds, and water vapor radiate ceaselessly, it is constantly available.

infrared thermography A remote-sensing technique using sensors and cameras sensitive to emitted thermal radiance to map the temperature of a surface; the technique can be used on an aircraft or on the ground. This technique has applications, for example, in producing images of the spatial variation of the temperature of a sea surface, or in images of the heat loss from an urban area.

insolation In climatology, the intensity of short-wavelength SOLAR RADIATION received over a given period of time on a horizontal surface, measured in watts per square meter (W m^{-2}). It may also refer to measurements on inclined planes not necessarily at the Earth's surface. Insolation is influenced by the time of the year, latitude, and atmospheric conditions, such as cloud amount. [From: Latin *insolatio* and *in*coming *sol*ar radi*ation*]

instability *See* absolute instability.

instantaneous field of view (IFOV) The size of the smallest PIXEL that can be practically imaged by the sensor system of a satellite, given its orbital elevation and the physical dimensions of the active part of the sensor on-board. For the AVHRR instrument on the NOAA satellites, the IFOV is an angle of 1.3 ± 0.1 milliradians, which equates, at its elevation of 850 km (530 miles), to a scan spot that is 1.1 km across at the SUBSATELLITE POINT.

instrument shelter (Stevenson screen) A box-like shelter, designed originally by the Scottish engineer Thomas Stevenson (1818–87), used to accommodate meteorological thermometers for the measurement of air temperatures unaffected by direct solar radiation, precipitation, and condensation. It consists of a white wooden box with a hinged door, which faces poleward, sited approximately 1.25 m (4 ft) above the ground, and allowing indirect ventilation through the bottom of the screen, louvered sides, and a double roof. The screen houses the wet- and dry-bulb thermometers and the maximum and minimum thermometer; larger versions also house a thermograph and hydrograph.

insular climate The climate of a small island or group of islands. Seasonal temperature change is low, especially in the low latitudes.

intensification A strengthening of the winds in association with a simultaneous strengthening of the PRESSURE GRADIENT within a weather-producing system such as a HURRICANE or a FRONTAL CYCLONE. *See also* deepening.

intensity The strength of the radiation signal emitted or received by a remote-sensing instrument. It is quantified in watts per square meter.

intensity of rainfall *See* rainfall intensity.

interception The process by which PRECIPITATION is captured by vegetation and prevented from reaching the ground. Some of the intercepted moisture may be evaporated directly back into the atmosphere from the plant surface. If the RAINFALL INTENSITY is high, the precipitation will overcome the carrying capacity of the plants and drip to the ground.

interglacial A period of warm climate between two glacial periods during which continental glaciers retreat. During the last 1,000,000 years interglacials have had a periodicity of 100,000 years, each lasting for approximately 10,000 years.

Intergovernmental Panel on Climate Change (IPCC) An international panel that was established in 1988 by the United Nations Environment Program (UNEP) and the World Meteorological Organization (WMO) to assess information on possible global warming. Scientific authors and editors produce periodic reports for the IPCC, which are then reviewed by international experts. After publication, many governments view reports as the official state of knowledge on climate change issues, and in most cases policy formulation revolves around this information. The IPCC produced its First Assessment Report in 1990, followed in 1995 by the Second Assessment Report.

International Global Atmospheric Chemistry Program (IGAC) A core project of the International Geosphere–Biosphere Program (IGBP). It aims to develop understanding of the processes determining atmospheric composition and of interactions between the chemical composition and physical and climatic processes. It also aims to predict the impact of natural and anthropogenic forcing on the chemical composition of the atmosphere. Among the projects areas of study are atmosphere–surface chemical exchanges, ozone in the troposphere, and aerosols and their radiative effects.

International Practical Temperature Scale *See* thermodynamic temperature.

interpluvial A period of geologic time characterized by a dry climate between PLUVIALS.

interstade (interstadial) A warmer interval during a glacial period where the temperatures are not sufficiently warm or prolonged to form an INTERGLACIAL.

Intertropical confluence A broadly east–west aligned zone in which the NORTHEAST and SOUTHEAST TRADE WINDS flow generally smoothly together from either side of a stretched-out asymptote. A CONFLUENCE refers only to the direction of the flow; it does not include changes of speed.

Intertropical Convergence Zone (ITCZ) The broadly east–west aligned narrow zone in which the NORTHEAST and SOUTHEAST TRADE WINDS flow generally smoothly together and converge. The ITCZ migrates into the northern hemisphere during the northern summer and into parts of the southern hemisphere during the southern summer, following the annual movement of the zenithal Sun with a time lag of about six weeks. The distribution of land and sea considerably modifies this simple pattern. Over most of the Pacific Ocean, the ITCZ remains in the northern hemisphere all year round. The ITCZ is most clearly developed over the tropical oceans (especially the eastern portions of the Pacific and Atlantic Oceans), where the northeast and southeast trade winds both slow down markedly as they flow smoothly in to the zone, known here as the DOLDRUMS. The marked convergence and the piling up of air that slows down in the lowest kilometer or so as it enters the ITCZ is connected to significant VERTICAL MOTION. Scattered along the zone are very vigorous CUMULONIMBUS clouds that produce heavy rainfall. The zone is sometimes also known as the *Intertropical Front* over the continents. *See also* South Pacific Convergence Zone.

Intertropical Front *See* Intertropical Convergence Zone.

intortus A cloud variety related uniquely to CIRRUS, the filaments of which are highly twisted or contorted into tangled shapes. *See also* cloud classification.

inversion (temperature inversion) An increase in temperature with increasing altitude. This is a reversal from the normal expected decrease of temperature with increasing height. A temperature inversion results in a warmer layer overlying a cooler layer. Inversions are a characteristic of the TROPOPAUSE and MESOPAUSE. *See also* inversion layer.

inversion layer (capping inversion, temperature inversion layer) A layer of the atmosphere in which there is an INVERSION of temperature, i.e. temperature increases with height, a reversal of the expected decrease of temperature with altitude. In the troposphere, an inversion layer is generally stable and tends to act as a lid or cap that inhibits vertical motion, resulting in an absence of turbulence and general atmospheric stability. If the air under an inversion layer is moist, the inversion can act as a cap to a layer of stratified cloud, restricting the development of convective cloud forms. Inversion layers can also trap dust, smoke, and other atmospheric pollutants; as a result, precipitation and visibility are greatly reduced below the layer. Inversions may occur with the passage of a cold front (*see* frontal inversion), the influx of relatively cold sea air via an onshore breeze (*see* surface inversion), the overnight radiative cooling of surface air (*see* radiation inversion), and through the presence of a high-pressure system (*see* subsidence inversion).

ionosphere The region of the upper ATMOSPHERE within which there is a high concentration of free electrons that affects the propagation of electromagnetic radio waves. It extends through the upper MESOSPHERE and THERMOSPHERE to indefinite heights, becoming most distinct beyond about 80 km (50 miles) above the Earth's surface. It is characterized by high concentrations of ions (hence its name) and electrons formed when solar X-ray and ultraviolet radiation ionize molecules or atoms into ion pairs (a negative electron leaves a neutral atom which then becomes positive while the electron attaches itself to another atom producing a negative ion). These ions and electrons are at sufficiently high concentrations to reflect and/or refract radio waves. The structure of the ionosphere is affected by the cycle of sunspot activity, as well as by geographical location and seasonal and diurnal changes.

Several layers or regions are identified within the ionosphere, based on the propagation of electromagnetic radio waves, with electron density increasing in general with altitude from the D-LAYER through the E-LAYER and with especially high ionization in the F-LAYERS. Records of the ionosphere show that it is an indicator of greenhouse effects. Increases in greenhouse gases in the lower atmosphere cause warming but in the upper atmosphere cooling and contraction takes place as less heat is being reflected from below. The height of the ionosphere is lowered by a small amount each year.

IPCC *See* Intergovernmental Panel on Climate Change.

iridescence The subtly tinted patches of red, green, and occasionally blue and yellow that appear around the edges of HIGH CLOUD within around 30° of the Sun. The colors are produced by the diffraction of sunlight by small cloud particles.

irradiance (radiant flux density) The rate of incident energy on unit area of a surface, measured in watts per square meter (W m^{-2}).

isallobar A line on a mean-sea-level weather map (synoptic chart) that connects points at which the time rate of change of atmospheric pressure is the same. A typical value (normally quoted as a rise or fall (or steady) over the three hours leading up to the observation hour) would be 2.0 hPa per 3 hours. A very rapidly deepening FRONTAL CYCLONE may, however, have falls of up to 10.0 hPa per 3 hours in extreme cases.

isanomal A line that connects places that experience the same anomaly of a particular variable within a particular period: for example, the difference (or anomaly) of one year's rainfall total from its long-term annual mean rainfall.

isentropic *See* isentropic surface.

isentropic analysis A technique in weather forecasting or research in which SYNOPTIC-SCALE air motions (e.g. within FRONTAL CYCLONES) are mapped across surfaces of constant dry-bulb or wet-bulb POTENTIAL TEMPERATURE, known as ISENTROPIC SURFACES. It is more realistic to do this because the ascending and descending circulation on such a scale is ADIABATIC (i.e. no heat is exchanged by the air parcel with its environment); such upslope and downslope motion glides three dimensionally along the surfaces of constant potential temperature. The air does not move *across* horizontal ISOBARIC SURFACES (i.e. surfaces at which all points are of the same atmospheric pressure, which are often used to depict upper-air features, such as the LONG WAVES), but instead, the air moves up and down *through* the isobaric surfaces. *See also*: analysis, isobaric analysis.

isentropic lift The lifting of air that follows an upward-sloping ISENTROPIC surface. Situations producing isentropic lift are associated with widespread stratiform clouds and precipitation.

isentropic surface (isentropic) A surface in space on which all points are at the same ENTROPY or constant POTENTIAL TEMPERATURE at a given time. *See also* isentropic analysis.

isobar (isobaric surface) A line on a chart connecting places with the same ATMOSPHERIC PRESSURE. Isobars are typically drawn on a MEAN-SEA-LEVEL PRESSURE (MSLP) chart at either 2 hPa, 4hPa, or 8 hPa intervals. Unlike the surface, where the meteorologist plots the spatial variation of pressure at a fixed height, upper-air charts show the variation in the height of a fixed pressure level, for example 850 hPa, 500 hPa. Lines on such charts are called isohypses or CONTOUR lines, rather than isobars.

isobaric analysis An ANALYSIS of atmospheric conditions for an ISOBARIC SURFACE. On surface isobaric charts, the isobars are traditionally plotted at 4 hPa intervals centered on 1000 hpa (i.e. 996, 1000, 1004, 1008 hPa, etc). Once plotted the centers of high and low pressure can be identified and are indicated by the letters H (high pressure) and L (low pressure). Other features, such as troughs of low pressure, are also indicated. *See also* isentropic analysis; surface weather chart.

isobaric chart (constant-pressure chart) A SYNOPTIC CHART that is produced for a particular ISOBARIC SURFACE (e.g. 1000, 850, 700, 500, 300, or 200 hPa). The atmospheric variables usually plotted on the isobaric chart include the height of the pressure surface, temperature, moisture content, and wind speed and direction.

isobaric surface (constant-pressure surface) A surface in space on which all points are at the same or constant ATMOSPHERIC PRESSURE at a given time. *See also* isobaric analysis.

isobront A line that connects the location of THUNDERSTORMS that exist at a specific time over a particular region.

isoceraunic line A contour joining places that have the same percentage frequency of days in the year on which THUNDER was heard.

isocheim A line on a map or chart that connects places that experience the same mean winter temperature.

isohaline In oceanography, a line on a chart connecting points of equal salinity.

isohel A line that connects places that observe equal durations of bright sunshine for a particular period (e.g. for one day, one week, or one month).

isohyet A line that joins places that observe equal rainfall totals over a particular period (e.g. one hour, one day, or one month).

isokinetic *See* isotach.

isoline (isopleth) A contour that joins places that have the same value of some variable, such as an ISOTHERM (DRY-BULB TEMPERATURE) or an ISOBAR (MEAN-SEA-LEVEL PRESSURE).

isomer **1.** In meteorology, a line that connects places with the same value of an individual month's rainfall, expressed as a percentage of their annual average rainfall. **2.** In chemistry, compounds with the same molecular formula but a different arrangement of the atoms, which may make them completely different compounds, e.g. CH_4N_2O may be either urea, $CO(NH_2)_2$, or ammonium cyanate, NH_4CNO.

isoneph A contour joining places of equal total CLOUD AMOUNT.

isonif A line connecting sites that have the same snow depth.

isopleth *See* isoline.

isopycnic A line (or a surface) of constant atmospheric density.

isoryme A contour connecting places that experience the same intensity of frost.

isotach (isokinetic) A line that joins places that have the same wind speed.

isothere A contour that connects places that have the same summer mean temperature.

isotherm A contour that connects places with the same DRY-BULB TEMPERATURE.

isothermal layer A layer in the atmosphere with zero LAPSE RATE of temperature, i.e. there is no change in temperature with height and it is thus characterized by a constant temperature.

isothermobath In oceanography, a line drawn to connect all points of equal temperature on a diagram showing the vertical temperature structure of the ocean.

ITCZ *See* Intertropical Convergence Zone.

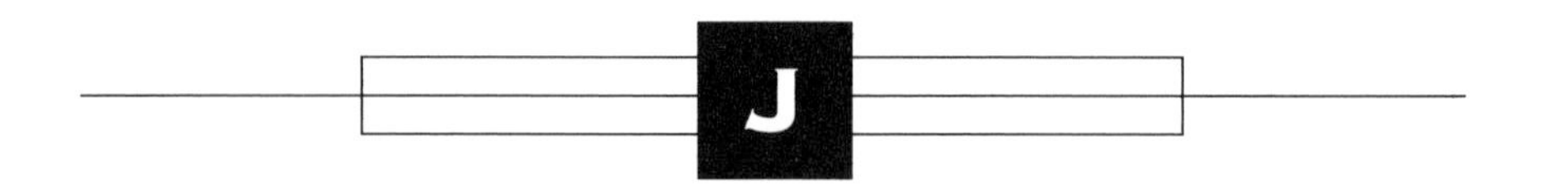

Jacob's ladder *See* crepuscular rays.

jet-effect wind *See* canyon wind.

jet maximum The elongated core of a JET STREAM, in which the wind speed is strongest. In the winter the maximum speed in the POLAR FRONT JET STREAM can reach 200 knots (100 m s^{-1} or 230 mph) or more on rare occasions. The air speeds up as it approaches the core in the ENTRANCE REGION, and slows down in the EXIT REGION.

jet streak A small-scale zone of maximum wind speed – a JET MAXIMUM – that runs through the snaking pattern of the JET STREAM. The jet maximum is most often a more-or-less stationary feature within the jet stream, through which the air blows very strongly. A jet streak is often associated with changes in surface-pressure developments, such as rapid deepeners (*see* explosive cyclogenesis).

jet stream A long snaking ribbon of fast flowing air, normally in the upper TROPOSPHERE. It is often thousands of kilometers long, a few hundred kilometers wide, and about one or two kilometers deep. It has marked WIND SHEAR above, below, and to the sides of the very elongated core, along which there are one or more speed maxima (*see* jet maximum). Significant CLEAR-AIR TURBULENCE is associated with the wind shear. The principal upper tropospheric jet streams are the SUBTROPICAL JET STREAM, POLAR FRONT JET STREAM, and TROPICAL EASTERLY JET STREAM. The subtropical jet is a geographically relatively fixed westerly flow feature over days and weeks, but migrates seasonally with the SUBTROPICAL ANTICYCLONES. The tropical easterly jet is geographically fixed and markedly seasonal, forming as part of the summer MONSOONs of southern Asia and West Africa (*see also* African monsoon; Asian monsoon). Its ENTRANCE REGION is above the Indonesian area while its EXIT REGION lies above sub-Saharan Africa. The polar front jet is strongest in winter when the POLAR FRONT's temperature contrasts are largest. It sinuous shape and day-to-day change in location across middle-latitude oceans means, unlike the previous two jets outlined, it does not show up as a distinct feature on monthly or seasonal mean maps of the upper tropospheric flow. *See also* Polar-Night Jet Stream.

jet stream cirrus An extensive sheet or elongated striations of cirrus associated with vertical motions around the core of a jet stream. Deep ascent often occurs on the warm low-latitude side of a jet stream with deep descent on the cold high-latitude flank. The ascent leads to CIRRIFORM cloud that exhibits a sharp edge coinciding roughly with the jet core. The cloud moves very quickly with the rapid flow at jet level.

JGOFS *See* Joint Global Ocean Flux Study.

Joint Global Ocean Flux Study (JGOFS) A program launched in 1987 to study the fluxes of carbon within the oceans (which are believed to contain 50 times as much carbon as the atmosphere), and exchanges of carbon between the atmosphere and oceans, on a global scale. It is a core project of the International Geosphere–Biosphere Program. The Program also aims to develop the capacity to predict the global-scale response to anthropogenic perturbations, especially those related to climate changes.

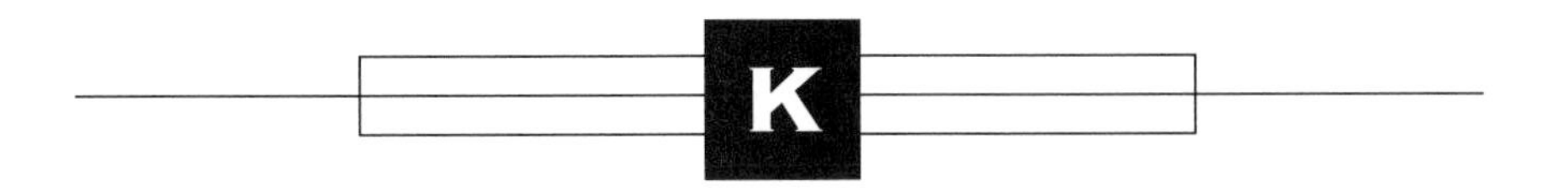

Kamchatka Current A cold ocean current that flows from the Bering Sea along the pensinsula of Kamchatka into the North Pacific Ocean and forms the OYASHIO CURRENT.

Kansan I and II A glacial period, the second of four glacial stages in North America, which started about 620,000 years ago, and lasted for about 60,000 years. During this period extensive areas of northern and central Europe and North America were covered in ice.

katabatic wind (drainage wind, gravity wind, mountain breeze) A local cold downhill wind that blows down valleys and mountain slopes (e.g. Norwegian fjords); the term is also applied to cold air that flows outward from icecaps, such as those in Antarctica and Greenland. A katabatic wind results from the radiative cooling of the upland ground surface at night, which in turn cools the air in contact with the slope. As the air becomes denser, it starts to move downhill by gravity flow beneath warmer less dense air, its direction being controlled almost exclusively by the orography. In a valley-wind system, in which cooling may occur over distances of 5 km (3 miles) or less, the katabatic flow's thickness rarely exceeds 10 m (30 ft); in Antarctica it could extend up to 500 m (1600 ft) into the atmosphere. *See also* bora.

kata-front A WARM FRONT or COLD FRONT that is overrun by drier air, or in which the warm air subsides, so that any cloud and precipitation tend to be suppressed. In this sense, they are generally inactive fronts. *Compare* ana-front.

kelvin (K) The SI unit of THERMODYNAMIC TEMPERATURE (absolute temperature); it is equal to 1/273.16 of the thermodynamic temperature of the triple point of water. The units of Kelvin and Centigrade degrees are equal. The unit is named for William Thomson, Baron Kelvin (1824–1907). *See also* Kelvin temperature.

Kelvin–Helmholtz instability A phenomenon that occurs occasionally in an atmospheric layer within which the wind speed increases rapidly with height. Kelvin–Helmholtz waves form in the layer of strong WIND SHEAR; the waves can break and are marked by a train of clouds that have the appearance of breaking ocean waves.

Kelvin temperature The THERMODYNAMIC TEMPERATURE (absolute temperature) of a substance measured in kelvins. The lowest temperature theoretically possible (*see* absolute zero) is 0 K, and the melting point of ice is 273.15 K, which is 0.01 K below the triple point of water.

Kelvin wave *See* El Niño Southern Oscillation.

Kew barometer A type of MERCURY BAROMETER. Designed in 1854 by P. Adie, this is a portable marine barometer with a graduated scale. The scale is designed to take account of any unsteadiness in the level of the mercury which may occur as a result of changes in air pressure through gusting winds or due to the rolling of a ship.

khamsin A LOCAL WIND in Egypt that can persist for several days. The khamsin is a hot and dry southerly wind, which often transports large quantities of dust from the

Sahara desert into the country's capital Cairo, sometimes pulling in air from Arabia and the Gulf of Aden. The khamsin occurs most frequently in April, May, and June; the visibility is often so poor that lights are sometimes required at midday. Synoptically, the khamsin is a product of the advection of warm air that occurs on the eastern limb of a low-pressure system that tracks either through the Mediterranean Basin or over North Africa from west to east.

knot A unit of WIND speed; 1 knot = 1.15 mph = 0.515 m s^{-1} = 1.85 km h^{-1}. The knot is equivalent to one nautical mile per hour, the word being derived from the method that sailors used to estimate a ship's speed in the days before accurate instruments became available. This consisted of a rope with regularly spaced knots, attached to a log-line. The ship's speed was proportional to the number of knots that were run off the reel in a fixed amount of time. The knot is used in weather observations and forecasting and is still widely used in shipping and aviation.

kona storm The local name for a storm in the Hawaiian Islands (North Pacific Ocean), characterized by strong southerly winds and heavy rain associated as a depression passes north of the islands. [Polynesian: 'leeward']

Köppen climate classification An empiric CLIMATE CLASSIFICATION system devised by the German botanist and climatologist, Wladimir Peter Köppen (1846–1940), first published in 1900, and subsequently revised and published in 1918. Further revisions were made by Köppen throughout his life and it has also been revised by other climatologists. The classification was based on the major vegetation zones and their climatic requirements. The climates of the world are divided into five major groups, each represented by capital letters (A, B, C, D, and E) and defined by temperature, with the exception of the B group, in which the controlling factor is dryness. Group A: *tropical moist (rainy) climate* with average temperatures above 18°C (64.4°F) and no real winter or summers seasons. Group B: *dry climates* in which evaporation exceeds precipitation. Group C: *warm temperate rainy climates* with the coldest month between –3°C (26.6°F) and 18°C (64.4°F) and the warmest month greater than 10°C (50°F). Group D: *cold boreal forest climates* with the coldest month less than –3°C (26.6°F); in North America this is modified to 0°C (32°F), and the warmest month greater than 10°C (50°F). Group E: *polar climates*, which are subdivided into *tundra climates* (ET) with temperatures in the warmest month between 0°C (32°F) and 10°C (50°F) *perpetual frost climate* (EF), with temperatures in the warmest month less than 0°C (32°F).

The major groups are further subdivided on the basis of seasonal precipitation and designated with the additional capital or lower-case letters. These include: f – absence of a dry season and adequate precipitation in all months; s – a dry summer season; w – a dry winter season; and m – monsoon climate with rain, except during short dry season (A group climates only). The arid climates (group B) are subdivided into: S – semiarid (steppe type), and W – arid (desert). Further additions to the code are lower-case letters designating temperature, which include: a – warmest month mean temperature is greater than 22°C (71.6°F); b – warmest month mean temperature is below 22°C (71.6°F); c – less than 4 months are over 10°C (50°F); d – coldest month is below –38°C (–36.4°F); h – mean annual temperature is greater than 18°C (64.4°F); and k – mean annual temperature is less than 18°C (64.4°F). The Köppen climate classification has been widely used, although it has been criticized. The main criticisms are that it excludes other meteorological factors, such as sunshine and wind, which are important to vegetation, and the poor correlation between the groups and actual vegetation distribution in the world. *See also* Flohn climate classification; Strahler climate classification; Thornthwaite climate classification.

kosava (kossava) A cold east to southeast wind experienced in central and eastern Europe. The kosava affects countries, such as Bulgaria and Serbia, usually blowing from the Aegean Sea region; for example, a kosava is a ravine wind blowing on the Danube to the southeast of Belgrade. *See also* canyon wind.

Kuroshio Current (Kurosiwo Current, Japan Current) A warm current in the Pacific Ocean that flows northeastward along the east coast of Japan. This current is a northward-flowing continuation of the Pacific north EQUATORIAL CURRENT between the Philippines and the east coast of Japan. Dense fog tends to develop at its boundary with the Oyashio Current due to the temperature contrast produced. The relative warmth from the Kuroshio Current moderates the climate of Taiwan and Japan as air is warmed as it flows toward the mainland.

Kyoto Protocol An international agreement, which has yet to be ratified, to reduce worldwide emissions of greenhouse gases, originally formulated in 1997 at the Third Conference of Parties (COP 3) in Kyoto, Japan. The agreement, if ratified, commits individual industrialized nations to specified reductions in six greenhouse gases or classes of gases: CARBON DIOXIDE (the most prominent), METHANE, NITROUS OXIDE, HYDROFLUOROCARBONS, PERFLUOROCARBONS, and SULFUR HEXAFLUORIDE. In addition, the protocol introduced the concept of emissions credits. In 2001, America indicated that it was unwilling to ratify the protocol and was considering counter proposals.

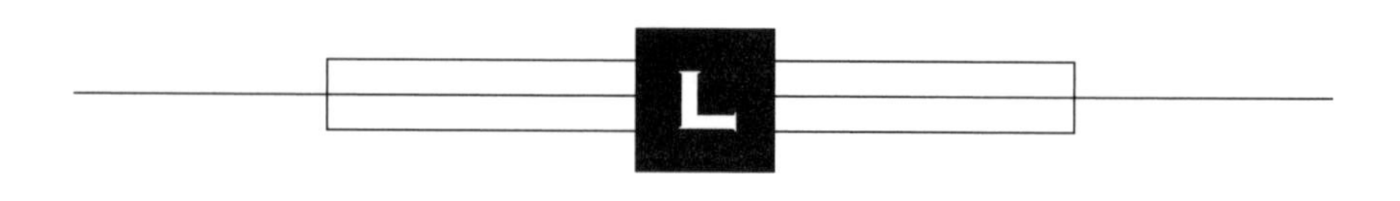

Labrador Current A cold current in the ATLANTIC OCEAN that flows south along the Labrador coast through the Grand Banks, where it splits to flow east and west. The eastward-flowing current joins the NORTH ATLANTIC DRIFT and the west branch flows into the Gulf of St Lawrence. The Labrador Current is a product of the convergence of the West Greenland and Baffin Island Currents and inflow from Hudson Bay.

lacunosus A variety of CUMULIFORM cloud in which the normally rather thin sheets, layers, or patches of cloud, most commonly CIRROCUMULUS and ALTOCUMULUS but also STRATOCUMULUS, display a more-or-less regular scattering of circular holes, often with fringed or ragged edges. The cloud 'units' and holes combined present a pattern like a honeycomb or a net. *See also* cloud classification.

lake breeze A phenomenon particularly evident on the Great Lakes of North America (although lake breezes occur globally) in calm summer conditions, when the surface temperature of the lake is cool relative to the land. A mesoscale anticyclone (high) develops over the lake; air consequently blows from the higher pressure over the lake to the lower pressure caused by heating over the land. The pressure gradient can sometimes be in excess of 2 hPa. The breeze's leading edge, the *lake breeze front*, a zone of atmospheric convergence and increased cloud development, may be felt up to 40 km (25 miles) inland from the lake.

Lamb's dust veil index *See* dust veil index.

laminar flow The very smooth flow of the air in which each particle follows exactly that of its predecessor. There is no mixing whatsoever of adjacent fluid.

land breeze The opposite of a SEA BREEZE, the land breeze is a diurnal LOCAL WIND that typically begins around midnight, and is a product of the more rapid cooling of the land surface relative to the sea. This offshore wind is most developed in calm conditions around dawn in tropical and mid-latitude locations when the geostrophic flow is slack.

Landsat A series of very high resolution Earth-surface monitoring satellites that grew out of the Earth Resources Technology Satellite 1 (ERTS–1) launched in July 1972 and renamed Landsat–1. The current version in the series is Landsat-7, launched in 1999, which images the Earth with a MULTISPECTRAL SCANNER, sensing in eight narrow spectral BANDS. It orbits at a height of 705 km (438 miles) with a period of 99 minutes and, unlike operational METEOROLOGICAL SATELLITES, has a repeat cycle (imaging the same region) of 16 days.

langley A unit of energy per unit area, equivalent to 1 calorie per square centimeter. It is named for Samuel Langley (1834–1906) of the Smithsonian Institute, a pioneer in solar radiation studies. The unit has been superseded by the SI unit W m^{-2} internationally but is still used in the US (1 langley = 698 W m^{-2}).

Langmuir theory of raindrop growth A variant of the COLLISION–COALESCENCE PROCESS of raindrop growth in clouds. Langmuir's experiments revealed that larger water droplets (19 μm) with their greater terminal velocities overtook and in some cases absorbed smaller droplets.

La Niña A climatic fluctuation in the equatorial Pacific in which there is an upwelling of cooler than normal waters off the coast of Peru and Ecuador (in contrast to EL NIÑO, which is characterized by unusually warm waters). It may follow an El Niño event but occurs with less frequency. The cold current affects the weather in the middle latitudes of the western Pacific Ocean and causes very hot summers in Japan. Whereas El Niño tends to reverse the normal climatic conditions of some regions (e.g. wet places become dry, and dry become wet), La Niña enhances normal conditions, for example, wet places become wetter. In the 1988–89 La Niña event, a build-up of exceptionally cold air in the Arctic produced North America's highest atmospheric pressure reading of 1075 hPa at Northway, Alaska. During the 1995–96 event, Winnipeg, Canada, which normally experiences cold winters had day-time temperatures of only –20°C (–4°F) and below with overnight lows of –30°C (–22°F) were experienced for 19 consecutive days. [Spanish: 'the little girl']

lapse rate The rate of change of a property (usually temperature) with height in the atmosphere. The overall average lapse rate is about 6.5°C km^{-1} up through the troposphere but there are widespread variations with height, season, and location. It is described as positive when the temperature decreases with height, and negative when the temperature increases with height (*see* inversion). *See also* dry adiabatic lapse rate; environmental lapse rate; saturated adiabatic lapse rate.

latent heat The amount of heat absorbed or released in the change of phase of a substance at constant temperature. In meteorology, latent heat is the amount of heat absorbed or released in the transformations between the three phases of water: ice, liquid water, and water vapor during the process of evaporation a change of phase from liquid (water) to gas (water vapor) takes place. The energy required for the transformation, the *latent heat of vaporization*, depends on temperature but is about 2.4×10^6 J kg^{-1} at 0°C. This energy, unlike some other transformations is recoverable through condensation in the atmosphere. The *latent heat of condensation* is the energy released during condensation from water vapor to liquid water. The *latent heat of fusion* is the energy released when water freezes to ice, or absorbed when ice melts to water. The *latent heat of sublimation* is the energy absorbed or released when water vapor changes directly to ice, or vice versa. Latent heat is transported around the world and plays an important role in maintaining the energy balance. *Compare* sensible heat.

latent heat flux The movement of water vapor (transporter of latent heat) from one location to another, for example from the tropics toward the poles, where there is an energy deficit. Globally, poleward latent heat flux reaches a maximum of 1.5×10^{15} watts at 38° N and at 40° S.

latent instability A vertical temperature profile (sounding) that is STABLE near the Earth's surface but unstable above, typically occurring at night. Although CONVECTION is absent at the time of the sounding, it may be released during daytime hours as a result of surface heating.

latitude A major factor, together with the tilt of the Earth on its axis, in the maldistribution of solar radiation over the globe. Indirectly, it produces a deficit of 130 W m^{-2} at the poles and a surplus of 30 W m^{-2} at the Equator. The climate system, through the GENERAL CIRCULATION, seeks to redress this imbalance.

layer cloud *See* stratiform.

layered haze A distinct layer of haze capped above and below by air with better visibility. Layering is associated with relatively thin slabs of STABLE air that may be a few hundred meters deep.

leader A preliminary electrical discharge that precedes the formation of a LIGHTNING channel, frequently propagating in steps until, in the case of a CLOUD-TO-GROUND DISCHARGE, it meets an upward-advancing streamer. A leader is faintly luminous; it is not associated with appreciable current flow, but its rapid propagation is essential to the formation of the final visible lightning channel through which the majority of the charge flows. *See also* dart leader.

lee *See* leeward.

lee depression (lee trough, orographic depression) A low-pressure feature that forms preferentially to the lee (or downwind) of a mountain barrier when air flow is at right angles to the barrier. The air column is 'squashed' by vertical contraction as it crosses the barrier. After crossing, the column of air expands to its original (upstream) depth and, as it does so, tends to develop a strong spin about the vertical axis that is related to the conservation of absolute vorticity. The spin to the lee of mountains is manifested as a low-pressure center. This process is linked to lows that occasionally form to the lee (east) of the Rocky Mountains with westerly winds (*see* Alberta clipper) and to the south of the western Alps under northerly flow.

lee trough *See* lee depression.

leeward (lee) The downwind side or side facing away from the wind. For example, the side of a range of mountains that is sheltered from the PREVAILING WIND, such as the Patagonian region in South America, which is located on the leeward side of the Andes, and hence is sheltered from the mid-latitude westerlies.

lee wave (mountain wave, hill wave) A mechanically induced circulation comprising a stationary (standing) wave or waves that occurs above and on the LEEWARD side of a mountain range or obstacle. Air flowing over the mountain or obstacle plunges downward and upward in a series of crests and troughs analogous to the standing waves sometimes seen in a stream. Lee waves only develop in a stably stratified atmosphere when there is a strong and constant horizontal wind flow of at least 15 knots (17 mph). Their wavelength, which can be up to 40 km (25 miles), although typically between 10 and 20 km (6 and 12 miles), is only dependent on the atmosphere and is independent of the orography, being calculated by means of the *Brunt–Vaisala frequency equation.* Lee waves may give rise to severe downslope winds and FÖHN conditions. WAVE CLOUDS form when the wave's crest reaches the cloud condensation level.

Examples include the Sierra wave that forms over the Sierra Nevada in California when a westerly airflow increases speed over the crest as a cold front approaches from the northwest; 'stationary' clouds over the leeward valleys indicate the presence of the lee wave. The helm wave of Cumbria, England, is another smaller-scale example of a lee wave.

lenticular cloud *See* lenticularis.

lenticularis (lenticular cloud) A type or species of cloud that has the shape of a smooth lens or almond, horizontally aligned, and frequently with well-defined outlines. Lenticularis occur quite often as WAVE CLOUDS in the crests of LEE WAVES generated downstream of hills and mountains – so can occur over a large region downwind of uplands. They can also occur in layers, sometimes resembling a stack of pancakes, in which the humidity, temperature, and wind profiles favor their development. ALTOCUMULUS lenticularis is the most common type although the form can occur with CIRROCUMULUS and STRATOCUMULUS too. *See also* cloud classification.

leste A dry hot southerly, southeasterly, or easterly wind that affects the Canary Islands, Madeira, and the North African mainland. Often laden with dust from north-central African desert regions, the leste blows as a result of the pressure gradient established between high pressure to the east and an approaching area of low pressure from the Atlantic. In the Canary

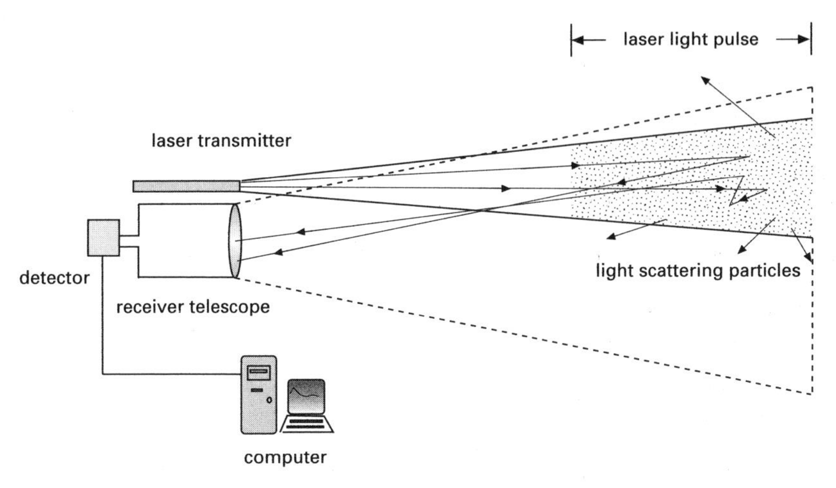

Lidar atmospheric sensing

Islands, the word *levanto* is used specifically for a hot dry southeasterly wind.

levanter (solano) An easterly wind that blows from the Mediterranean Sea through the Straits of Gibraltar between Spain and Morocco. It occurs most frequently between July and November. The levanter generates LEE WAVES and a BANNER CLOUD on the western flank of the Rock of Gibraltar. As the wind velocity increases, this cloud lifts and disperses; under such exceptional easterly storm conditions the levanter is referred to as the *levantades*.

leveche A hot dry south or southeast sand- and dust-laden SCIROCCO-type wind that is experienced in southeastern parts of Spain, for example in the Alicante district, extending only a few miles inland. It originates over the deserts of North Africa and Saudi Arabia. Its properties and synoptic climatology are similar to those of the LESTE wind, occurring most frequently during the summer.

libeccio A LOCAL WIND experienced in the central Mediterranean Basin. The libeccio blows from the west or southwest, advecting moist Atlantic air masses into the region. Occurring most frequently during the winter, the libeccio can bring stormy conditions and even snow when the air stream is forced to ascend, such as when it approaches the west coast of Corsica.

lidar A remote-sensing technique, similar to RADAR. It consists of a transmitter from which a laser pulse is emitted into the sky. The light particles (photons) scattered back (e.g. from aerosols or water vapor within the atmosphere) are received by a telescope and measured at the receiver of the lidar instrument (see diagram). Lidar has a number of applications in meteorology and climatology; these include cloud-base recorders; detecting and studying pollutants, aerosols, and water vapor within the atmosphere; and studies of cloud physics (e.g. the airborne cloud physics lidar operated by NASA). [From: *li*ght *d*etection *a*nd *r*anging]

lifetime *See* atmospheric lifetime.

lifted index (LI) A measure of atmospheric INSTABILITY. The lifted index is determined by calculating the temperature that air near the Earth's surface would have after it has been lifted to a higher level (usually the 500 hPa level) and then comparing it to the actual temperature observed at that level. Negative lifted index values indicate instability; positive values indicate STABILITY. High negative lifted index values indicate a very UNSTABLE atmosphere, often leading to strong updrafts within any developing THUNDERSTORM. *See* elevated convection.

lifting condensation level The level in the atmosphere at which an unsaturated air parcel that is lifted at the DRY ADIABATIC LAPSE RATE would become saturated.

lightning A transient electrical phenomenon leading to substantial exchange of charge between two regions of the atmosphere. A lightning path is typically kilometers long with considerable luminosity along the path; it may generate THUNDER. Lightning is commonly associated with THUNDERSTORMS and conveys charge from the cloud to the surface (CLOUD-TO-GROUND DISCHARGE), between clouds (CLOUD-TO-CLOUD DISCHARGE), to the air (CLOUD-TO-AIR DISCHARGE) and above clouds (*see* sprite). *See* ball lightning; beaded lightning; forked lightning; heat lightning; sheet lightning. *See also* lightning conductor.

lightning conductor An electrical connection to an elevated spike mounted at the highest point of a building, structure, ship, or aircraft intended to carry lightning currents harmlessly or away from critical regions.

line scanner An instrument used in remote sensing in which a mirror is used to sweep the ground surface normal to the flight path of the platform to produce an image. For example, it is used on-board the NOAA and TIROS meteorological satellites to produce cloud imagery. As the satellite orbits the Earth, the spinning mirror of the line scanner samples extremely rapidly along one line swiped at right angles to the track of the satellite. As it orbits, the rate of spin of the mirror is organized in such a way that the next scan line just touches its predecessor so that a continuous image is built up line-by-line. The line scanner on these satellites produces 360 lines every minute.

line source The linear origin of pollutants typically associated with transport routes, such as roads (e.g. the emissions from a line of traffic along a highway) and flight paths of aircraft. *Compare* area source; point source.

line squall *See* squall line.

liquid-in-glass thermometer *See* thermometer.

Little Climatic Optimum *See* climatic optimum.

little dry season A break in the summer or monsoon rains of the tropics and subtropics, sometimes known by local names (*see* veranillo). The poleward movement of the Intertropical Convergence Zone (ITCZ) is followed by the poleward migration of the main rains in some parts of Asia, Africa, and Central America. As the zenithal Sun begins its retreat back toward the Equator, there may be a noticeable lessening in the rains along the 10°–15° N latitude.

Little Ice Age A period between about 1550 and 1850 AD during which the temperatures in Europe, North America, and Asia were cooler than at present and conditions were harsher. During this period there was increased glaciation around mountainous areas, such as the Alps and Sierra Nevada, as well as Alaska. Rivers, such as the Thames in London, froze over.

local climate The climate of a small area as distinguished from the climate of the surrounding area. *See also* microclimate.

local storm report (LSR) In the US, a storm report produced by local National Weather Service (NWS) offices to inform users of reports of severe and/or significant weather-related events.

local wind A wind that affects a limited geographical area and is usually seasonally specific. There are two main causes of local winds: (1) A regional wind-flow pattern that is modified by the disposition of relief, an example being the MISTRAL. (2) The FÖHN effect resulting from the adiabatic compression of descending air to the LEEWARD of a mountain range, one example being the CHINOOK. Despite having similar generic origins, a vast plethora of names exist for local winds dependent on the regional setting.

lofting In pollution studies, a pattern of flow that occurs when the top of a plume from a chimney stack disperses into slightly turbulent or neutral airflow conditions, while the lower part of the plume is prevented from dispersing down toward the surface by a stable boundary layer, especially at night.

long-range weather forecasting Weather predictions that are for periods significantly longer than those handled by the MEDIUM-RANGE FORECAST. It usually refers to periods of one month or longer ahead, overlapping into the realm of SEASONAL PREDICTION.

long wave (Rossby wave) A very large-scale wave on an upper-air isobaric analysis of the middle and upper TROPOSPHERE. Long waves consist of a sequence of RIDGES and TROUGHS, with a wavelength that is typically a few thousand kilometers. They stretch right around the Earth, principally in middle latitudes, with typically four or five waves around a hemisphere. Long waves are strongly linked to surface weather patterns and the POLAR FRONT JET STREAMS are located along the central axis of the waves in the upper troposphere.

longwave radiation (terrestrial radiation, infrared radiation) ELECTROMAGNETIC RADIATION, in the INFRARED part of the spectrum, that is emitted from the surface of the Earth and the atmosphere in the absolute temperature range 200–300 K. It has a maximum intensity at about 10 μm with a range of 3–100 μm. *Compare* shortwave radiation.

low A usually extensive region, from several hundred to several thousand kilometers across, of relatively low barometric pressure, linked to cyclonic inflow of air, its large-scale ascent, and often cloudy and wet conditions. *See also* cyclone.

low cloud In CLOUD CLASSIFICATION, cloud with a base not more than 2000 m (6500 ft) above the surface. The class includes CUMULUS, CUMULONIMBUS, STRATUS, and STRATOCUMULUS.

low-latitude climates The group of five climates of the low-latitude regions of the world identified in the STRAHLER CLIMATE CLASSIFICATION: wet equatorial climate, trade wind littoral climate, tropical desert and steppe climates, west-coast desert climate, and tropical wet-dry climate.

low-level jet A maximum of wind speed in the lower TROPOSPHERE, associated with a marked horizontal temperature gradient. Examples occur just ahead of some strong surface COLD FRONTS and along the coast of Somalia in northeast Africa. This latter Somali Jet is a component part of the summer ASIAN MONSOON; it turns from southeasterly, to southerly, to a southwesterly current along its axis as it blows across the Arabian Sea toward the Indian subcontinent. Its JET MAXIMUM lies about 1 km (0.6 mile) above the surface and although the JET STREAM width is narrow, it is responsible for about one half of the air's mass transported across the entire Equator into the northern (summer) hemisphere.

low-precipitation storm (LP storm) A THUNDERSTORM, occurring along or near a dry line, that seldom produces significant precipitation. These storms have a prominent bell-shaped main tower and are capable of producing tornadoes and very large hail and much lightning. They commonly occur over the High Plains of the US. *Compare* high-precipitation storm.

low-pressure system *See* cyclone.

low Sun The time of year, usually winter, when the elevation of the Sun at noon (solar time) is relatively low in the sky and when the receipt of radiation is likely to fall below about 4 MJ m^{-2}day.

luminous night cloud *See* noctilucent cloud.

lysimeter An instrument used to study aspects of the hydrologic cycle, including the assessment of EVAPOTRANSPIRATION losses from soils and the soil moisture content. It consists of a suitable container placed below ground surface to intercept and collect water moving down through the soil.

mackerel sky A complete or very extensive cover of high ALTOCUMULUS or CIRROCUMULUS cloud, with ripples and patches separated by small areas of blue sky, that resembles the pattern on the skin of a mackerel. It applies particularly to the species of cloud VERTEBRATUS, but the term is commonly applied to other clouds that exhibit similar patterns.

macroburst *See* downburst.

macroclimate The climate of an extensive geographic region, such as part or all of a continent, hundreds to thousands of kilometers across. *Compare* mesoclimate; microclimate.

macrometeorology The study of large-scale meteorological processes, such as those occurring over a large region, continent, or the whole Earth. These include monsoons, long-wave troughs, and the general circulation of the atmosphere.

macroscale Denoting the largest-scale areas or processes in meteorology, from an extensive region to continental to global. *Compare* mesoscale; microscale.

maestro A cold northwesterly wind that affects the central Mediterranean Basin. It is experienced in the Tyrrhenian, Adriatic, and Ionian seas, and along the west coast of Corsica and Sardinia. It blows when a low-pressure system is located to the east over the Balkan Peninsula, the result of this being that this region is no longer in the warm sector of the low.

magnetopause The boundary that surrounds the MAGNETOSPHERE, separating the flow of the solar wind from the charged particles of the magnetosphere.

magnetosheath The intervening layer between the MAGNETOPAUSE and the shock front caused by the solar wind.

magnetosphere The region of the atmosphere, encompassing the EXOSPHERE and most of the IONOSPHERE in which the Earth's magnetic field influences the motion of ionized particles. The magnetic field acts as a shield for the Earth deflecting and preventing solar winds from reaching the surface. The shape of the magnetosphere is distorted by the influence of the solar wind: on the daylight side facing the Sun the magnetic field lines are compressed by the solar wind; on the night side (away from the Sun) they are elongated to form a long *magnetotail.*

Mai-yu season (Bai-u season) The time that the 'plum rains' occur across much of central China, after the cessation of the northeast trade airflow from the Pacific Ocean. The rains tend to occur once the southwest airstream has become established. By mid-June the warm humid air lies across much of central China. Small disturbances produce heavy rainfall of a non-frontal nature, these are called the plum rains, or Mai-yu (Mei-yu), since they arrive as the plums are ripening. Weekly rainfall may exceed 60 mm (2.36 in).

Maloja wind A warm föhn-type wind that blows down the Upper Engadine valley by day from the summit of the Maloja Pass in Switzerland; at night the wind blows up valley.

mamma (mammatus) A supplementary

cloud feature consisting of hanging rounded udder-like protuberances on the underside of another cloud, perhaps most commonly observed on the underside of laterally spreading anvils on top of CUMULONIMBUS clouds. Their appearance can be dramatic when illuminated from below by the setting Sun. Mamma clouds also occur beneath ALTOCUMULUS, ALTOSTRATUS, STRATOCUMULUS, CIRRUS, and CIRROCUMULUS. *See also* cloud classification.

mammatus *See* mamma.

manometer An instrument consisting of a liquid column gauge used for measuring pressure differences. The MERCURY BAROMETER, in which mercury is balanced against atmospheric pressure, is a type of manometer.

mares' tails A common name for a type of CIRRUS in the form of elongated wispy streaks.

marin A warm SCIROCCO-type wind that blows across the Gulf of Lions into southeastern France. Usually oppressive, it can be accompanied by overcast conditions and heavy rain.

marine climate *See* marine west coast climate; maritime climate.

marine meteorology The study of the atmospheric phenomena over ocean and coastal areas and the interaction of the atmosphere and fluid surface of the ocean. It includes wind and wave measurement, wind systems, boundary-layer processes, mesoscale coastal phenomena, sea breezes, and tropical cyclones.

marine west coast climate The climate of the west coastal regions of the continents, between 35° and 60° N and S. Precipitation is abundant all year with a winter maximum, and prevailing westerlies and cyclonic storms can occur throughout the year. Annual temperature ranges are low with mild winters and generally moderate summers. In western Europe these climates can extend inland into central Europe; in North and South America, Australia, and New Zealand mountain ranges aligned generally north–south confine the climates to the coastal margins. The climate is designated Cfb and Cfc in the KÖPPEN CLIMATE CLASSIFICATION.

maritime air An AIR MASS that is generally moist and mild or warm from its SOURCE REGION over an ocean, or through modification after passage across an ocean surface.

maritime climate (marine climate) The climate of a region that is strongly influenced by its location in relation to the ocean. The moderating influence of the ocean gives these climates their characteristically low diurnal and annual temperature range, while the availability of moisture is responsible for comparatively high precipitation and humidity.

maritime polar air mass (mP) A cool and moist AIR MASS that has swept across many hundreds to a few thousand kilometers of middle- or higher-latitude ocean from a cold continental SOURCE REGION. The process of flowing across an ocean after leaving the source region means that these air masses are strongly influenced by huge SENSIBLE HEAT and LATENT HEAT FLUXES.

maritime tropical air mass (mT) A warm and moist AIR MASS that originates over the subtropical oceans beneath the SUBTROPICAL ANTICYCLONES, out of which the air spirals toward higher and lower latitudes. Maritime tropical air tends to be laden with stratiform cloud, and brings mild and humid conditions followed generally by light precipitation by the time it arrives in middle latitudes.

marsh gas *See* methane.

Maunder minimum A period of relative solar inactivity from 1645 to 1715 during which it is believed there were fewer sunspots than have occurred after this period, according to evidence put forward in the 1890s by the British astronomer E.

Walter Maunder (1851–1928). It coincided with the period of colder than average temperatures of the LITTLE ICE AGE. Evidence suggests that other periods of similar inactivity have occurred in the more distant past.

maximum and minimum thermometer A THERMOMETER used to measure both maximum and minimum temperatures over a given time period. *See also* maximum thermometer; minimum thermometer.

maximum temperature The highest temperature achieved during a specific period, typically over 24 hours. It is usually recorded with a standard MAXIMUM THERMOMETER.

maximum thermometer A THERMOMETER designed to record the highest temperature reached over a given time period between settings. This is usually a mercury-in-glass thermometer in which a constriction above the bulb allows the mercury to expand and flow past it on expansion with a rise in temperature but stops it flowing back in the opposite direction upon contraction. Resetting is carried out by shaking the thermometer vigorously until the mercury is below the constriction. *See also* maximum and minimum thermometer; minimum thermometer.

mb *See* millibar.

MCC *See* mesoscale convective complex.

MCS *See* mesoscale convective system.

mean sea level (MSL) The average level of the surface of the sea for all stages of the tide. The level of the sea is constantly changing through tidal oscillations, waves, swell, and atmospheric pressure changes. The mean sea level is calculated from observed heights taken at hourly intervals, generally over a 19-year period, using methods that take account of the fluctuations in the surface level. The world mean sea level is calculated from many observations made globally and can change in relation to eustatic (*see* eustasy) changes and also changes in the height of land surfaces (isostatic changes).

mean-sea-level pressure The value of ATMOSPHERIC PRESSURE after it has been reduced from its station value (i.e. the actual observation) to the common datum of mean-sea-level. This eliminates the very marked influence of elevation on the value of recorded pressure; near sea level the pressure decreases by about 1 hPa for every 10 meters of ascent.

mean temperature The average temperature taken over a specified period. For example; mean daily temperature is the average temperature observed at regular time intervals throughout the solar day.

mediocris A species of CUMULUS cloud showing moderate vertical extent and small protuberances on the tops. *See also* cloud classification.

Mediterranean climate A climatic type that is distinguished by its unique precipitation cycle, with hot dry summers and mild moist winters. It occurs around latitude 30°–45°N and S along the west coasts of continents and poleward of the eastern side of the subtropical anticyclones (highs). During the summer the poleward migration of the anticyclones into the Mediterranean climate regions brings dry continental tropical air. In winter the climate is influenced by westerly airflow bringing moist air, cyclonic storms, and abundant rainfall. Total annual rainfall amounts vary considerably depending on the location, decreasing over continental locations and toward the tropics.

The largest extent of this climate type is around the Mediterranean Sea, in Europe, extending from Portugal and Morocco in the west to Eurasia in the east. Mediterranean climates also occur in central and southern California in North America; in the southern hemisphere it occurs along the coast of Chile, South Africa, and southwestern Australia. It is denoted by Csa and Csb in the KÖPPEN CLIMATE CLASSIFICATION.

Mediterranean front That part of the POLAR FRONT when located over the Mediterranean–Caspian Sea areas in winter. It separates Mediterranean air to the north from the continental tropical air flowing northeastward from the Sahara in the south. Temperatures may vary by as much as 12–16°C across the front in late winter.

Mediterranean water A region of deep dense water in the Mediterranean. It is formed by the excessive evaporation of surface water and the resultant instability, which culminates in the sinking and spreading out of water at the bottom of the Mediterranean Basin. Mediterranean water also overflows across the submarine sill at the Straits of Gibraltar where it sinks to depths of about 1000 m (3000 ft) and spreads out on the bottom of the Atlantic Ocean.

medium-range weather forecasting The process of predicting the weather on a global scale from three to seven or even up to ten days ahead. The EUROPEAN CENTRE FOR MEDIUM-RANGE WEATHER FORECASTS (ECMWF), for example, specializes in this period; other centers, including the US NATIONAL WEATHER SERVICE (NWS), fulfil the same role.

megathermal climate A climate type with high temperatures. In the KÖPPEN CLIMATE CLASSIFICATION it is the A group of climates (humid subtropical or tropical), with monthly mean temperatures above 18°C (16.4°F). The same name is also used in the THORNTHWAITE CLIMATE CLASSIFICATION.

meltemi *See* etesian wind.

mercury barometer (mercurial barometer) A type of BAROMETER used for measuring atmospheric pressure. It consists of a vertical column of mercury contained in a glass tube that has been evacuated and sealed at the top and marked with a scale. The weight of the mercury in the tube is balanced by the atmospheric pressure acting on the exposed cistern of mercury at the base of the column, the height of the column varying with changes in atmospheric pressure. Readings from the mercury barometer need to be corrected to take into account air temperature and local value of the acceleration of free fall and adjusted to mean-sea-level. Although this type of barometer is still widely used it is being superseded in many weather stations by solid-state barometers.

The mercury barometer, historically the first instrument used to measure atmospheric pressure, was invented in 1643 by the Italian mathematician and physicist Evangelista Torricelli (1608–47). *See also* Fortin barometer; Kew barometer.

mercury-in-glass thermometer A THERMOMETER containing mercury, which acts as a heat-measure medium. The mercury is contained within a glass bulb or reservoir attached to a thin tube within a glass sheath. With temperature changes the mercury expands as it is heated or contracts as it is cooled and moves up or down the attached tube. Changes in the height of the mercury column can be read from a graduated scale etched onto the side of the thermometer. The mercury-in-glass thermometer is unsuitable for recording exceptionally low temperatures because of mercury's freezing point of –38.9°C (–38°F); a thermometer containing alcohol, which has a freezing point of –114.4°C (–174°F), is used instead to record very low temperatures.

meridional circulation The component of the large-scale atmospheric motion that is parallel to meridians or lines of longitude and thus shows large north–south movement. *Compare* zonal flow.

meridional temperature gradient The temperature gradient in the direction of the geographic meridian, for example the difference in temperature between a location in the north and one on the Equator.

mesoclimate The climate of a small geographical area created, for example, by differences in the vegetation cover (e.g. forest, crops), differences in elevation (e.g. upland areas), differences in slope aspect (e.g.

in valleys), water bodies, or urban areas. Such climates occur on the scale of tens to hundreds of kilometers and within the broader extensive climate regions of the MACROCLIMATES. *Compare* microclimate.

mesocyclone A cyclonic circulation consisting of a column of rapidly rotating air associated with some deep convective storms. The mesocyclone, which is a feature of all SUPERCELL thunderstorms, occurs within the cumulonimbus cloud and is usually up to 10 km (6 miles) in diameter. A TORNADO may develop from the mesocyclone's rotation and extend below the cloud to the ground.

mesohigh (mesoscale high) A MESOSCALE high-pressure feature that is associated with the cold dense current of air that flows out of a THUNDERSTORM as a DOWNDRAFT. The cold air radiates horizontally away from underneath a thunderstorm, leading to a sudden jump in surface pressure (apart from the temperature drop and sudden onset of gusty winds).

mesolow (mesoscale low, subsynoptic low) A MESOSCALE low-pressure region.

mesometeorology The study of middle-scale weather and atmospheric phenomena, such as tropical cyclones and polar lows, thunderstorms, mesoscale convective systems (MCS), and tornadoes. These features are intermediate between microscale and synoptic scale or macroscale. *Compare* macrometeorology; micrometeorology.

mesonet A network of environmental monitoring stations designed to measure the environment at the size and duration of mesoscale weather events, i.e. at the scale of few kilometers to several hundred kilometers and of a few minutes to several hours duration.

mesopause The boundary that lies about 80 km (50 miles) above the Earth's surface and caps the MESOSPHERE, at which point temperatures stop decreasing.

mesoscale Denoting meteorological phenomena or weather systems that range in size from about 80 to several hundred km (50 to several hundred miles) horizontally. Such features include squall lines, mesoscale convective complexes, and mesoscale convective systems. They are smaller than the SYNOPTIC-SCALE weather systems but larger than MICROSCALE. Mesoscale is sometimes applied to smaller-scale systems, for example, between 10 and 100 km (6–60 miles). *See also* macroscale.

mesoscale convective complex (MCC) A large and persistent MESOSCALE CONVECTIVE SYSTEM, which establishes inertially stable circulations and does not readily decay. It is classified in terms of the areal extent of the cloud shield as observed on infrared satellite imagery: the area of cloud top with temperature below –32°C must exceed 100,000 km^2; that with temperature below –52°C must exceed 50,000 km^2. In addition, it must last at least 6 hours.

mesoscale convective system (MCS) A deep convective system of a number of thunderstorms that can extend for up to around 500 km (300 miles) in horizontal extent and last for 6–12 hours. It is distinguished in synoptic observations by an extensive middle- or upper-level stratiform/cirrus shield. There is frequent CLOUD-TO-CLOUD LIGHTNING within the system. *See also* mesoscale convective complex.

mesoscale system (mesosystem) *See* mesoscale; mesoscale convective complex; mesoscale convective system.

mesosphere The atmospheric layer that lies above the STRATOSPHERE, separated from it by the STRATOPAUSE at about 40–50 km (25–30 miles) above the Earth's surface, and below the thermosphere, from which it is separated by the MESOPAUSE at approximately 80 km (50 miles) above the Earth's surface. It is the coldest atmospheric layer with temperatures falling rapidly with height to about –100°C at 80 km (50 miles). This fall is a result of the lack of water vapor, dust, and ozone, which absorb incoming solar radiation.

mesothermal climate A climate type with moderate temperatures. In the KÖPPEN CLIMATE CLASSIFICATION it is the C group of climates (warm temperate rainy climates), in which the monthly mean temperature for the coldest month is between –3°C (27°F) and 18°C (64°F) and for the warmest month is greater than 10°C (50°F). The same name is used in the THORNTHWAITE CLIMATE CLASSIFICATION.

METAR A special SYNOPTIC CODE used for routine weather reports from airfields and airports that are made quickly available to aviation.

METEOR A long-lived series of Russian near polar-orbiting meteorological satellites.

meteoric water Water on the surface of the Earth that is derived from atmospheric PRECIPITATION.

meteorological computer model A computer-based MODEL used to simulate meteorological processes.

Meteorological Operational Satellite *See* METOP.

meteorological satellite (weather satellite) An artificial satellite in orbit around the Earth, most commonly in either a POLAR ORBIT or a GEOSTATIONARY ORBIT, with the specific operational aims that include mapping cloud distribution and sea-surface temperature, and providing vertical temperature and humidity profiles together with cloud motion vectors.

Meteorological Service of Canada A federal organization, a part of ENVIRONMENT CANADA, that undertakes day-to-day weather prediction and research into future climate. It also monitors water and air quality, and the extent of sea ice around Canadian waters.

meteorology The study and science of all the phenomena and processes of the atmosphere, and the application of such knowledge to the forecasting of weather. It includes the physical and chemical processes that take place within the atmosphere, the motions (dynamics), and the interactions of the atmosphere with the oceans and surface of the Earth. *See also* climatology; dynamic meteorology; physical meteorology; synoptic meteorology; weather forecasting.

Meteosat The European community geostationary satellite series, operated by EUMETSAT. A Meteosat is positioned above Europe and Africa at approximately 0° longitude as one of the operational meteorological GEOSTATIONARY SATELLITES (currently five or six) stationed above the Equator at 0° to provide coverage of the whole Earth. The first satellite of the series, Meteosat–1, was launched by the European Space Agency (ESA) in 1977. The Meteosats scan using the wavelengths of: visible (0.5–0.9 µm); thermal infrared (10.5–12.5 µm); and water vapor (5.7–7.1 µm).

The *Meteosat Second Generation* (MSG) series is under development and scheduled to come into operation during the early years of the first decade of the 21st century.

methane (CH_4, marsh gas) A colorless, odorless, and flammable hydrocarbon gas, which is a TRACE GAS in the Earth's atmosphere. Emissions of methane are produced through the venting of natural gas and production of petroleum, decomposition of landfill and animal waste, animal digestion, rice cultivation, and coal production. Methane is a potent GREENHOUSE GAS, and concentrations within the atmosphere have been rising at the rate of about 0.6% per year.

Met Office The national weather service in the UK, currently based in Bracknell but scheduled to move to Exeter in 2003. The organization can be traced back to 1854, when it originated as a small department of the Board of Trade. Its role has expanded recently from one focusing primarily on the weather to a broader service that includes the impact of weather on the environment with interests in the other environmental sciences, such as hydrology and oceanography.

METOP (Meteorological Operational Satellite) A new series of polar-orbiting meteorological satellites under development, which will carry instruments for the EUROPEAN SPACE AGENCY (ESA), EUMETSAT, NOAA, and the French space agency (CNES). METOP–1, due to be launched in 2003, is scheduled to fly at 835 km (520 miles) in a five-day repeat Sun-synchronous orbit. The satellite series is intended to eventually form part of a NOAA–EUMETSAT system.

microburst A narrow DOWNBURST that is either dry or wet. Microbursts appear in association with shafts of precipitation that fall from the base of CUMULONIMBUS clouds. Microbursts are of short duration (less than five minutes) and affect a small area of less than 4 km (2.5 miles). A *wet microburst* occurs when the heavy rainfall drags a column of air down with it, hitting the surface and spreading out horizontally as a very gusty current of air. A *dry microburst* is one that is common across the High Plains in the US, where the cloud base is high and the air between the base and the ground is normally dry. In these conditions, the rain evaporates before reaching the surface, so cooling the air it falls through. This leads to a microburst that also splays out sideways on reaching the ground. Microbursts are particularly hazardous to aircraft, especially when landing or taking off.

microclimate The climate of a small distinct area, usually in the lowest layer of the atmosphere above the ground surface, in which features of the surface, such as vegetation and buildings, influence temperature, humidity, and wind. Different microclimates may occur at different layers within a plant canopy, or on either side of a building, for example. Such climates typically occur at scales of less than 100 meters. *Compare* macroclimate; mesoclimate.

microclimatology The study of a MICROCLIMATE, encompassing temperature, moisture and wind profiles within the lowest level of the atmosphere, including the effects of vegetation, soil moisture content, and urbanization.

micrometeorology The study of small-scale atmospheric phenomena that occur in the lowest levels of the atmosphere, in the PLANETARY BOUNDARY LAYER; it includes the processes of exchanges of heat, moisture, and momentum with the surface of the Earth; the SURFACE BOUNDARY LAYER; and TURBULENCE. *Compare* macrometeorology; mesometeorology.

micrometer (μm) The SI unit of length equal to 10^{-6} meter (formerly known as the *micron*). It is encountered in meteorology, for example, in the diameters and radii of aerosols, water droplets, and particles, and in the wavelengths of electromagnetic radiation.

micron A former name for the micrometer.

microscale Denoting the smallest scale of meteorological phenomena, ranging in size from a few centimeters up to several kilometers. The term is also applied to small-scale phenomena of short duration of a few minutes. *Compare* macroscale; mesoscale.

microthermal climate A climate type with low temperatures, short summers, and long cold winters. In the KÖPPEN CLIMATE CLASSIFICATION it is the D group of climates (cold boreal forest climate), with mean monthly winter temperatures of less than –3°C (27°F). The same name is also used in the THORNTHWAITE CLIMATE CLASSIFICATION.

middle clouds In CLOUD CLASSIFICATION, cloud with a base between approximately 2000 and 7000 m (6500 and 23,000 ft) above the surface. These values vary seasonally and latitudinally such that they are higher in summer and in low latitudes. The class includes ALTOCUMULUS, ALTOSTRATUS, and NIMBOSTRATUS.

middle-latitude steppe and desert climate The climates of the cool deserts and steppe in North and South America and Asia, which are located within the continental interiors far from sources of moisture and often bounded by mountain barriers on the windward side. Cool deserts extend from about 35° latitude as far north as 50° latitude N and S; the cool steppes extend nearly to 60°N in Canada. Winters are cold with little precipitation; summer precipitation tends to be cyclonic. The climates are denoted by the letters BSk and BWk in the KÖPPEN CLIMATE CLASSIFICATION.

mid-latitude climate The group of five climates of the middle-latitude regions of the world identified in the STRAHLER CLIMATE CLASSIFICATION: humid subtropical, marine west-coast, Mediterranean, mid-latitude desert and steppe, and humid continental, which are controlled by both tropical and polar air masses.

mid-latitude cyclone *See* cyclone.

mid-latitude depression *See* cyclone.

mid-level cooling Localized cooling of air in the middle TROPOSPHERE between 2400 and 7500 m (8000 and 25,000 ft). This cooling process is able to destabilize the entire atmosphere, for example when a mid-level cold pool approaches a region.

midnight Sun The phenomenon that occurs in the polar regions, as a result of the Earth's tilt about its axis, in which the Sun remains visible above the horizon for over 24 hours. The Sun remains above the horizon from about March 20 through September 23 at the North Pole, and from about September 23 through March 20 at the South Pole. The length of time for which the midnight Sun occurs decreases toward the Arctic and Antarctic Circles. Along the Arctic Circle (66°32′ N) the Sun remains above the horizon on June 21, the northern hemisphere summer solstice; along the Antarctic Circle (66°32′ S) this occurs on December 22, the southern hemisphere summer solstice.

Mie scattering A theory of the SCATTERING of ELECTROMAGNETIC RADIATION by particles in the atmosphere, proposed by G. Mie in 1908. The degree of scattering of direct radiation depends upon particle size. Whenever particles of a similar size to the wavelength of the incident radiation are present, Mie scattering will occur; this is predominantly in the forward direction in all wavelengths, especially in a turbid (dusty) atmosphere. This produces a grayish sky as no wavelengths in the visible band are preferentially scattered. The particles causing Mie scattering are spherical and include water vapor, dust, smoke, and pollen. It occurs primarily in the lower atmosphere *See also* non-selective scattering; Rayleigh scattering.

Milankovitch cycles (astronomical theory of climate change) A theory of the cycle of long-term climate change, and the ICE AGES, correlated with the distribution of insolation that is determined by variations in the Earth's orbit around the Sun. The theory was suggested during the mid 1800s and subsequently formulated and published in 1920 by the Serbian astronomer and mathematician Milutin Milankovitch (1879–1958).

There are three main variables that influence the amount and distribution of the solar radiation received by the Earth: (1) Changes in the tilt of the Earth's axis, which affects the severity of seasons. The tilt or inclination varies between about 21.8° and 24.4° on a cycle of some 41,000 years. (2) The eccentricity of the Earth's orbit about the Sun, which varies on a primary periodicity of approximately 100,000 years and a secondary periodicity of 400,000 years. (3) Changes in the Earth's axis of rotation, which affects the timing of the equinoxes – the precession of the eqinoxes – and the timing of the APHELION and PERIHELION, and occurs on a period of approximately 22,000 years. This alters the timing of the aphelion and perihelion in relation to the seasons. The combinations of these cyclical variations affect the severity of the seasons.

millibar (mb) The unit of ATMOSPHERIC PRESSURE that is frequently used by meteorologists. The millibar is defined as one thousandth (10^{-3}) of one BAR. A pressure of 1 mb is equivalent to 0.7501 mm of mercury. The millibar has been largely replaced in scientific work by the HECTOPASCAL (hPa) but remains in general use for weather forecasting. 1 mb is equal to 1 hPa.

minimum temperature The lowest temperature achieved during a specified period of time, typically over a 24-hour period. It is usually recorded with a standard MINIMUM THERMOMETER within an instrument shelter; additional measurements of minimum temperatures are made over concrete, over soil, and over grass.

minimum thermometer A THERMOMETER used for measuring the lowest temperature attained over a given time period (usually a day). These thermometers are usually alcohol-in-glass thermometers; the alcohol has a lower freezing point of –114.4°C (–174°F) compared to that of –38.9°C (–38°F) for mercury. An index marker is contained within the thin bore of the thermometer. As the alcohol cools and contracts the marker is carried down by the alcohol's meniscus toward the bulb of the thermometer, but as it warms and expands the marker remains unmoved. The end of the marker furthest from the bulb indicates the lowest temperature reached since it was last reset. The thermometer is reset by tilting it until the marker is returned to position at the meniscus. *See also* maximum and minimum thermometer; maximum thermometer.

mirage An optical effect of the lowest atmosphere in which images of objects appear displaced from their actual position and/or distorted as a result of the differential refraction of light rays passing from one density of air to another. A range of effects occurs according to the variations in density of the air, which are determined by temperature and humidity. An *inferior mirage* occurs when the image appears below its actual position. This results when the air closest to the ground surface is strongly heated, for example, above the surface of an asphalt road on a hot day or above a desert surface. Light rays near the ground surface are strongly refracted upward and produce the visual effect of shimmering pools of water. A *superior mirage* occurs when a layer of warm air overlies a much cooler layer directly above the ground surface, as occurs when warmer air flows above an ice surface. The light rays are strongly refracted downward resulting in the effect that the image appears to be above its actual position. In polar regions, for example, distant objects may appear to hang above the snow or ice surface. *See also* fata morgana.

mist A condition in which the horizontal visibility is reduced by small suspended water droplets or wetted hygroscopic nuclei such that it is greater than 1000 m (3000 ft) but less than 2000 m (6500 ft); if the visibility is less than 1000 m it is defined as FOG. The obscuring particles have typical diameters of a few tens of micrometers. *See also* Scotch mist.

mistral A strong cold dry LOCAL WIND that can be experienced along the Mediterranean coast from Barcelona to Genoa, although it is most frequently observed in the Rhône delta region. It occurs most often during the winter season, for example, Marseille, France, is subject to the mistral on 100 days in an average year. Synoptically, the mistral blows as a result of the pressure gradient established by high pressure over central Europe and/or the western Mediterranean and a deep low-pressure area situated in the Gulf of Lions or Gulf of Genoa. The resulting north to northwesterly wind is funneled down the Rhône Valley between the Massif Central and Alpine ranges, transferring cold polar maritime air southward. The mistral often attains speeds at the surface in excess of 30 knots (35 mph), sometimes being strengthened by KATABATIC effects; the sky is typically devoid of clouds at the time and the wind can reduce the temperature to below 0°C (32°F). WINDBREAKS, such as thick hedges of cypress trees, have been planted throughout the region to protect crops, orchards, and gardens from the damaging force and coldness of the mistral. Occasionally the mistral has crossed the Mediterranean Sea and been felt in such North African countries as Tunisia.

mixed cloud A cloud composed of both liquid water droplets and ice crystals. ALTOSTRATUS, CUMULONIMBUS, and NIMBOSTRATUS are examples.

mixed layer (turbulent boundary layer) That part of the PLANETARY BOUNDARY LAYER that is dominated by turbulent diffusion caused by eddies generated by friction with the surface and thermals arising from surface heat sources (*see* eddy diffusion). Surface heating during the day and the absence of temperature inversions allow components of the air within the planetary boundary layer to exhibit mainly random vertical movements. Such movements may become more organized into gusts of wind and dust devils during the afternoon. Despite being random, the turbulent movements allow the transfer of atmospheric properties, such as heat, water vapor, momentum, and air pollutants, from the near surface up through the planetary boundary layer.

mixing condensation level The lowest level at which condensation occurs when two air masses possessing different initial temperature and DEW-POINT profiles are mixed together. The process of vertical mixing has the effect of averaging the properties of the two original air masses, resulting in over-saturation and therefore condensation.

mixing depth (mixing layer) The vertical extent of the atmospheric layer near the Earth's surface in which convection and turbulence are pronounced, causing the mixing of air and any pollutants present.

mixing fog *See* evaporation fog.

mixing ratio (humidity mixing ratio) A measurement of atmospheric moisture content, usually expressed as the dimensionless ratio of mass of WATER VAPOR in a parcel of air to the unit mass of dry air (typically grams of water vapor per kilogram of dry air).

mizzle *See* Scotch mist.

mock Moon *See* paraselene.

mock Sun *See* parhelion.

model A representation of a system or process. A tool used to simulate or forecast the behavior of the atmosphere–ocean system. Such models are used to produce weather forecasts from hours to days or long-term climate conditions over periods of years to thousands of years. Examples of models include CLIMATE MODELS, ENERGY BALANCE MODELS, GENERAL CIRCULATION MODELS, RADIATIVE-CONVECTIVE MODELS, and STATISTICAL-DYNAMICAL MODELS.

moist adiabatic lapse rate *See* saturated adiabatic lapse rate.

moisture advection The predominantly horizontal movement of atmospheric water vapor from one location to another. This water vapor flux is important for maintaining a location's moisture budget, a rapid flux being required if substantially high precipitation totals are to be generated.

moisture index A measure of the annual water balance for an area, incorporated in a new CLIMATE CLASSIFICATION emphasizing the concept of EVAPOTRANSPIRATION that was devised by the American climatologist, Charles Warren Thornthwaite, in 1948, and slightly revised in 1955. The moisture index included evapotranspiration and is expressed by the formula:

$$I_m = 100 \times (S - D)/PE,$$

where I_m is the moisture index, S is the water surplus in months when precipitation exceeds evapotranspiration, D is the water deficit in months when evapotranspiration exceeds precipitation, and *PE* is potential evaporation. It has a value of zero when precipitation is equal to potential evapotranspiration; a positive value indicates a moist climate, and a negative value indicates a dry climate. *See also* Thornthwaite climate classification.

monsoon The seasonal change in wind direction associated with differential heating of land and sea areas. The word, from the Arabic *mawsim*, season, was originally applied to the seasonal reversal of winds in the Arabian Sea that blow for about six months from the northeast and six months from the southwest. The monsoon is associated with the seasonal changes that take place in the pressure systems together with the positions of the upper wind patterns and jet streams. In winter the intense high-pressure system, the Siberian anticyclone, is located over central Asia, and winds flow out from this (the northeast monsoon). During summer it is absent, low-pressure is centered over the Middle East (the Asiatic low), and the southwest monsoon is dominant. The Indian subcontinent is especially affected by the monsoon but monsoons also occur in Africa, northern Australia and to a limited extent in the southwestern US. The word is now more generally used in the tropics and subtropics for any region affected by a reversal of wind, particularly when accompanied by copious amounts of rainfall. *See also* African monsoon; Asian monsoon. *See also* burst of the monsoon.

monsoon climate *See* tropical monsoon and trade-wind littoral climate.

monsoon current The humid southwest flow of air that crosses the southern Arabian Sea and heralds the onset of the southwest MONSOON in India. The airstream usually reaches Kerala state in early June, and proceeds northeast toward Calcutta, before moving erratically northwest up the Indo-Gangetic Plain along the foothills of the Himalayas.

monsoon depression A low-pressure system that normally forms in the Bay of Bengal north of 18°N latitude and moves west-northwest across the central and northern parts of India producing widespread rains, especially in the southern sector of the traveling low. Between five and seven every year form in this way. These depressions rarely develop into devastating cyclonic storms.

monsoon trough An elongated area of low pressure positioned on average ENE–WNW across the plains of north India during the summer MONSOON season. It may be regarded as the most northerly position of the Intertropical Convergence Zone (ITCZ) in India and marks the approximate line along which the majority of MONSOON DEPRESSIONS travel during June to late September.

month degrees In climate classifications, a parameter that is sometimes used to indicate the conditions suitable for vegetation growth, in which the excess of mean monthly temperatures greater than 6°C (43°F) are summed and used as accumulated temperature.

Montreal Protocol (Montreal Protocol on Substances that Deplete the Ozone Layer) The international agreement drawn up in Montreal in 1987 by the United Nations Environment Program (UNEP), and ratified by most of the industrialized nations in 1989, under which ozone-depleting substances are regulated. The original agreement sought a 50% reduction in the use of chlorofluorocarbons (CFCs) by 1992, relative to 1986 levels. Subsequent amendments expanded the list of substances to be regulated, and brought forward the dates for these to be phased out. The Protocol uses several international bodies to report on the science of ozone depletion and provide information for policy development. *See also* Kyoto Protocol.

morning glory A very long spectacular narrow ROLL CLOUD that sweeps across part of the Gulf of Carpentaria in northern Australia. It can stretch up to 1000 km (600 miles) in length and to 3000 m (10,000 ft) in height, and may sometimes be as low as 50 m (160 ft). It is oriented NNW–SSE and advances from east to west at speeds of 25–45 knots (30–50 mph). The morning glory brings squall-like winds and fine mist, but little precipitation. One theory for its formation is that it is somehow related to the double sea breeze penetration that occasionally occurs from either coast of the Cape York peninsula on the eastern flank of the Gulf of Carpentaria.

mother cloud The type of cloud from which another one develops. For example, anvil cirrus can be associated, as a mother cloud, with MAMMA.

mother-of-pearl cloud *See* nacreous cloud.

mountain breeze *See* mountain wind.

mountain gap wind *See* canyon wind.

mountain meteorology The study of atmospheric phenomena in mountainous regions and areas of complex terrain. The scale of the phenomena and processes vary from a few meters to thousands of kilometers. It includes gravity waves, convection cells, and mesoscale mountain circulations.

mountain wave *See* lee wave.

mountain wind (mountain breeze) A KATABATIC WIND that is most developed in valley systems during the middle of the night when it flows down from the mountains, filling the valley in the late night and early morning period.

Compare valley wind.

Mozambique Current A warm west Indian Ocean current, that flows between Madagascar and Africa, influencing the climate of the island and the mainland. South of Madagascar, the Mozambique Current feeds into the AGULHAS CURRENT.

MSL *See* mean sea level.

multicell storm A thunderstorm that contains more than one cell, i.e. more than one updraft and downdraft system. There may be a number of updrafts and downdrafts within the storm in close proximity. *See also* squall line.

multiple-vortex tornado (multivortex tornado) A TORNADO that possesses two or more condensation funnels or debris clouds.

multispectral imaging The process of remotely sensing a natural scene from a satellite or a planetary surface in a large number of narrow wavebands. Each BAND is normally selected to map a different feature from other bands. *See also* remote sensing.

multispectral satellite imagery A set of simultaneous satellite images of the same scene that are recorded in different narrow wavebands, usually to stress the nature of Earth surface features.

multispectral scanner An on-board satellite instrument that is capable of imaging the Earth in a number of narrow wavebands simultaneously, to build up a line-by-line image of the surface. LANDSAT, for example, uses such a scanner.

mutatus Denoting a MOTHER CLOUD changed in such a way that the entire cloud is affected.

nacreous cloud (mother-of-pearl cloud) A rare kind of cloud that occurs in the stratosphere, mainly at high latitudes, and at elevations around 24 km (15 miles). They are normally of lenticular form and exhibit brilliant IRIDESCENCE at angles within some 40° of the Sun soon after sunset or before sunrise. They appear to be orographic in origin and are likely to be forced by flow over mountains that produces wavy undulations at great heights. Their formation is thought to require a temperature of –80°C to –90°C (–112°F to –130°F) at cloud level.

nadir The point on the celestial sphere diametrically opposite to the ZENITH.

NAO *See* North Atlantic Oscillation.

National Ambient Air Quality Standards (NAAQS) In the US, the targets set for an acceptable concentration of specific pollutants in air. Under the 1970 Clean Air Act and amendments, national *Primary Ambient Air Quality Standards*, set by the Environmental Protection Agency (EPA), define levels of air quality designed with an adequate margin of public health safety. The EPA is required to set primary standards for the major air pollutants at levels that are necessary to protect the public health with an adequate margin of safety. *Secondary Ambient Air Quality Standards* are defined with a view to protecting the public welfare from any known or anticipated adverse effects of a pollutant.

National Centers for Environmental Prediction (NCEP) A US organization, part of the NATIONAL OCEANIC AND ATMOSPHERIC ADMINISTRATION's (NOAA's) National Weather Service, that is made up of nine national centers with the aim of providing timely and accurate forecasts. The centers include the TROPICAL PREDICTION CENTER, STORM PREDICTION CENTER, and CLIMATE PREDICTION CENTER, as well as others, such as the Aviation Weather Center in Kansas City, Missouri. It was established in 1958 as part of the former National Meteorological Center.

National Environmental Satellite, Data and Information Service (NESDIS) In the US, an organization that is part of the NATIONAL OCEANIC AND ATMOSPHERIC ADMINISTRATION (NOAA); it manages its environmental and operational weather satellite program, including the processing, storage, and distribution of the huge volume of data provided by the satellite network.

National Fire Danger Rating System (NFDRS) A method of rating the potential danger of FIRE for large areas across the US. Operational, since 1972, this system uses daily fire weather observations and forecasts as well as integrating and interpreting seasonal weather trends to determine the risk of fire.

National Hurricane Center *See* Tropical Prediction Center.

National Oceanic and Atmospheric Administration (NOAA) A major organization, within the US Department of Commerce. Its role is to conduct research and to gather data about the global oceans, atmosphere, space, and the Sun. It aims to describe and predict changes in the Earth's environment and to conserve and wisely manage coastal resources. Two of its many constituent services are the NATIONAL WEATHER SERVICE (NWS) and the NATIONAL ENVIRONMENTAL SATELLITE, DATA AND INFORMATION SERVICE (NESDIS).

National Severe Storms Laboratory (NSSL) In the US, a research laboratory in Norman, Oklahoma, sponsored by the NATIONAL OCEANIC AND ATMOSPHERIC ADMINISTRATION (NOAA); it has the role of 'enhancing NOAA's capabilities to provide accurate and timely forecasts and warnings of hazardous weather events (e.g. blizzards, ice storms, flash floods, tornadoes, lightning, etc.)'. Its work includes undertaking research to improve understanding of weather processes associated with such severe phenomena, as well as improving the techniques of prediction and warning.

National Weather Service (NWS) The US operational weather service, forming a part of the NATIONAL OCEANIC AND ATMOSPHERIC ADMINISTRATION (NOAA); it is made up of a headquarters, national centers, and regional offices, which together provide expertise in the fields of meteorology, hydrology, and climate science. The US regions and their HQ offices are: Alaska (Anchorage), Pacific (Honolulu), Western (Salt Lake City), Central (Kansas City), Southern (Fort Worth) and Eastern (Bohemia). There are also 13 regional river forecast offices. The five centers, including the STORM PREDICTION CENTER and the TROPICAL PREDICTION CENTER have the role of delivering analysis and forecast products for all regional offices and other US government agencies.

natural hazard A natural process or event that causes harm to people and/or property. Natural hazards connected with the weather and climate include drought, fire, floods, tropical cyclones, tornadoes, severe thunderstorms, and snow or ice storms. These, however, only constitute natural hazards if they pose a threat, risk, danger, or peril to human activity, otherwise they are seen as extreme natural events.

natural-siphon rainfall recorder *See* tipping-bucket rain gauge.

NCEP *See* National Centers for Environmental Prediction.

nebulosus A type or species of cloud that appears as a nebulous layer or veil with no distinct detail. It applies only to STRATUS and CIRROSTRATUS. *See also* cloud classification.

negative tilt trough The angular displacement of a trough line from a meridian in which the axis of the trough is rotated clockwise from the meridian.

neoglacial A series of relatively small glacier readvances during the last few years of the Holocene geologic epoch.

nephanalysis A technique to produce an analysis of cloud type from studying satellite cloud images or synoptic charts. The subjective method is no longer used, though there are ways to assess the type and height of cloud automatically from such data.

nephoscope An instrument used to measure the speed and direction of cloud movement relative to a point on the ground directly below it.

neritic Denoting the region of shallow water lying directly above the continental shelf, usually less than 200 m (660 ft) deep.

NESDIS *See* National Environmental Satellite, Data and Information Service.

net radiation (Q^* net all-wave radiation) The net value of downward incoming shortwave and upward outgoing longwave radiation fluxes at the Earth–atmosphere interface. This is the most important surface energy exchange measurement or parameter since it represents the limit to the available energy either as source or as sink. For a daytime surface energy budget:

$$Q^* = (K_{down} - K_{up}) + (L_{down} - L_{up}) = K^* + L^*,$$

where Q^* is net (all-wave) radiation, K is shortwave radiation, and L is longwave radiation, $_{down}$ is incoming radiation and $_{up}$ is outgoing radiation. At night K^* equals zero. *See also* radiation budget.

net radiometer An instrument that measures net upward and downward long- and short-wave radiation flux; i.e. the difference in intensity between radiation entering and leaving the Earth's surface.

neutral stability (indifferent equilibrium) A state of atmospheric STABILITY when it is neither STABLE nor UNSTABLE. It occurs either when the ENVIRONMENTAL LAPSE RATE (ELR) of a layer of unsaturated air is equal to the DRY ADIABATIC LAPSE RATE (DALR), or when the environmental lapse rate of a saturated air layer is the same as the SATURATED ADIABATIC LAPSE RATE (SALR). If an air parcel is perturbed in a neutral stability state, it will remain stationary, neither moving up or down. Neutral stability rarely occurs in the atmosphere, usually only prevailing for short periods of time.

nevados A local name for an intensely cold KATABATIC WIND experienced in the valleys of Ecuador, South America. The nevados is caused partly by nocturnal cooling of air by radiation and also by its contact with snow and ice over the Andean mountain range.

névé *See* firn.

NEXRAD (Next Generation Weather Radar) A network of DOPPLER RADARS, operated by the NATIONAL WEATHER SERVICE (NWS) that virtually covers the whole of the contiguous US. The radar provides instantaneous maps, which are produced every 5–6 minutes, of the accurate location and intensity of most PRECIPITATION within about 120 km (75 miles) of the system, and of intense precipitation within some 200 km (124 miles). Additionally, the radial component of the wind toward or away from the radar is mapped. [From: *nex*t generation *rad*ar]

Next Generation Weather Radar *See* NEXRAD.

nightglow The weak radiance that is present in the sky at night, excluding light from the Moon and stars, that results from the from radiation emitted by atmospheric reactions occurring in ionized particles.

nimbostratus One of the cloud genera (*see* cloud classification); a type of LOW CLOUD that may extend into the middle levels and from which precipitation falls. It is often a deep gray ragged sheet or layer of cloud that obscures the Sun. It is the typical cloud on a very wet dull day and is associated with virtually continuous light to moderate precipitation of rain or snow. Low, ragged FRACTUS clouds may form in the SUBCLOUD LAYER. It usually follows ALTOSTRATUS in the sequence of clouds ahead of an approaching WARM FRONT.

Nimbus A series of Sun-synchronous METEOROLOGICAL SATELLITES that were operated by NASA, the first of which (Nimbus 1) was launched in 1964 and the last (Nimbus 7) in 1978. The series was one of second-generation meteorological satellites, following on from the original TIROS series. *See also* POES.

nitrogen (N) The most abundant gas in the Earth's atmosphere constituting approximately 78.08% of the volume of dry AIR; it is colorless and odorless.

nitrogen cycle The global-scale exchange of nitrogen and its compounds between different reservoirs. Nitrogen gas in the atmosphere finds its way into the soil in the form of nitrates as a result of *nitrogen fixation* by bacteria and by lightning. From the soil it passes to plants in protein and other molecules and thence to animals, who consume the plants. From dead plants and animals it returns by *nitrification* back into the soil, where some of it returns to the atmosphere as a result of the activities of denitrifying bacteria.

nitrogen oxides Gases consisting of nitrogen combined with oxygen, formed during the high-temperature combustion of fossil fuels. The two principal products of this are *nitrogen dioxide* (NO_2) and *nitric oxide* (NO). Their concentration in urban areas is typically ten to a hundred times that in the rural areas because they are produced, for example, by motor vehicles and electric power stations. Nitrogen dioxide can react with water vapor to produce nitric acid, adding to the ACID RAIN problem; if its concentration is great enough, it can produce a reddish-brown tinge to the sky in dry conditions.

nitrous oxide (N_2O) A colorless TRACE GAS within the atmosphere that acts as a powerful GREENHOUSE GAS. It is produced by the burning of vegetation, agricultural practices (such as the use of commercial and organic fertilizers), combustion of fossil fuels, and industrial emissions in nitric acid production.

NOAA *See* National Oceanic and Atmospheric Administration.

NOAA meteorological satellite *See* GOES; POES.

noctilucent cloud Literally, clouds that are bright at night, against a black sky. These rare very high-altitude clouds are observed occasionally poleward of about 50° latitude around the midnight hours in the summertime when they appear with a bluish-white glow. In North America they may be seen from mid-Canada and further north. They occur at an altitude of 80–85 km (50–53 miles) and move extremely rapidly at between 100–300 knots (115–350 mph). They are composed of minute ice particles with diameters of 0.1–0.3 μm, possibly composed of water that has entered the atmosphere on meteoric dust.

nocturnal radiation Radiative exchanges during the hours of darkness; since there is no solar radiation available this involves only longwave radiation. Nocturnal radiation represents the excess radiation emitted by the Earth's surface at night over that received by the Earth's surface from the atmosphere (clouds and greenhouse gases). It is indicated by negative values in NET RADIATION.

non-attainment level The reduced level of emissions or of ambient pollution concentration aimed for but that is unlikely to be achieved given current economic policies or best available techniques not entailing excessive costs (BATNEEC).

non-frontal depression A synoptic-scale low-pressure system that is not associated with fronts and therefore has no significant AIR MASS changes within its circulation. A HEAT LOW and a CUT-OFF LOW are examples of non-frontal lows.

non-selective scattering The SCATTERING of photons of ELECTROMAGNETIC RADIATION, usually by molecules and particles in the atmosphere that are larger than the wavelength of the incident radiation, for example water droplets and dust particles. This scattering takes place in all directions and may be upward or downward with respect to the Earth's surface, and primary, secondary, or tertiary. *See* diffuse radiation. *See also* Mie scattering; Rayleigh scattering.

nor'easter (northeaster) A cyclonic storm that affects the east coast of North America, especially during winter, with maximum intensities near New England. It develops in the lower middle latitudes, 30°–40°N, some 160 km (100 miles) off the coast. Wind gusts can be in excess of 64 knots (i.e. hurricane force) and given its North Atlantic source region, a nor'easter is usually cold, generally moist in character, and is frequently accompanied by much cloud bringing precipitation in the form of heavy snow and/or rain.

normal In meteorology and climatology, the average value of an element (e.g. temperature or precipitation), over a given period. The timescale of 30 years is the most frequently used.

norte (el norte) **1.** An outbreak of cold polar air, a continuation of the North

American NORTHER, experienced during the winter in Central America causing a drop in temperature of the order of 6–9°C. It is particularly evident on the narrow 250-km (150-mile) wide Isthmus of Tehuantepec in southeast Mexico, where such a wind can reach GALE force and blow incessantly throughout the winter season. It is generally accompanied by heavy rain, but the airflow becomes markedly drier on the Pacific side of the Isthmus.
2. A cold northerly wind that affects eastern Spain during the winter, blowing when high pressure is resident over the Iberian Mesata.
3. A warm and humid northerly wind experienced in Argentina.

North Atlantic Drift (North Atlantic Current) The relatively warm current that flows northeastward in the North Atlantic Ocean from approximately 40°N, 45°W off the Grand Banks of Newfoundland, Canada, toward Europe. The North Atlantic Drift is frequently called the GULF STREAM but is more correctly an extension of this latter current. Off the coast of the British Isles it splits, the southerly current becomes the CANARIES CURRENT and the northerly current flows along the coast of northwestern Europe.

North Atlantic Oscillation (NAO) A large-scale oscillation in the pressure systems over the North Atlantic that particularly affects winter weather, especially of Europe. The oscillation centers on the relationship between two pressure systems: the low-pressure system located over Greenland and the high-pressure system located over the North Atlantic west of the Azores, between about 35° and 40°N. During the North Atlantic Oscillation's positive (high) index phase, the ICELANDIC LOW across the high latitudes of the North Atlantic is deeper than normal, while the subtropical anticyclone (high) over the central North Atlantic – the AZORES HIGH – is stronger than normal. As a result of the increased pressure difference, the winter storms are strong and take a more northerly track. Milder wetter conditions with above normal temperatures occur across northern Europe and the eastern US; cold dry winters occur over Canada and Greenland. Below normal precipitation occurs across southern and central Europe. During strong negative (low) index phases of the North Atlantic Oscillation, the Azores high and the Icelandic low are weak and there is a correspondingly reduced pressure gradient. The storms follow a more west–east track bringing moist air across to southern Europe, the Mediterranean, and into the Middle East, while Greenland is milder and a northerly and easterly flow of Arctic air brings cold winter conditions to northwest Europe. The winter North Atlantic Oscillation tends to show interannual and interdecadal variability. *See also* El Niño Southern Oscillation.

northeaster *See* nor'easter.

northeast trade winds (northeast trades) The markedly persistent and strong TRADE WINDS that are best developed across the subtropical oceans of the northern hemisphere. They flow toward the INTERTROPICAL CONVERGENCE ZONE (ITCZ) and are most pronounced between 5° and 15°N. They flow smoothly toward the SOUTHEAST TRADE WINDS into this major region of CONVERGENCE and associated ascent in the ITCZ.

North Equatorial Current *See* Equatorial Current.

norther A northerly wind that causes a sudden plunge in temperature in the southern US, sometimes of the order of 20°C in only 24 hours. It is caused by polar air, which sweeps southward on the western side of a low-pressure system. The norther can lead to very squally conditions, and is often accompanied by hail and thunderstorms. Wind speeds may exceed 60 knots (70 mph) and it is a hazard for farmers and growers. The norther is not always moist, however. In the Central Valley of California, for instance, descent of the airflow from the mountains often produces a dry and sometimes dusty wind. In light of their similarities, the terms norther and NORTE are sometimes used interchangeably.

northern lights *See* aurora.

North Pacific Current (North Pacific Drift) A warm ocean current that flows eastward across the North Pacific Ocean, as an extension of the KUROSHIO CURRENT.

North Pacific high (North Pacific anticyclone, Hawaiian high, Hawaiian anticyclone) The SUBTROPICAL ANTICYCLONE over the North Pacific Ocean, often in the vicinity of the Hawaiian Islands and beyond. During the winter it is weak and shrinking and migrates to lower latitudes around 20°N, in summer the anticyclone intensifies, expands, and migrates northward to around 40°N.

nor'wester (northwester) **1.** A moderate to strong northwesterly wind that brings cool to cold conditions to regions east of the North American Rocky Mountains.
2. A warm, dry, and often strong FÖHN-type wind that descends the LEEWARD slopes of the Southern Alps in South Island, New Zealand, to affect regions to the south and east, such as the Canterbury Plains.
3. A type of squall, often generating violent thunderstorms, that affects regions of Northern India, including Assam and Bengal, and Myanmar (Burma). It usually occurs during March, April, May, and June and accounts for the majority of the rainfall in these regions during this season.
4. A GALE-force northwesterly wind that often affects the Cape Region of South Africa, especially during the winter.

nowcasting A modern method of predicting the weather from some few minutes to up to about six hours ahead. Special schemes have been developed for such very short-term prediction, involving the use of radar observations of precipitation, satellite cloud and water vapor images, wind profilers, and relatively dense and frequent surface weather observations (in some areas, such as the Oklahoma Mesonet). One thrust of the nowcasting method is to draw together these various types of data into a user-friendly display system for forecasters, enabling them to monitor very frequently the evolution of SYNOPTIC and smaller-scale features. The essence of the method is, however, a basic extrapolation of pre-existing features of the meteorological fields.

NSSL *See* National Severe Storms Laboratory.

nuclear winter A scenario in which a series of nuclear explosions raises dust into the TROPOSPHERE and lower STRATOSPHERE, leading to long periods of dense overcast skies with very little light penetrating to the Earth's surface. The thermal structure of the atmosphere would be completely altered by such an event.

nucleus (*plural* nuclei) In meteorology, a tiny particle suspended in the atmosphere. The smallest and most plentiful particles are known as AITKEN NUCLEI and have a radius of less than 0.1 µm. *Large nuclei* have radii of 0.2–1 µm and the largest, the *giant nuclei*, have radii of over 1 µm. *See also* cloud condensation nucleus; freezing nucleus.

numerical modeling A MODEL that uses mathematical techniques to represent and investigate physical processes. Such models are usually computer simulations, although they may range from simplistic balance models (e.g. estimating the input and output of radiation through the atmosphere) or sophisticated GENERAL CIRCULATION MODELS.

numerical weather prediction (NWP, numerical weather forecasting) The method, using complex computer models, of forecasting the evolution of the atmosphere and weather over periods that vary from less than one hour to some two weeks. It is numerical because the procedure involves the mathematical processing of an extremely large number of meteorological measures in order to predict the weather. The pioneer of numerical weather prediction was Lewis Richardson who, during World War I, undertook the first ever mathematically based prediction of a pressure field across parts of western Europe. He founded the system of assimilating scattered values onto a regular network of latitude/longitude grid points, which formed the essential framework for calculating future changes of temperature, wind, etc. Modern numerical weather forecasting features feeding a large range of different observations into the world's fastest computers in order to calculate the forecast as quickly as possible.

nutrient cycle *See* biogeochemical cycle.

NWP *See* numerical weather prediction.

NWS *See* National Weather Service.

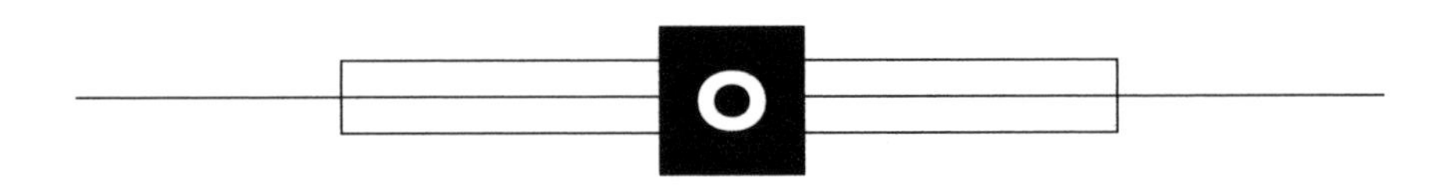

objective analysis A critical technique used in WEATHER FORECASTING that produces the best possible fields of meteorological variables given that the original observations are unevenly scattered around the globe. It is a completely automatic mathematical method that is used to set the initial conditions from which the weather forecast is run. As part of the process, erroneous observations are eliminated from these starting conditions.

observations (weather observations) A wide range of meteorological measurements made at the surface by eye or by using instruments every hour, on the hour, at busy airports for example (*see* station model). UPPER-AIR STATIONS provide routinely timed observations from the RADIOSONDES they release, while other data from commercial aircraft and POLAR-ORBITING SATELLITES are produced at asynoptic times (i.e. not exactly on the hour). Weather RADAR and satellites in GEOSYNCHRONOUS ORBITS also observe aspects of the weather at routine times.

occluded front (occlusion) A complex front that occurs during the later stage of the evolution of a FRONTAL CYCLONE (frontal depression). It indicates that the warm, MARITIME TROPICAL air at the surface in the WARM SECTOR has been lifted away from the surface into the lower and/or middle TROPOSPHERE by the colder denser air of the cold front. At the surface, the front separates polar air on either side of it and is often a region of deep cumulonimbus cloud and precipitation, sometimes with fierce squalls and hail and thunder. *Warm occlusions* are those with relatively cooler air ahead of the warm front compared to the cold air behind the cold front, while *cold occlusions* exhibit the opposite thermal contrasts. *See also* cold front; warm front.

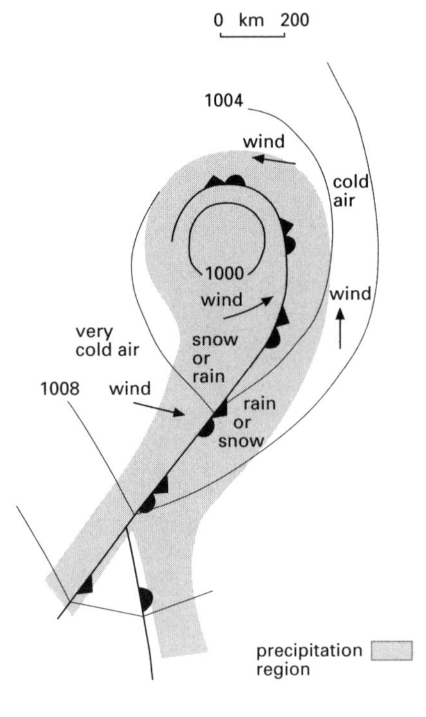

Occluded front

occluded mesocyclone A MESOCYCLONE in which air from the rear-flank downdraft has cut off the inflow of warm unstable low-level air.

occlusion *See* occluded front.

occult deposition The interception of acidic cloud droplets by vegetation, chiefly in hilly areas. This may constitute about 10% of the total deposition. Concentration of sulfur oxides (SOx) and nitrogen oxides (NOx) in such droplets may be much higher than in washout (rainfall) conditions. *See also* acid rain.

ocean The mass of salt water that covers nearly 71% of the Earth's surface. Although only one basin, the ocean is split, geographically, into major oceans; these include the ATLANTIC, PACIFIC, and the INDIAN OCEANS. To the north and south of the globe are the ARCTIC and ANTARCTIC OCEANS, respectively.

ocean climate *See* maritime climate.

ocean current The movement or flow of ocean water along a defined path. This movement can be permanent or semi-permanent. The currents are generally wind-driven or thermohaline, i.e. produced by the sinking of high-density water created by temperature and/or salinity differences. The major ocean currents have an important influence on the climate due to the high specific heat capacity of water; they transport heat from the tropics to higher latitudes, and return cool water. In the northern hemisphere, those flowing generally northward, toward the North Pole, are warm and act to raise coastal temperatures, for example the GULF STREAM and the NORTH ATLANTIC DRIFT; those flowing southward, toward the Equator, are cold and lower than the temperatures along coastal regions, for example, the OYASHIO and CALIFORNIA CURRENTS.

oceanicity The degree to which the climate of a location is influenced by the sea, as opposed to land. *Compare* continentality.

oceanography The scientific study of the seas and oceans. It is mainly divided into geologic oceanography, physical oceanography, chemical oceanography, and marine biology.

ocean weather ship An ocean-going vessel equipped with meteorological instruments to carry out and record meteorological observations.

offshore current 1. A current in the sea or ocean that flows parallel to the coast.
2. Any current that flows away from shore.

okta The internationally used basic unit for assessing cloud cover for both total cloud amount and the cover associated with individual cloud layers. One okta is the same as one eighth of the celestial dome. Eight oktas means the sky is overcast, 4 oktas means that it is half covered, 0 oktas means the sky is clear, and 9 oktas means that the sky is obscured – by blowing snow or fog for example – so that the observer cannot see the sky to report the cloud cover.

opacus A cloud variety, a sheet or layer of cloud that is sufficiently thick or opaque to blot out the Sun or the Moon. It is used in conjunction with STRATUS, STRATOCUMULUS, ALTOCUMULUS, and ALTOSTRATUS. *See also* cloud classification.

open pan evaporimeter *See* evaporimeter.

orbital forcing The influence exerted by variations in the orbit and rotation of the Earth on climate. *See also* Milankovitch cycles.

orographic Of or relating to mountains; for example weather phenomena, such as clouds and rainfall, that are generated as a result of a mountain barrier interfering with the physics and dynamics of the atmospheric flow.

orographic cirrus HIGH CLOUD produced by the flow in the LEE WAVE of moist air in the middle and/or upper troposphere across a mountain barrier, normally at a high angle to the alignment of the line of mountain peaks and ridges. The moist air is sometimes forced to 'jump' up and over the barrier, which can lead to the production of extensive CIRRUS and CIRROSTRATUS that streams away downwind from the line of the watershed. The cloud persists for as long as the conditions pertain, which may be some hours or half a day to a day and can stretch many hundreds of kilometers downwind, producing overcast skies a very large distance from the mountains.

orographic cloud Cloud produced by the forced uplift of a flow of moist air up and over hills and mountains. The reduced pressure as the air ascends is associated with adiabatic cooling and, if conditions are right, to the condensation of myriads of cloud droplets.

orographic depression *See* lee depression.

orographic lift (orographic uplift) The forced ascent of air over a barrier of hills or mountains. If the air is moist the uplift can give rise to adiabatic cooling leading to saturation and condensation and the formation of OROGRAPHIC CLOUD.

orographic precipitation (orographic rainfall, relief rainfall) PRECIPITATION caused by the vertical uplift of a moist air stream as it flows over a topographic barrier, such as a mountain range. The air stream cools to DEW-POINT as it ascends, with a layer of OROGRAPHIC CLOUD forming if the condensation level is reached. If sufficient water vapor is present, considerable amounts of rain are deposited on the high ground. The generation of orographic precipitation is favored when: (1) the mountain range is extensive and oriented at right angles to the direction of the air flow (e.g. the Southern Alps of New Zealand) and (2) cyclonic or convective processes are already active in the atmosphere (the orographic component is normally weak and merely acts as a trigger mechanism).

Compression and consequently warming of the air stream on the leeward slopes of a mountain range results in the formation of a RAIN SHADOW. For example, the eastern slopes of the Western Ghats in India are sheltered from the southwest monsoon and in some years receive less than 600 mm (24 in) of rainfall, whereas totals on the windward slopes can exceed 5000 mm (200 in) per annum.

oscillation In meteorology, a variation from the mean value for a meteorological parameter or phenomenon that takes place on a cyclical basis. *See also* North Atlantic Oscillation; Southern Oscillation.

outer convective band *See* feeder band.

outflow boundary The boundary between the cooled outflow air from a THUNDERSTORM and the surrounding environmental air, similar in characteristics to a cold weather front. New thunderstorms often develop along these outflow boundaries.

overcast The condition of cloud cover, for any level of cloud or mixture of levels of cloud, when the sky is completely covered.

overrunning The action of one layer of air, often relatively warm, moving over another, more dense layer at the surface, as in the case of a WARM FRONT.

overshooting top In the cumulonimbus cloud of a THUNDERSTORM or a SUPERCELL thunderstorm with a vigorous updraft, a bulge at the top of the ANVIL produced as the updraft rises above the top of the cloud. Overshooting tops are generally short-lived but those that persist are a strong indication that the storm's updraft is strong enough to produce severe weather.

oxygen (O) The second most abundant gas in the Earth's atmosphere. It exists in two forms: the diatomic molecule (O_2), constituting approximately 20.95% of the volume of dry AIR, and triatomic oxygen, known as OZONE (O_3), which occurs in trace amounts but is of considerable importance in protecting the Earth from the harmful effects of the Sun's ultraviolet radiation.

oxygen-isotope analysis In paleoclimatology, the technique of obtaining evidence of past climates, including ice ages, from stable oxygen–isotope concentrations in cores from sources, such as marine sediments, ice sheets, tree trunks, and stalactites and stalagmites within caves. It is used, for example, in estimating the temperature of water at the time the shells of marine organisms were deposited from the oxygen retained in the calcium carbonate of the shells. This is based on the ratio of the stable oxygen isotopes, ^{18}O and ^{16}O, which is temperature-dependent in water; ^{18}O concentrations increase with a lowering of temperature as a result of the lighter ^{16}O being preferentially evaporated. Stable oxygen isotopes are also used to extract past climatic information from ice cores. To a large degree the isotope concentration can be considered as a function of the temperature at which the water condensed. *See also* oxygen-isotope ratio.

oxygen-isotope ratio The ratio of the two main stable isotopes of oxygen: ^{16}O and ^{18}O. Variations of the ratio are used as an indicator of paleotemperatures (the temperatures of past climates). Differences in the mass of the two isotopes result in the lighter ^{16}O being evaporated from water bodies in preference to the heavier ^{18}O; atmospheric water thus has a deficiency of ^{18}O.

Oyashio Current (Oya Current, Kuril Current) A cold current in the Pacific Ocean that flows southwest from the Bering Sea and along the east coast of Japan, where it meets the KUROSHIO CURRENT.

ozone (O_3) A TRACE GAS consisting of the triatomic form of OXYGEN. It is found in its highest concentrations in the STRATOSPHERE, where it is constantly created and destroyed naturally by the interaction between ultraviolet solar radiation and oxygen (*see* Chapman cycle of ozone formation). In the stratosphere it provides an important shield, protecting the Earth from the Sun's harmful ultraviolet rays. Within the upper troposphere, ozone acts as a GREENHOUSE GAS but the contribution it makes to the enhanced GREENHOUSE EFFECT has yet to be fully established. Ozone occurs also in polluted air within the lower TROPOSPHERE as a chemical oxidant and a major constituent of photochemical SMOG above urban and industrial areas. This is particularly apparent on sunny days because solar radiation is required for its formation. As a pollutant it is an irritant affecting the human respiratory system. *See also* ozone depletion; ozone layer.

ozone-depleting substance (ODS) *See* ozone depletion.

ozone depletion The artificial destruction of stratospheric OZONE above the Antarctic and surrounding regions by the interaction of ultraviolet solar radiation with *ozone-depleting substances* (ODSs). These compounds contain chlorine and bromine, such as the CHLOROFLUOROCARBONS (CFCs), that have ascended from the Earth's surface. The destruction occurs annually, beginning in the spring of the southern hemisphere when solar radiation begins to illuminate the highest latitudes. The intensely cold conditions in the Antarctic stratosphere generate POLAR STRATOSPHERIC CLOUDS, the ice crystals of which provide the platform on which the chemical reactions take place to destroy the ozone. The result is the formation of the *ozone hole*, a large area of the stratosphere with extremely low levels of ozone. The size of the ozone hole is affected, in the long term, by the extent of the emissions of ozone-depleting substances and by temperature variations in the upper atmosphere. Colder conditions contribute to the freezing of more ice crystals and hence more chemical activity. Ozone depletion is not restricted solely to Antarctica; it also occurs to a lesser extent over parts of North America, Europe, Asia, Africa, Australia, and South America.

The ozone destruction enables damaging ULTRAVIOLET RADIATION to reach the surface of the Earth, where it is harmful to living organisms. In humans, for example, exposure to increased levels of UVB radiation can cause skin cancers. Measures taken to prevent ozone depletion included the MONTREAL PROTOCOL of 1987, which led to decisions to phase out the use of HALONS and CFCs.

ozone layer (ozonosphere) The layer in the atmosphere at an altitude of about 10–50 km (6–30 miles) above the Earth's surface in which ozone concentrations are at their greatest. Ozone is generated by photochemical reactions between 20–60 km (12–35 miles) above the Earth's surface. There is a marked contrast between ozone levels in the TROPOSPHERE and STRATOSPHERE, which is the result of little mixing between the two layers. The stratosphere contains about 90% of the atmosphere's ozone. At the Earth's surface, ozone is destroyed through reactions with plants and water whereas in the stratosphere it is very stable and may survive for several months.

ozonesonde An instrument attached to an air-borne balloon (*see* balloon sounding) used to measure the vertical variations of OZONE concentration in the atmosphere. It consists of an ozone sensor connected to a RADIOSONDE, which transfers data to a ground station and measures humidity, pressure, temperature, wind speed and direction, and geopotential height simultaneously with ozone sampling. *See also* ozone layer.

ozonosphere *See* ozone layer.

Pacific Ocean The largest and deepest of the Earth's oceans. It covers approximately 168 million sq km (65 million sq miles) and occupies 33% of the Earth's surface. This ocean extends from the Arctic to the Antarctic regions and between North and South America on the east and Australia and Asia on the west. Its maximum length is about 14,500 km (9000 miles) and its maximum width about 17,500 km (11,000 miles) between the Isthmus of Panama and the Malay Peninsula. Its average depth is around 4000 m (13,000 ft) with its greatest depth in the Mariana Trench, approximately 11,000 m (36,000 ft) deep.

paleoclimatic indicator Any one of a number of sources from which information and evidence about past climates in geologic history (paleoclimates) may be derived. These include microfossils (e.g. foraminifera), macrofossils (e.g. mollusks and corals), plants (e.g. tree rings and pollen), as well as glacial, periglacial (e.g. ice-wedge casts), and pluvial deposits, which provide morphological evidence for past climates.

paleoclimatology The study of past CLIMATES. Increasingly this study involves looking at causes of the variations among climates of past geologic ages and relating them to changes in more recent climate conditions. Methods to provide information and data about past climates include dendrochronology, palynology and pollen analysis, ice-core analysis, deep-sea sediment core analysis, as well as the study of glacial deposits, fossils, and rock records.

Palmer Drought Severity Index (PDSI) A meteorological drought severity index that was developed by Wayne C. Palmer in 1965 and now used by many US government agencies. The index shows long-term abnormal dryness or wetness, reflecting the supply-and-demand concept of the regional water balance equation. The PDSI is calculated based on precipitation and temperature data, potential and actual evapotranspiration, runoff, as well as the local available water content of the soil. Drought conditions are shown in the index by minus numbers: 0 is normal, –2 is moderate drought, –3 is severe drought, and –4 or less is extreme drought. The index is also used to indicate moist conditions: 2 is unusually moist, 3 is very moist, and 4 or more is extremely moist.

palynology The study of living and fossil pollen grains and spores, together with some other microfossils. It has applications in the study of past climates (paleoclimatology) and climate change. *See also* pollen analysis.

pampero The local name in Argentina and Uruguay for a line squall with a rapid increase in wind speed; it is sometimes accompanied by rain, thunder, and lightning. It typically includes an arched cloud formation along the leading edge and is followed by a cool southwesterly wind.

pancake ice Small thin disks or cakes of ice, roughly circular in shape with raised rims, that form on the surface during the freezing of freshwater lakes and ocean saltwater. Pancake ice and ice rind may eventually freeze together to form larger floes or an extensive, unbroken sheet of ice.

pannus A type of ACCESSORY CLOUD that occurs below another main cloud and is occasionally attached to it, in the form of a more-or-less continuous sheet of ragged shreds. The main cloud can be ALTOSTRATUS, NIMBOSTRATUS, CUMULUS, or CUMULONIMBUS. *See also* cloud classification.

papagayo A strong, cold, and dry northerly wind that blows across the Gulf of Papagayo into northwestern districts of Costa Rica, Central America. It is also experienced on the Mexican Plateau and is similar to the NORTE and NORTHER.

paraselene (mock Moon) An optical phenomenon, a type of HALO, in which a bright area of light appears on a level with the Moon. It is formed in the same way as the solar PARHELION but is less bright.

parcel of air *See* air parcel.

parhelion (mock Sun, Sun dog) An optical phenomenon, a type of HALO, in which a patch of bright light is visible at the same level as the Sun. It is formed by the refraction of sunlight by hexagonal plate-like ice crystals with diameters less than 30 μm (0.001 in) that are present, for example, within cirrostratus or cirrus cloud. A parhelion generally occurs on the main 22° halo around the Sun (the halo itself is not always visible), and the phenomenon often occurs as a pair of parhelia with one either side of the Sun. More rarely it occurs at other points on the parhelic circle (e.g. 46° or 120°). *Compare* paraselene.

partial drought In the UK, a period of at least 29 consecutive days during which mean daily rainfall over that period is less than 0.25 mm (0.01 in). This definition is inappropriate for dissimilar climates; it was originally defined in 1887 by the former British Rainfall Organisation but is no longer much used. *See also* absolute drought.

pascal (Pa) The SI unit of PRESSURE defined as 1 newton per square meter. The multiple HECTOPASCAL is used in scientific meteorology.

passive remote sensing The means of imaging atmospheric and other phenomena in which the sensing instrument measures the radiation being emitted by the source. Unlike ACTIVE REMOTE SENSING, it does not itself emit radiation to enable the measurement to be made. INFRARED SATELLITE IMAGERY is an example of passive remote sensing in which the instrument measures the INFRARED RADIATION radiated up to the satellite from the Earth.

past weather Part of the synoptic report from surface stations that summarizes not more than two types of weather that have occurred in the past three or six hours depending on the observation hour. The observer can select from 10 basic weather types, including RAIN, DRIZZLE, SNOW, SHOWER, FOG, and THUNDERSTORM. The past weather symbols form part of the STATION MODEL.

PDSI *See* Palmer Drought Severity Index.

pearl-necklace lightning *See* beaded lightning.

Penman formula In climatology, a method of estimating POTENTIAL EVAPOTRANSPIRATION devised by H. L. Penman. Evaporation loss is expressed in terms of the solar radiation, mean air temperature, mean atmospheric relative humidity, and mean wind speed. The formula and its modifications have been widely used by agricultural and soil scientists.

pentad In meteorology, a period of five successive days. It divides conveniently into the number of days (365) of a normal year and is used in preference to the week in the analysis of meteorological data.

perenially frozen ground *See* permafrost.

perfluorocarbon (PFC) An artificial compound composed of carbon and fluorine, the two main compounds being perfluoromethane (CF_4) and perfluoroethane (C_2F_6). The PFCs were introduced, together with the hydrofluorocarbons

(HFCs) as alternatives to ozone-depleting substances, such as the CHLOROFLUOROCARBONS (CFCs). They do not harm the ozone layer but as GREENHOUSE GASES possessing long atmospheric lifetimes (up to 50,000 years) they were one of the six gases listed for limitation in the KYOTO PROTOCOL. They are emitted as by-products in industrial processes (notably aluminum smelting).

perigee In remote sensing, the point along a satellite's orbit at which it is nearest to the Earth or the point that it is orbiting. More generally in astronomy any Earth satellite, including the Moon, is said to be in perigee when it is closest to the Earth. *Compare* apogee.

periglacial climate Originally defined as the climate that was associated with regions adjacent to and affected by the Quaternary ice sheets, the term has also been expanded to cover contemporary climates adjacent to ice sheets, ice caps, and glaciers. It may also include TUNDRA and cold climates in which frosts and permafrost processes occur even if they are not in close proximity to ice sheets. Periglacial climates are arid and temperatures may be below freezing for six months or more.

perihelion The point in a planet's orbit at which it is closest to the Sun. For the Earth, this occurs about January 3 when the Earth–Sun distance is about 1.5% less than the mean Earth–Sun distance. The effect of this proximity to the Sun on the Earth's climate is to lessen the severity of winters in the northern hemisphere.

perlucidus A variety of cloud; a widespread sheet or patch of cloud that has marked breaks between the cloud elements such that the Sun, Moon, clear sky, or overlying cloud can be seen. It is applied to STRATOCUMULUS and ALTOCUMULUS. *See also* cloud classification.

permafrost (perenially frozen ground) Ground that remains below 0°C (32°F) and persists for at least two consecutive years. Around 25% of the Earth's surface is underlain by permafrost, predominantly in Russia, Canada, and Alaska (where it occupies around 80% of the land area). It can extend to depths in excess of 500 m (1640 ft) in Canada and Alaska and to over 700 m (2300 ft) in Siberia. Recent studies of permafrost in Alaska and Siberia have revealed warming within some permafrost areas; for example, in Fairbanks (Alaska) and Yakutsk (Siberia) the permafrost warmed by 1.5°C in the 30 years leading up to 2000.

permanent snow line *See* snow line.

perpetual frost climate One of the climate subgroups in the KÖPPEN CLIMATE CLASSIFICATION, designated as EF. This is an ice-sheet climate with mean monthly temperatures below 0°C (32°F) throughout the year.

persistence In meteorology, weather conditions that continue for longer than the average or expected time.

perturbation In meteorology, a disturbance in a field. An example is a traveling disturbance in the pressure field, such as a LOW. At one site, a time series of pressure will illustrate the passage of a low because the graph will be perturbed as the low moves across. Perturbations can also occur in other fields, such as temperature, wind, or humidity, and on a range of space scales. A THUNDERSTORM can be viewed as a perturbation of atmospheric properties within a larger scale environment.

Peru Current (Humboldt Current) A cold current in the southeast Pacific Ocean that flows north along the coasts of Chile and Peru. The EKMAN EFFECT tends to deflect the flow in a westerly direction generating upwelling of nutrient-rich cold water, which supports diverse ecosystems and important fisheries, from deeper levels to replace the flow. During an EL NIÑO event the upwelling is reduced resulting in a rise in sea surface temperature and a decline in fish stocks and other marine life.

PFC *See* perfluorocarbon.

PGF *See* pressure gradient force.

pH A quantitative measure of the degree of acidity or alkalinity of a solution. It is defined as the negative logarithm to the base ten of the hydrogen-ion concentration, expressed in gram ions per liter of solution:

$$pH = -\log_{10} [H^+] = \log_{10} 1/[H^+].$$

The *pH scale* is 0–14: a neutral medium (e.g. pure water) has a pH of 7 (i.e. a hydrogen ion concentration of 10^{-7} mole/dm^3, which equals the hydoxyl ion concentration); acidic solutions have a pH below 7; alkaline solutions have a pH greater than 7. Because the pH scale is logarithmic, a decrease in pH from, for example, 4 to 3 indicates a ten-fold increase in the acidity of the solution. Rainwater is slightly acidic (but see ACID RAIN, in which the pH is below 5). Water in saline environments may have a pH value of 9 or higher; the pH of ocean water typically falls in the range 8.1–8.3.

photochemical smog *See* smog.

photodisintegration The splitting of molecules by PHOTODISSOCIATION in such a way that the recombining of the molecular fragments is unlikely.

photodissociation The splitting of a molecule into two or more smaller molecules or atoms by absorption of radiation, particularly of high-energy ultraviolet radiation. This process contributes to the production of photochemical SMOG and to the production of OZONE in the atmosphere. The *photodissociation rate* is the number of molecules dissociated by radiation per unit volume per second.

photogrammetry The branch of remote sensing concerned with the plotting and accurate measurement of features visible on aerial photographs or satellite images.

photometeor Any one of a number of luminous phenomena produced in the atmosphere as a result of diffraction, refraction, reflection, and/or interference from the Sun or Moon, such as a halo, corona, glory, green flash, rainbow, fogbow, shimmer, scintillation, mirage, and crepuscular rays.

photometry The branch of physics concerned with the measurement of the intensity of light.

photosphere That part of the Sun's gaseous atmosphere from which continuous emission of radiation takes place. Known as the surface or visible disk of the Sun, it consists of plasma about 1000 km (621 miles) thick at a temperature of about 6000 K.

photosynthesis The complex process by which organic compounds are synthesized by chlorophyll-containing cells within green plants and some bacteria, from carbon dioxide and a source of hydrogen, such as water, in the presence of radiant energy from the Sun. During the process oxygen is simultaneously released. Photosynthesis is a vital process for life on Earth as plants, either directly or indirectly, provide the food source for other forms of life. It is, in addition, the primary source of oxygen in the atmosphere and the absorption of carbon dioxide is vital in the carbon cycle.

physical meteorology The branch of meteorology concerned with the structure and composition of the atmosphere. Areas of study include aerosols and cloud physics, precipitation processes, atmospheric electricity, radiation, atmospheric thermodynamics, the boundary layer, turbulence, and gravity waves.

physiological drought **1.** A temporary daytime state of drought in plants due to the losses of water by transpiration being more rapid than that taken up by roots although the soil may have an adequate supply; it is usually followed by the night-time recovery of the plant.
2. The condition of plants that suffer the loss of a physiologically suitable water source and are unable to utilize the water

that is present, for example frozen water or salt water.

pibal *See* pilot balloon.

Piché evaporimeter *See* evaporimeter.

pileus (cap cloud) A small ACCESSORY CLOUD that sometimes forms above CUMULUS or CUMULONIMBUS clouds. It appears as a smooth shallow 'cap', and is formed by air flowing over the top of the deep cloud, which acts somewhat like a hill, forcing the moist environmental air around and over it. This is similar to the way in which a lenticular cloud is formed. Occasionally several layers of pileus may be observed. [Latin: 'cap'] *See also* cloud classification.

pilot balloon A small meteorological balloon, inflated with hydrogen or helium, and used for visually observing the upper wind speed and direction above a station. The balloon, which carries no instruments, ascends at a constant rate: the ascension rate is approximately determined by careful inflation to a given total lift. A *pilot balloon observation* (*pibal*), the measurement and calculation of the wind speed and direction, is made by reading the elevation and azimuth angles of a theodolite while visually tracking the pilot balloon, or the balloon can be tracked by radar. *See also* balloon sounding; radiosonde.

pitot tube anemometer An ANEMOMETER that uses the build up of pressure in a *Pitot tube* to measure wind speed. The pitot tube, invented by Henri Pitot (1695–1771), consists of an L-shaped tube with an opening at one end facing into the flow. The difference in pressure between the inside and the outside of the tube is measured and converted to a wind speed. This type of anemometer is useful in strong steady streams of air.

pixel In remote sensing, the smallest element or building block of a remotely sensed image. The best pixel dimension for all AVHRR images is at the SUBSATELLITE POINT, where it is a 1.1 km square. [Short for: picture element]

Planck's law (Planck radiation formula) A law of radiation stating that the wavelength of propagation of radiation depends on the temperature of the emitting body. It is named for the German physicist Max Planck (1858–1947), who discovered it. Planck's law describes the distribution of energy across the wavelengths of the electromagnetic spectrum for a given temperature of a perfect radiator (black body). It introduced the concept of energy quanta, upon which the quantum theory is based.

planetary boundary layer (friction layer) The atmospheric boundary layer that occupies approximately the lowest kilometer of the troposphere, but height varies considerably according to location and time from around 500 m (1600 ft) above ocean surfaces, or on calm nights above land surfaces, to 1.5 km (1 mile) during daytime. It is the layer that is influenced by the frictional effects of the Earth's surface, and above the SURFACE BOUNDARY LAYER is also known as the EKMAN LAYER since the effects of friction decrease with height. The top of the planetary boundary layer is the level at which there is no longer any influence of friction on airflow, which then approximates to the GEOSTROPHIC WIND. The top of the planetary boundary layer also rises during the day as turbulence increases and falls at night when turbulence is dampened and a temperature inversion may be present.

planetary vorticity *See* vorticity.

planetary wind One of the major wind systems of the Earth that results from a combination of solar radiation and the rotation of the Earth, including the TRADE WINDS and WESTERLIES. *Compare* local wind.

plan position indicator (PPI) A standard type of RADAR image display in which the antenna is at the center of a circular map. The location of the phenomenon being sensed, such as precipitation, is displayed in terms of its plan position, in which the variables are the radial distance from the antenna and the AZIMUTH ANGLE.

plate tectonics The theory that the Earth's lithosphere is divided into mobile oceanic and continental plates, which are in motion relative to each other. The relative positions and extent of the major land masses and ocean areas have altered over geologic time with consequent affects on climate.

plume blight In the US, an identifiable coherent pollution plume that is visually intrusive in an area in which good visibility is regarded as an important characteristic.

plum rains *See* Mai-yu season.

pluvial A period of greater precipitation and a wetter climate than preceding or succeeding periods. The evidence of past pluvials is provided in the geologic record, especially in tropical and subtropical latitudes. During full glaciation in polar regions a number of the semidesert areas of the tropics experienced pluvials, in response to the southward movement of the circulation belts. *Compare* interpluvial.

pluviometric coefficient In climatology, a measure of the degree of climatological seasonality in the RAINFALL receipt of a location. Pluviometric coefficients are usually calculated on a month-by-month basis and then averaged. For example, suppose the thirty-year mean annual rainfall at a location (X) is 1200 mm and the thirty-year mean January rainfall is 50 mm, then the pluviometric coefficient for January is calculated as follows:

The amount of rainfall occurring in January if the total is evenly distributed throughout the year is $(31/365.25) \times 1200 = 101.85$ mm. Then January's pluviometric coefficient (PC) is $50/101.85 = 0.49$.

The more that the pluviometric coefficient diverges from 1, the more that the distribution of rainfall at X is becoming less uniform in time, with marked drier and wetter periods evident throughout the year. A line joining all points in space that have identical pluviometric coefficients is an *equipluve*.

POES (Polar-orbiting Operational Environmental Satellite) The current series of US POLAR-ORBITING SATELLITES operated by the National Oceanic and Atmospheric Administration (NOAA), comprised at any one time of two satellites that circle the Earth at right angles to each other. The first series of polar-orbiting satellites to provide global meteorological data began with TIROS (Television and Infrared Observation Satellite), launched by NASA in 1960. NOAA has been flying its own satellites since the launch of NOAA–6 in 1979; prior to this the NOAA series (TIROS 2 to 5) was also operated by NASA. A new series of five satellites was launched in 1998 with NOAA–15, which carry more advanced instruments. NOAA–16, the most recent of the series, came into operation in 2001 to replace NOAA–14. The POES satellites provide meteorological data that is transmitted globally and used by the National Weather Service (NWS) in 3–10 day forecasts and 30–90 day climate outlooks. Images are provided of cloud cover, vertical temperature and humidity profiles, and surface parameters, including sea surface temperature, snow, and ice cover.

point rainfall The amount of RAIN that falls at one location. It does not necessarily provide a good indicator of rainfall receipt over the surrounding area.

point source In air pollution studies, any industrial unit that discharges pollutants into the atmosphere. A chimney stack of a factory that emits at least 100 tonnes a year is subject to a special emissions inventory and regulation. In the US, such stacks are known as *special emission sources*. *Compare* area source; line source.

polar Denoting either of the Earth's poles or the region within the ARCTIC or ANTARCTIC Circles.

polar-air depression *See* polar low.

polar air mass An extensive region of air (AIR MASS) that has a source in ANTICYCLONES over wintertime higher- and middle-latitude continents. CONTINENTAL

POLAR (cP) air is very cold, dry, and stable but if it flows out across adjacent oceans the air mass is significantly modified through being strongly warmed and moistened to become MARITIME POLAR (mP) air, which is unstable.

polar cell A weak cellular circulation that occurs in the north–south vertical plane within the high-latitude TROPOSPHERE. It consists of sinking motion over the poles, easterlies in the lower troposphere that flow toward middle latitudes, ascent in frontal zones, and a return flow in the upper troposphere back toward the pole. It is the weakest of the cells in the THREE-CELL MODEL.

polar climate The climate of the regions that lie poleward of the Arctic or Antarctic Circles. Mean monthly temperatures are below 0°C (32°F) throughout the year with extremes of cold temperatures during the winter; annual temperature ranges are large. During summer there are long periods of continuous day (*see* midnight Sun), while during winter there are long periods of continuous night. Precipitation is generally low with most occurring on the coastal margins. High winds occur around the margins of ice caps of Greenland and Antarctica. The climate is denoted by the letter E in the KÖPPEN CLIMATE CLASSIFICATION.

polar easterlies The relatively weak east winds that flow out across the highest latitudes from the ANTICYCLONES (highs) that sometimes occur over polar areas.

polar front The gently sloping semipermanent zones that separates warm moist MARITIME TROPICAL air from cold MARITIME POLAR air along the middle latitudes of each hemisphere. Over the North Atlantic and North Pacific Oceans it can often be traced as a continuous line over several thousand kilometers, generally aligned southwest to northeast. It is the front along which cyclone waves (*see* frontal wave) form and run, and from which most CYCLONES evolve. The polar front is intimately linked to the POLAR FRONT JET STREAM. *See also* Atlantic polar front.

Polar Front Jet Stream The marked upper tropospheric core of very fast flowing air associated with the POLAR FRONT. It is usually many thousands of kilometers long, perhaps a hundred wide, and a kilometer or two deep. It is roughly twice as strong in the winter than in the summer because of the increased frontal thermal contrasts; wind speeds in the winter can reach 200 knots (230 mph). *See also* jet stream.

polar high (polar anticyclone) An extensive region of high pressure across the highest latitudes, the polar regions; a source of very cold and generally dry air. The anticyclone over Antarctica (the *Antarctic high*) is semi-permanent; that over the Arctic (the *Arctic high*) is generally seasonal.

polar low (polar-air depression) A relatively small-scale non-frontal LOW or low-pressure system that characterizes the highest latitudes; it is a SECONDARY DEPRESSION that forms over oceans poleward of the polar front and can produce blustery snowy conditions, particularly in the wintertime, when they are most common.

Polar-Night Jet Stream The westerly wind speed maximum that occurs in the lower STRATOSPHERE during the prolonged periods of darkness at the highest of latitudes. The darkness leads to deeply cold air over the poles and, consequently, a westerly jet. The situation is reversed in the long daylight summer in high latitudes when the poles are warmer (in the lower stratosphere) than further toward the Equator, and as a result the flow evolves into an easterly current. *See also* jet stream.

polar orbiter *See* polar-orbiting satellite.

Polar-orbiting Operational Environmental Satellite *See* POES.

polar-orbiting satellite A satellite with a near-circular north–south trajectory, such as that of the POES series of satellites operated by the US National Oceanic and Atmospheric Administration (NOAA), which crosses over or near the polar regions on every orbit. With the POES series elevation, the orbital period is about 100 minutes, so the satellite crosses near a pole approximately every 50 minutes.

polar outbreak The rapid incursion of much colder and drier polar air, preceded by a cold front, over a high- or middle-latitude site. In the winter, such outbreaks can sometimes reach the subtropics, ending up occasionally as CUT-OFF LOWS, and bring storms followed by cold clear weather. They are not normally seen as a summer phenomenon.

polar stratospheric cloud (PSC) Ice crystal clouds found within the lower STRATOSPHERE when temperatures fall below –85°C (–121°F). Such a depth of cold is found usually only above the Antarctic during the long polar winter. The ice crystals that compose these polar stratographic clouds provide the crucial platform on which the chemical reactions responsible for OZONE DEPLETION occur during the late winter and early spring.

polar vortex *See* circumpolar vortex.

polar zone *See* frigid zone.

pole of cold *See* cold pole.

pollen analysis The study of fossil pollens and spores in sediments, such as peat bogs and lake sediments. As pollen is particularly resistant to decay, these studies enable the plant species to be ascertained, which in turn indicates the temperature and moisture conditions of past geologic times for an area. Pollen analysis can yield important evidence for past climates as different species of plants vary in their growing requirements. *See also* palynology.

pollutant A natural or synthetic substance or effect that causes deterioration in the biosphere by changing environmental conditions, such as altering growth rates of species, interfering with the food chain, causing problems with human health, well-being, and property.

pollution The introduction into the environment of substances (solid, liquid, or gas), that are likely to have an adverse effect on susceptible components of the biosphere, including soil and water habitats, ambient air, and all or any forms of life. *See also* air pollution.

pollution dome The accumulation of pollutants in an urban or industrial zone over a period of at least a few days due to a lack of ventilation in the PLANETARY BOUNDARY LAYER caused by persistent high pressure. Pollutants accumulate just beneath the temperature INVERSION level.

pollution plume An elevated point-source emission from a chimney stack with pollutants capable of being transported many kilometers downwind. Characteristics include initial plume rise, Gaussian distribution of pollutant concentrations in the horizontal, and vertical and dry deposition in a zone 10 to 100 km (6.2–62 miles) downstream.

ponente A westerly wind that blows across the Mediterranean Basin. The word is most commonly applied to a SEA BREEZE felt along the western coast of Italy.

potential energy The energy possessed by a body in the gravitational field of the Earth. It can be expressed as the energy required to bring the body (e.g. a parcel of air) from a fixed datum, usually mean sea level, to a given position above sea level.

potential evapotranspiration The maximum amount of WATER VAPOR that is capable of being released to the atmosphere by the processes of EVAPORATION and TRANSPIRATION in a given weather situation from a surface covered by green vegetation that has a continuous supply of water and does not completely shade the ground. It is a measure of the ability of the atmosphere to remove water from the surface through the processes of evaporation and transpiration assuming no control on water supply. Three conditions are important in this process: energy (often through sunlight), eddy diffusion through wind, and the water-vapor gradient between the ground surface and atmosphere. Formula for the estimation of potential evapotranspiration include the PENMAN FORMULA and THORNTHWAITE'S INDEX OF POTENTIAL EVAPOTRANSPIRATION.

potential instability of an air mass *See* convective instability.

potential temperature The temperature that an air parcel would have if brought from its initial state to the standard pressure of 1000 hPa along the DRY ADIABATIC LAPSE RATE.

powder snow A type of SNOW in which the crystals are dry, loose, and unconsolidated. It occurs at very low temperatures, being common in polar regions and mountain areas. Powder snow is ideal for skiing.

PPI *See* plan position indicator.

praecipitatio A supplementary cloud feature in CLOUD CLASSIFICATION that includes all PRECIPITATION that falls from the cloud base to reach the ground (for example, rain, snow, sleet, hail, graupel, and drizzle); it excludes VIRGA. It is associated with ALTOSTRATUS, CUMULONIMBUS, NIMBOSTRATUS, STRATOCUMULUS, CUMULUS, and STRATUS.

precession The motion of the Earth's rotation about its axis, which is continually changing its direction in space. The *precession of the equinoxes* refers to the westward drift of the location of the PERIHELION such that the orbital ellipse traced out by the Earth slowly moves around the Sun with periodicities of 23,000 years and 18,800 years. It is one of the factors contributing to variations in the amount of solar energy received by the Earth over long periods.

precipitable water (total column moisture) The amount of RAINFALL that would result if all of the water vapor present in a standard column of air above a specified pressure level was condensed and precipitated on to a horizontal surface of unit area. The equation for precipitable water (M_w), expressed in grams, is:

$$M_w = (1/g).\int_{p_1}^{p_2} r dp$$

where g is the acceleration of free fall (9.8 m s^{-2}), p_1 is the pressure (hPa or mb) at the top of the column and p_2 is the pressure at the bottom, r is the humidity MIXING RATIO. M_w is summed over a number of layers. As precipitation-generation processes are not completely efficient, the amount of rainfall deposited in reality is usually significantly lower than the theoretical maximum M_w value.

precipitation The deposition of water, in either liquid or solid form, derived from atmospheric sources that falls onto the Earth's surface. Precipitation encompasses RAIN, DRIZZLE, FREEZING RAIN, FREEZING DRIZZLE, SNOW, SLEET, HAIL, GRAUPEL, ICE PELLETS, and other forms. It is collected and measured primarily in a RAIN GAUGE. Any solid precipitation (e.g. snow) is melted before the total depth of water collected is measured in millimeters or inches. As rain is usually the main component of precipitation, the terms RAINFALL and precipitation are used interchangeably.

Precipitation is initiated within clouds by the BERGERON–FINDEISEN PROCESS, by the COLLISION–COALESCENCE PROCESS, or by a combination of these processes. Three types of precipitation are recognized according to the mechanism of uplift in their formation: CONVECTIVE PRECIPITATION, FRONTAL PRECIPITATION, and OROGRAPHIC PRECIPITATION.

precipitation-efficiency index (precipitation-effectiveness index, precipitation-evaporation ratio) A measure of the usefulness of rainfall input that remains following evaporation losses (EFFECTIVE PRECIPITATION) for human activities (e.g. water supplies and agriculture). A precipitation-efficiency index (P/E) was devised in 1931 by the American climatologist C. W. Thornthwaite (1889–1963) to define objectively the Earth's major climatic regions. His formula is:

$$P/E = 11.5[p/(T - 10)]^{10/9}$$

where p is the mean monthly precipitation (inches) and T is the mean monthly temperature (°F). As T (and so evaporation) increases, the denominator of the above expression increases and hence P/E falls.

precipitation gauge *See* rain gauge.

precipitation radar A RADAR designed specifically to sense the location and intensity of RAINDROPS. *See also* Doppler radar.

precipitation variability The variations in precipitation totals through time or over space. In climates where CONVECTIVE PRECIPITATION is dominant, for instance, rainfall totals can vary significantly between two places that are only 3 miles (5 km) apart. Over time, precipitation input can vary at hourly, daily, weekly, monthly, seasonal, yearly and decadal timescales. The degree of precipitation variability can be quantified by means of any standard statistical measure such as the standard deviation or the coefficient of variation.

present weather Part of the synoptic report that relates to the presence of weather features, such as fog, rain, drizzle, snow, or showers, at the time that the observation is being made. There is also the possibility of reporting one of these phenomena that did not occur at the time of observation, but did so in the past hour, since the previous observation. There are 100 possibilities (00–99 in the SYNOPTIC CODE) for present weather, of which only one must be reported. In general, low numbers relate to less important features, such as increase in cloud or decrease in cloud, while the highest numbers refer to THUNDERSTORMS of various intensities. The present weather symbol is plotted as a part of the STATION MODEL.

pressure The force per unit area. By Newton's 2nd law, force is equal to the product of mass (m) and acceleration. Weight (W) is a force, defined as the product of an object's mass and the acceleration of free fall: $W = mg$. The SI unit of force is the newton (N); 1N being the magnitude of the force that gives an acceleration of 1 m s^{-2} to a body of mass 1 kg. Pressure is then defined as the number of newtons (N) acting per square meter (m^2). The SI unit of pressure is called the PASCAL (Pa), which is defined as 1 N m^{-2}.

pressure altimeter *See* altimeter.

pressure gradient The horizontal or vertical rate of change of barometric pressure. It is expressed in hPa per meter, for example; the horizontal wind speed is directly related to the magnitude of the horizontal GRADIENT of pressure. The pressure gradient in the vertical is very steep indeed, but the PRESSURE GRADIENT FORCE associated with it is balanced by the force of gravity.

pressure gradient force The acceleration experienced by a unit mass of air in response to a change of barometric pressure in either the horizontal or vertical plane, i.e. a gradient of pressure. A strong pressure gradient force leads to strong winds, while a weak gradient is associated with weak flow in the atmosphere. For horizontal motion, this force is equated to the Coriolis force in the geostrophic balance (*see* Coriolis effect). In the case of vertical (up or down) motion, it is equated to the product of the acceleration of free fall and the air density in the hydrostatic balance.

pressure gradient wind *See* gradient wind.

pressure-plate anemometer An ANEMOMETER that uses the build up of pressure on a flat plate to determine wind speed. In this case the depression of a flat plate placed into the wind is measured. This type of anemometer is used to measure atmospheric turbulence.

pressure system A synoptic-scale feature on mean-sea-level charts. A LOW, CYCLONE, RIDGE, TROUGH, and an ANTICYCLONE (high) are all pressure systems.

pressure tendency (barometric tendency) In synoptic meteorology, the change in ATMOSPHERIC PRESSURE that has occurred at a given location on the Earth's surface, usually over a period of three hours. For example, suppose at point X the atmospheric pressure at 3.00 pm is 1021 hPa, and at 6.00 pm it is 1018 hPa, then the pressure tendency over this 3-hour period is –3 hPa. An ISALLOBAR joins locations with the same pressure tendency over a given period. Isallobar maps are often used to deduce the location of FRONTS. Pressure tendencies are reported for the last 24 hours rather than the last three in tropical latitudes. This is because there is a marked atmospheric tidal change of pressure that dominates pressure variations through the day and swamps the normally subtle pressure changes that occur in relation to weather disturbances.

pressure-tube anemometer An ANEMOMETER similar to the PITOT TUBE ANEMOMETER but using a tube which is open at both ends and has a larger diameter at the ends than at the middle. Wind speed is calculated by measuring the pressure in the middle of the tube, at the constriction.

prevailing wind The geostrophic-scale wind direction that occurs most frequently in a particular area; for example, in southern Chile the prevailing wind is from the west, i.e. westerly. The prevailing wind is thus distinct from the DOMINANT WIND. The prevailing wind depends on the site's position in the global atmospheric circulation system (*see* general circulation). In many regions (e.g. India) the prevailing wind varies on a seasonal basis, being affected as the world's pressure belts change their latitudinal positions.

Prevention of Significant Deterioration In the US, a pollution control program inserted into the 1977 Clean Air Amendments. It seeks to identify three classes of areas: (1) to afford a level of protection to non-polluted areas beyond the provisions of the National Ambient Air Quality Standards (NAAQS); (2) to allow areas of moderate economic growth by relaxing air-quality constraints; and (3) to redefine some areas from the second class of area to permit pollution levels to rise to federal standards.

Primary Ambient Air Quality Standards *See* National Ambient Air Quality Standards.

process lapse rate *See* lapse rate.

profiler An instrument that can be either satellite- or surface-based to provide instantaneous vertical sections of variables, such as dry-bulb temperature, humidity, and wind speed. A *wind profiler* is a surface-based DOPPLER RADAR instrument that can take instantaneous high-resolution measurements of the wind, including velocity, direction, and turbulence, in the vertical column above the radar site. Some wind profilers also measure temperature using a radio acoustic sounding system (RASS).

propeller anemometer An ANEMOMETER consisting of propellers (frequently helicoidal in shape) mounted on a horizontal pivotal axis. The propellers rotate in the wind generating an electric current proportional to the wind speed, which is read from an appropriately calibrated meter. The *vane-oriented propeller anemometer* consists of a two, three, or four bladed propeller that is turned into the wind by an attached vane. *See also* cup anemometer.

PSC *See* polar stratospheric cloud.

pseudoadiabatic chart *See* Stüve diagram.

psychrometer A type of HYGROMETER used to determine relative humidity of the atmosphere. It consists of two THERMOMETERS, one a WET-BULB THERMOMETER in which the bulb is kept wet by a surrounding wet-cloth wick and the other a DRY-BULB THERMOMETER. The wet-bulb temperature is lower than that of the dry-bulb in unsaturated air at a specified relative humidity. This lower temperature is a result of the release of latent heat, which is needed for the evaporation of water from the muslin, from the air surrounding the wet bulb. The psychrometer works on the basis that the difference between the two thermometers will be greater in drier air. Simultaneous readings are taken from the thermometers and relative humidity and dew-point determined with the use of psychometric tables. *See also* sling psychrometer.

pulse length The duration of the extremely short burst of radiation that is emitted by some ACTIVE REMOTE SENSING instruments, such as RADAR and LIDAR.

pulse storm A THUNDERSTORM that produces brief but strong UPDRAFTS. These are common across the Deep South of the US during the summer where conditions are very humid. The main threat of damage comes from sudden down rushes of colder denser air from within the storm, and from very large, and continually growing, hailstones, which are repeatedly moved up and down through the storm. Flash flooding is also a threat, as these storms can remain stationary for long periods of time.

pyranometer An instrument with a hemispherical field of view used to measure diffuse radiation and/or total (beam plus diffuse) radiation incident on a horizontal surface. It generally consists of a THERMOPILE sensor connected to a recorder, encased in a protective optically ground hemispheric glass cover. A shading disk on a tracking arm or a shadow band may be used to block direct radiation to obtain measurements of diffuse radiation only. *See also* pyrheliometer.

pyrheliometer An instrument with a narrow field of view used for measuring direct solar radiation incident on the collection unit, perpendicular to the Sun's rays. To obtain continuous readings the pyrheliometer is mounted on a Sun-following power-driven tracker to ensure that it is always aimed at the Sun. *See also* pyranometer.

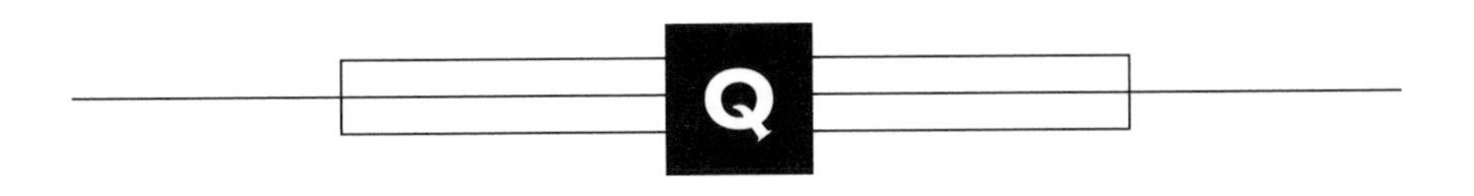

quasi-biennial oscillation (QBO, stratospheric oscillation) A marked reversal of the winds in the lower part of the tropical STRATOSPHERE in which the direction changes gradually from westerly to easterly and back to westerly with an average period of 26 months, or roughly two years.

quasi-stationary front *See* stationary front.

quickflow The most rapidly responding hydrologic processes and parts of the catchment area. An arbitrary cut-off line is drawn on a stream hydrograph, anything above this level represents the quickflow.

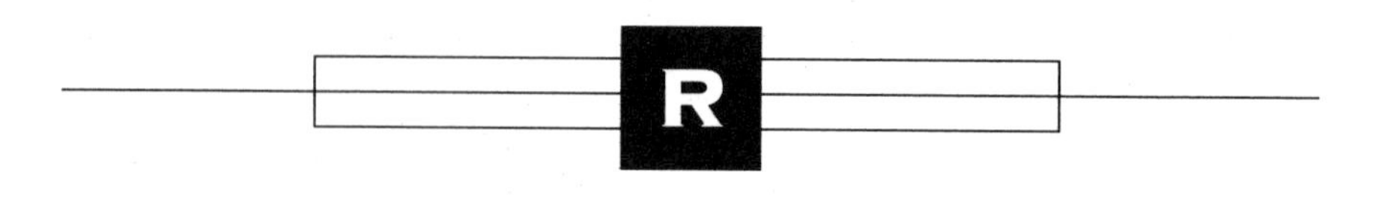

radar A REMOTE-SENSING instrument, developed before World War II, used to detect distant objects as a result of their ability to reflect or scatter a beam of microwave electromagnetic radiation. In meteorology, radar is used in weather forecasting to detect clouds and precipitation, and severe weather phenomena, such as thunderstorms and tornadoes; the latter especially with the use of DOPPLER RADAR. Radar is also used in the analysis and study of such meteorological phenomena. *See* NEXRAD. *See also* lidar. [From: *Ra*dio *d*etection *a*nd *r*anging]

radar altimetry A method, using RADAR, of monitoring the distance from a satellite or an aircraft to a surface (e.g. a sea or ice surface) by sending a pulse of microwave radiation from the altimeter down to the surface – and measuring its travel time to cover that distance and return to the receiver on the satellite. The instrument can estimate the distance to within a few centimeters, permitting the routine mapping of the height of wind-driven waves.

radar echo The signal returned to a radar antenna from specific objects, such as PRECIPITATION particles. The BACKSCATTER produced by the objects is also known as the echo.

radar imaging A method, using RADAR, of mapping the location and characteristics of selected environmental phenomena by emitting a pulse of microwave radiation and analyzing the image partly scattered back by them. Images of the intensity and distribution of rain, or the height and orientation of wind-driven ocean waves are examples of radar imaging.

radar meteorology The branch of meteorology concerned with the use of (primarily) ground-based RADAR in both the analysis and prediction of atmospheric phenomena over a wide variety of distances. It includes research into the structure and evolution of TORNADOES, TROPICAL CYCLONES, and EXTRATROPICAL CYCLONES. This work enables the occurrence of these phenomena to be predicted.

radar scatterometer An instrument that illuminates the Earth's surface from a satellite or aircraft in order to deduce environmental variables from the BACKSCATTER, which varies with the roughness of the land, ice, and sea surfaces.

radar winds Atmospheric motion sensed by tracking a radar target attached to a RADIOSONDE or the motion monitored by a DOPPLER RADAR.

radiance The radiant flux density per unit solid angle, of a source of radiation. It is measured by satellites, for example, and is expressed as a flow of energy in watts per square meter (per unit solid angle).

radiation The transmission of energy principally by electromagnetic waves but also by particulate cosmic radiation through space or a material medium. Electromagnetic waves do not need air as a medium through which to propagate.

radiation balance The resultant flux of incoming solar and outgoing terrestrial radiation through a horizontal surface. It is regarded as positive if the incoming flux exceeds the outgoing flux. *See* net radiation.

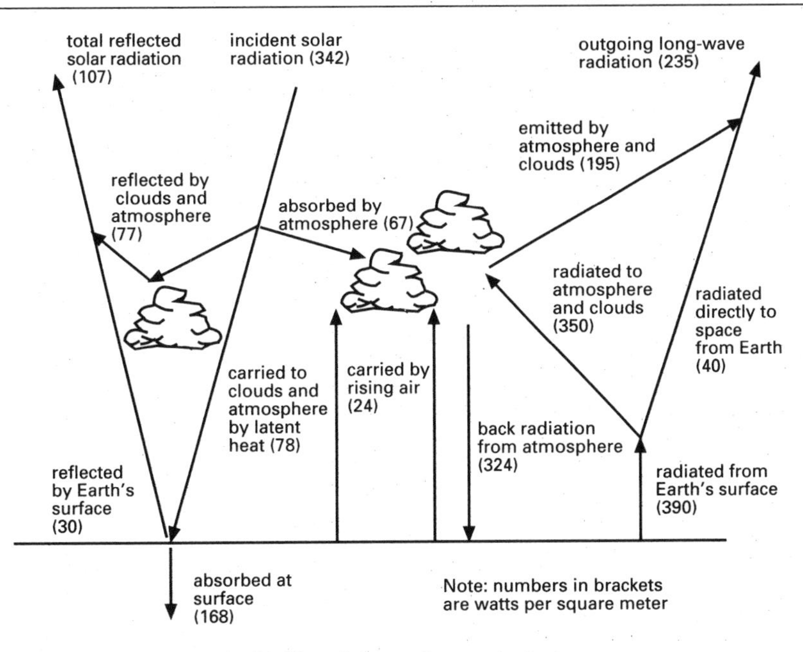

Earth's radiation and energy budget

radiation budget The flux of individual components of the incoming solar and outgoing longwave (terrestrial) radiation; it includes ABSORPTION and REFLECTANCE.

radiation densimeter An instrument used to measure the density (per unit area or per unit volume) of radiation at a specific location.

radiation fog (ground fog) FOG formed over land, generally at night under clear skies when conditions are calm or very nearly calm in moist air. It relates to the cooling mechanism being provided by RADIATIVE COOLING of the Earth's surface and lowest layers of the atmosphere in which the temperature of the air near the ground is reduced to below its dew-point. The fog is often deepest at sunrise but usually disperses soon after dawn when heated by solar radiation; it may persist for longer over cold wet surfaces in winter. Radiation fog occurs most commonly in the autumn and winter when cooling is strong and the hours of darkness are more extensive. Low-lying river valleys are particularly susceptible to radiation fog as a result of the high moisture levels and cooler air sinking into the valley.

radiation inversion A temperature INVERSION in the layer of the atmosphere in contact with the ground surface caused by the dominance of outgoing terrestrial radiation at night.

When NET RADIATION is negative, as occurs during the night, the air in contact with the ground is cooled so that the normal decline in temperature with height is reversed or inverted.

radiation night A night on which the lowest layers of the atmosphere are without cloud or wind allowing strong radiational cooling of the ground and adjacent air to take place. Such nights occur most frequently in winter in air masses with low absolute humidity.

radiative-convective model (RCM) A simple thermodynamic MODEL used to simulate the equilibrium temperature profile of the atmosphere and surface using known solar radiation at the top of the atmosphere and information about the atmospheric composition and surface albedo. Composition of the atmosphere may include information about cloud heights, depth, and cloud water amount, as well as water vapor and other gas concentrations.

radiative cooling A negative flux of radiation from the Earth's surface leading to cooling in the adjacent layer of air during night under clear sky conditions. *See* radiation night.

radiative forcing A change in the RADIATION BALANCE at the TROPOPAUSE in response to some externally applied alteration in radiation to the climate system, such as an increase in GREENHOUSE GAS concentrations (excluding water vapor), forcing an adjustment in the RADIATION BUDGET.

radiative warming *See* global warming.

radiatus A variety of cloud that occurs in parallel bands, aligned such that the lines of cloud appear to converge toward a point on the horizon (the radiation point) as a result of the effect of perspective. It is applied to CUMULUS, STRATOCUMULUS, ALTOSTRATUS, ALTOCUMULUS, and CIRRUS. *See also* cloud classification.

radio altimeter *See* altimeter.

radiocarbon dating *See* carbon dating.

radiometer An instrument used for detecting and measuring electromagnetic radiation. Radiometers can respond to a wide band of wavelengths (e.g. solar radiation) or to very narrow bands (e.g. infrared radiation). *See also* pyranometer; pyrheliometer.

radiometry The science of measuring radiation, especially that associated with infrared radiation.

radiosonde (radio-sounding device) An instrument package, attached to a meteorological balloon filled with a lighter-than-air gas such as helium or hydrogen, and carried aloft through the atmosphere to send back meteorological readings including pressure, temperature, and humidity via a radio transmitter. The data collected is transmitted to a ground station. The balloon is tracked by a radiotheodolite or by radar to determine wind speed and direction. Radiotheodolites used to track the balloon and receive the radio signals are being replaced by the use of the GLOBAL POSITIONING SYSTEM (GPS). Balloon-borne radiosondes can reach altitudes as high as 30,000 km (100,000 ft) before the balloon bursts and the radiosonde is returned to the ground carried by parachute. *See also* balloon sounding; ozonesonde; pilot balloon.

rain PRECIPITATION in the form of liquid water droplets, with diameters greater than 0.5 mm (0.02 in). It is formed by condensation of water vapor in the atmosphere and falls under the influence of gravity from clouds to the Earth's surface.

rainbow A well-known optical phenomenon that takes the form of an arc of the colors of the spectrum, which occurs when the Sun's rays are refracted and reflected by water droplets in the atmosphere. It appears when the Sun is shining and it is raining in the opposite direction to the Sun; it is seen when the observer is between the Sun and the rain. The angle of reflection determines the size of the rainbow. A *primary bow* appears with an angular distance of about 42° centered on the antisolar point; a *secondary bow* is often also visible with an angular distance of 51°. The bow is produced when sunlight is refracted as it passes through the individual water droplets and is reflected from the rear of the droplets. In the primary bow the color separation resulting from the refraction produces a spectrum with red on the outside of the bow going through to violet on the inner edge. The secondary bow, if visible, is dimmer and the colors appear in the bow in reverse order (i.e. red is on the inside and violet on the outside) as the light is reflected twice within the raindrop before it exits.

rain day 1. In the US, any day with measurable rainfall.
2. In the UK, an obsolete term for a period of 24 hours from 0900 UTC during which at least 0.2 mm (0.008 in) of rain is recorded. In other countries, the threshold of 1 mm (0.04 in) is sometimes used.

raindrop A droplet of water with a diameter of at least 0.5 mm (0.02 in) and usually in the range 1–2 mm (0.04–0.08 in) that falls to the Earth's surface from the atmosphere; DRIZZLE droplets have diameters less than 0.5 mm (0.02 in). Very small droplets evaporate in the unsaturated air below the cloud base; large droplets, with diameters around 5 mm (0.2 in), frequently disintegrate under aerodynamic forces before reaching the ground.

raindrop size spectrum The size distribution of RAINDROPS in a PRECIPITATION event. In an intense rainfall event, drops with diameters over 5 mm (0.2 in) can occur; in an event of moderate intensity the droplets are typically less than 0.75 mm (0.03 in) in diameter.

rain factor In climatology, an index that quantifies the degree of aridity of a location. Devised in 1920 by R. Lang, the rain factor (r_f) is expressed as mean annual precipitation (mm)/mean annual temperature (°C). As precipitation decreases and temperature increases, r_f decreases and the degree of aridity increases.

rainfall The total amount of PRECIPITATION of all forms. It is usually the total depth of water (including any melted solid forms of precipitation that have occurred) collected in a rain gauge at a particular location over a period of time, and is usually measured once every 24 hours. Rainfall and precipitation are also used interchangeably.

rainfall intensity (precipitation intensity) The amount of RAINFALL or PRECIPITATION that occurs over a particular period; a measure of the rate of rainfall receipt. Rainfall intensity (rainfall amount/duration) is typically expressed in millimeters or inches per unit time, usually per hour (mm/h or in/h). High rainfall intensities often lead to localized flooding, especially when they occur following periods of heavy rainfall.

rain-free base The low dark cloud base from which no rain is seen to be falling. It usually extends away from the core, where air is rising in the updraft, in a THUNDERSTORM.

rain gauge (precipitation gauge) An instrument used for collecting and measuring precipitation (rainfall and solid precipitation, e.g. snow). The standard rain gauge consists of a 50 cm (20 in) high cylinder with a 20 cm (8 in) diameter funnel and a narrow tube to collect the rain and minimize evaporation losses. The depth of water is usually measured once a day by an observer; either directly from the graduated tube, or with a standard ruler, or the collected water is poured into a graduated measuring vessel. Any water that has overflowed from the inner tube into the outer cylinder is also measured. Rainfall amounts are determined to the nearest 0.1 mm or 0.01 in. Some rain gauges also record rainfall intensity. Self-recording rain gauges include the HYETOGRAPH, TIPPING-BUCKET RAIN GAUGE, and the WEIGHING RAIN GAUGE.

rain making Attempts to generate rain by artificial means, primarily through CLOUD-SEEDING experiments. *See* artificial rain.

rain shadow An area to the lee of a mountain or range of mountains or hills that is sheltered from the prevailing rain-bearing winds and so receives lower amounts of rainfall than the windward side. A rain shadow is a product of the OROGRAPHIC PRECIPITATION mechanism: forced ascent of moist air results in heavy rainfall on the windward slope, but this air stream is compressed, and so warms and dries out, as it descends the lee slope. For example, the desert areas of eastern California and Nevada form a rain shadow, being sheltered from the prevailing westerlies by the Coastal Range and Sierra Nevada.

range The straight-line distance from a RADAR's antenna to the target. It is not the distance projected vertically onto the surface, but the true value along the inclined radar beam.

Rankine scale An absolute temperature scale that is sometimes used in the US. The zero point of the scale is at absolute zero but it differs from the Kelvin scale (*see* Kelvin temperature) in that it is based on the units of the FAHRENHEIT SCALE, so one Rankine degree (1°R) is 5/9 of one kelvin degree (1 K). The freezing point of water on the Rankine scale is 491.67°R. It is named for W. J. M. Rankine (1820–70).

Raoult's law A physical chemistry law that describes the relationship between the VAPOR PRESSURE of a solution and the SATURATION VAPOR PRESSURE of a component over a solution, and the mole fraction of the component in the solution. The law states that the partial pressure of a solute in equilibrium with a solution is equal to the mole fraction of the solute times the vapor pressure of the pure solute. The presence of a solute will lower the saturation vapor pressure. The law was discovered by the French scientist François Raoult (1830–1901).

rapid deepener *See* explosive cyclogenesis.

rare gas *See* inert gas.

ravine wind A wind generated as a result of a pressure difference (i.e. pressure gradient) between two ends of a narrow valley. The ravine wind blows from higher to lower pressure, its velocity being increased by the FUNNELING effect created by the presence of the ravine itself.

rawinsonde A RADIOSONDE, tracked by radar or radiotheodolite to measure wind speed and direction aloft.

Rayleigh scattering The SCATTERING of ELECTROMAGNETIC RADIATION produced in the atmosphere when the scattering particles, such as air molecules, are small compared to the wavelength of the incident RADIATION. The amount scattered is inversely proportional to the fourth power of the wavelength: the shorter the wavelength of the incident light, the more light is scattered. The Rayleigh scattering dominates the shorter end of the visible spectrum, hence the blue color of the daytime sky. Sunsets and sunrises appear reddish because the light travels a longer path through the atmosphere than at midday and the scattering of the shorter blue wavelength is more complete leaving the remaining reddish colors of the spectrum. *Compare* Mie scattering; non-selective scattering.

RCM *See* radiative-convective model.

Réaumur scale A temperature scale developed in 1731 by the French physicist and inventor, René-Antoine Ferchault de Réaumur (1683–1757). The freezing point of water is represented by 0 degrees and the boiling point by 80 degrees, at normal atmospheric pressure. The scale is now little used.

recurrence interval The average number of years that separate a RAINFALL INTENSITY of a given magnitude at a particular location. The recurrence interval is used to define the event's *return period.* For example, suppose that there have only been two occasions over the last 100 years when the rainfall intensity at a location X has equaled or exceeded 75 mm/h (3 in/h). The return period of such an event is thus once in every 50 years. (This does not mean that events of this magnitude occur at prescribed 50-year intervals, e.g. 1900, 1950, 2000, but that 50 years is statistically the most likely interval between such events.) An analogy can be drawn with a bag containing 98 black balls and 2 white balls. While the most likely outcome is for 1 in every 50 balls drawn from the bag to be white, it is possible that the 2 white balls could be drawn in rapid succession, or even consecutively. The occurrence of two 75 mm/h (3 in/h) events in the course of only two years is thus not necessarily inconsistent with such events having a return period of 1 in 50 years. Recurrence interval/return period estimation is most reliable when based on long time series (>100 years); estimates made using short record lengths are frequently unreliable. Return periods are often given following major flood events in order to quantify the rarity of such events.

red rain *See* blood rain.

reflectance (reflection, reflectivity) The return into space of solar radiation falling on a surface that is the boundary between two media (such as between clear air and a cloud). Total reflection occurs when all the incident radiation is returned. About 4% of the incoming solar radiation reaching the Earth's surface is directly reflected back to space, with no change to the wavelength. *See also* albedo.

refraction The deflection in the direction of propagation of a wave (e.g. electromagnetic radiation) when it passes obliquely from one medium to another in which the wave velocity is different. This is described by SNELL'S LAW. In the atmosphere the different mediums may be, for example, air layers of different density or temperature, or the air and water droplets. Refraction is responsible for a number of phenomena in the atmosphere, including HALO phenomena, RAINBOWS, and MIRAGES.

regime A recurring pattern, which can occur with climate.

relative humidity The amount of WATER VAPOR in the atmosphere at a given temperature usually expressed as a percentage of the amount of moisture that can exist in the same volume of air if it is saturated at the same temperature. Warm air can contain more water vapor than cold air, so the same amount of ABSOLUTE HUMIDITY or SPECIFIC HUMIDITY in a parcel of air will have a higher relative humidity if the air is cooler and a lower relative humidity if the air is warmer. *See also* dewpoint.

relative vorticity *See* vorticity.

relief rain (relief rainfall) *See* orographic precipitation.

remote sensing The range of methods used in measuring environmental variables by monitoring, quantifying, and processing signals either emitted by natural features (PASSIVE REMOTE SENSING) or scattered back to an artificial signal source (ACTIVE REMOTE SENSING). The method is described as remote as the measuring instruments used are never in direct physical contact with the medium being measured. SATELLITE imagery, RADAR and LIDAR, and aerial photography are all examples of remote-sensing methods that have extensive applications in meteorology and climatology.

reshabar 1. A strong northwesterly wind that blows across the Caucasus Mountains from the Black Sea in the west to the Caspian Sea in the east.
2. A LOCAL WIND that affects northern Syria, northern Iraq, western Iran, and southeastern Turkey, corresponding roughly to the Kurdish homeland. It is cold in winter and hot in summer.

residence time The length of time for which a particular constituent persists within a reservoir, such as the atmosphere. Residence times of gases in the atmosphere vary greatly; for ammonia it is around 10 days, for nitrous oxide approximately 20 years, while for nitrogen it is about 16 million years. *See also* atmospheric lifetime.

resolution In remote sensing, the size of the smallest element in the satellite-viewed scene that can be resolved by an instrument, which determines the sharpness of the image received. The theoretical limit of what can be resolved by an optical system depends on the wavelength of the radiation being observed (the smaller the better), the size of the sensor's instantaneous field of view (IFOV; the smaller the better), and the distance from the Earth (the smaller the better).

retrogression (retrograde motion) In meteorology, motion that goes against the normal flow. For example, a situation in which large-scale ROSSBY WAVES (long waves) move westward, contrary to the generally westerly winds flowing through the pattern. More generally, it is the condition in which such waves move in the opposite direction to the air flowing through them. It can also refer to a complete change in direction of a weather system, such as a LOW.

return flow 1. A part of an eddy that occasionally forms to the lee of a hill top or mountain peak. Under certain conditions a rotor or roll-like motion about a horizontal axis can form to the lee; the return flow is that component of the rotor that flows back toward the top or peak, in opposition to the general air flow across the barrier.
2. Winds that flow on the west side of an eastward-moving anticyclone (high); from the south in the northern hemisphere and from the north in the southern hemisphere.

return period *See* recurrence interval.

return stroke (main stroke, return streamer) The final phase of a CLOUD-TO-GROUND DISCHARGE stroke in which intense luminosity propagates upward from the Earth's surface to the base of a thundercloud, and along which the major charge transfer (~10 C at ~30 kA) occurs. It takes less than 100 microseconds. The return stroke follows the formation of a suitable lightning channel by a STEPPED LEADER and ascending streamers.

revolving storm *See* tropical cyclone.

ridge (wedge) An elongated area of relatively high pressure, characterized by a generally anticyclonic type of weather (i.e. generally settled fair weather), that may be an extension of an ANTICYCLONE or high-pressure region and is bounded by low pressure. It is similar in the pattern of mean-sea-level isobars to a ridge on a land-surface topography map. *Compare* trough.

rime (rime ice) A deposit of white, opaque, rough-textured ice crystals that forms when a cloud or fog of supercooled water droplets comes into contact with a solid surface in which the temperature is below 0°C (32°F), the freezing point of water. *Soft rime* forms from the water droplets in fog, *hard rime* from the freezing of water drops in DRIZZLE. Rime ice accumulates on the windward side of objects (e.g. a fence) to form frost feathers. It occurs mainly in upland areas in winter and is uncommon at low elevations where supercooled fogs are rare.

roaring forties The strong westerly winds of the temperate belt of the southern hemisphere, usually defined as being between 40° and 50°S and thus encompassing part of the desolate Southern Ocean, where there are few landmasses to disrupt the atmospheric flow. Consequently, this region is affected by strong and persistent westerly winds of over 40 knots (46 mph), with depressions tracking from west to east with great regularity; they bring rain but also relatively mild conditions. Its stormy conditions and isolation make this a hazardous zone, which is avoided by most shipping. *See also* brave west winds.

rocketsonde A package of instruments transported into the thermosphere of the upper atmosphere by a meteorological rocket before being ejected and descending to the Earth's surface borne by parachute. It provides instantaneous vertical profiles of a number of variables, such as temperature, pressure, ozone levels, and wind speed and direction as it descends. For example, temperature is measured by a thermistor, the results being transmitted to a ground station, and the rocketsonde may be tracked by radar or carry a transponder that can be tracked by a radio-direction finder and ranging system to determine altitude and wind speed. *See also* radiosonde.

roll cloud An elongated low-level ACCESSORY CLOUD in the shape of a horizontal tube that appears to spin slowly about its horizontal axis and is completely detached from the cumulonimbus base. A relatively rare cloud, it occurs along the GUST FRONT that flows out across the surface from a thunderstorm and typically is observed along the leading edge of thunderstorms or squall lines. It is sometimes associated with a COLD FRONT. The formation of roll clouds is not fully understood but may result from the motions of warm air riding up and over the cool air from the downdraft that is moving down and under creating a swirling motion or eddy.

rope cloud Long thin lines of low cloud that have been observed on satellite images to be associated with rapidly deepening FRONTAL CYCLONES.

rope stage The decaying stage of a TORNADO indicated by a narrowing and shrinking of its condensation funnel.

Rossby wave *See* long wave.

rotating cups anemometer *See* cup anemometer.

rotation of the Earth The rate at which the Earth rotates, an important factor in many of the equations that are used to explain the motions of the atmosphere. It is 2π radians per day (or in 86,400 seconds), i.e. 7.27×10^{-5} radians per second.

rotor cloud A small ragged cloud that sometimes occurs to the lee of hill tops or mountain peaks in association with an eddy that forms just downwind of such features, or underneath an outbreak of lenticular cloud (*see* lenticularis). In the orographic case, the eddy rotates about a horizontal axis so that the upper part of the rotor blows toward the top or peak, contrary to the general flow across the barrier. Rotors can be cloud free, but if the humidity conditions are right, they can contain a slowly rotating cloud. Underneath lenticular clouds to the lee of a barrier, the rotor clouds mark the presence of an overturning cell of air that is often very turbulent.

rotor streaming Air motion in a vertical plane that is a product of streamline difluence to the lee of a topographic barrier. It occurs downwind of the highest velocities, which occur in the immediate lee of the mountain range. The mesoscale circulation cell established can often lead to the local wind direction (the DOMINANT WIND) being diametrically opposed to the geostrophic-flow direction (the PREVAILING WIND). The phenomenon is chiefly associated with LEE WAVES. The ascent of moist air in the eddy circulation often produces ROTOR CLOUD, such as that seen to the east of the New Zealand Alps.

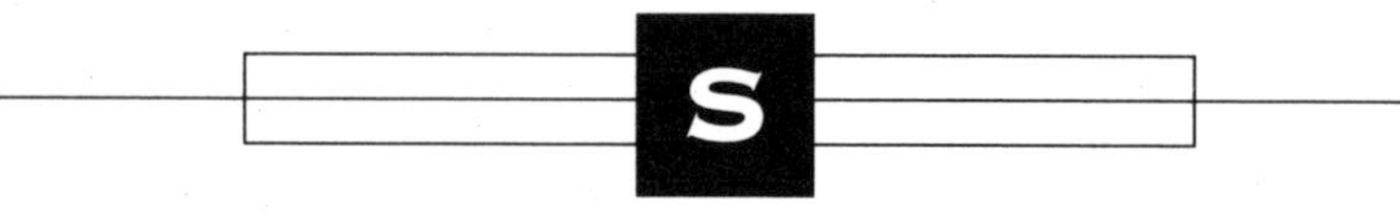

Saffir–Simpson Hurricane Scale A reference scale to classify the strength of hurricanes, based on central pressure, wind speed, and the magnitude of an associated STORM SURGE. It is used by the US National Weather Service to assess the potential damage to property and flooding from a hurricane landfall. The scale is also now used elsewhere in the world to assess the strength of all TROPICAL CYCLONES with wind speeds greater than 64 knots (75 mph). All winds in the scale use the US one-minute average. The scale was developed in the early 1970s by Herbert Saffir (1917–), a consulting engineer, and Robert Simpson (1912–), then director of the US National Hurricane Center.

Saint Elmo's fire (corposant) The visible luminous glow of an electrical discharge that may be seen, for example, on the masts of ships, aircraft propellers and wing tips, lightning conductors, and metal towers; it can also often be heard. It occurs during thunderstorms when there is high voltage within the atmosphere between the cloud and the ground and the electrical charge affects the air molecules, tearing them apart, and causing them to glow.

Saint Swithin's day In UK weather folklore, it is said that if it rains on St Swithin's day (July 15), rain will also fall on the 40 days following. There are no grounds for this in reality.

SALR *See* saturated adiabatic lapse rate.

saltation An EOLIAN transport process, in which a turbulent airflow lifts particles steeply from the surface into a highly localized zone possessing higher velocities. The particles are then transported a short distance downstream, before being gently dropped to the surface under the force of gravity.

samoon (samun) A characteristically hot and dry FÖHN-type wind in Iran; it results from the ADIABATIC compression of an airflow descending from the mountains of Kurdistan.

samuel *See* simoom.

sandstorm A desert storm in which fine or coarse sand is levitated by strong winds. Visibility is reduced, sometimes to only a

SAFFIR–SIMPSON HURRICANE SCALE

strength	*central pressure*		*wind speed*		*storm surge magnitude*	
	inches Hg	*hPa*	*knots*	*mph*	*m*	*ft*
1	28.91	980	64–82	75–95	1.2–1.5	4–5
2	28.50–28.91	965–979	83–95	96–110	1.8–2.4	6–8
3	27.91–28.49	945–964	96–113	111–130	2.7–3.7	9–12
4	27.17–27.90	920–944	114–135	131–155	4–5.5	13–18
5	<27.17	<920	>135	>155	>5.5	>18

few meters. Sandstorms are usually vertically shallow, rarely exceeding 30 m (100 ft) in depth. They may last from minutes to hours and can progress across desert areas. *See also* dust storm.

Santa Ana (Santa Anna) A LOCAL WIND with FÖHN characteristics experienced in southern California when a north to east wind blows during the passage of a low-pressure system off the Californian coast in late fall, winter, and early spring. The name derives from the town of Santa Ana, located to the southeast of Los Angeles, California. The wind blows from the Great Basin high, which lies usually over S Idaho, Utah, Nevada, and eastern California, then crosses the coastal ranges, descending through narrow passes, such as the Cajon and Santa Ana, to reach the southwest Californian coast as a hot, dry (with a consequent lowering of humidity), and sometimes dusty wind. The desiccative power of the Santa Ana can cause much damage to crops, particularly when it occurs during the spring growing season, and can also create dangerous wildfire conditions, especially in late fall.

satellite 1. (natural satellite) A body that is constrained to orbit a larger celestial body; for example, the Moon is a satellite of the Earth, which in turn is a satellite of the Sun.
2. (artificial satellite) A manufactured spacecraft launched into orbit around the Earth or another celestial body. METEOROLOGICAL SATELLITES provide considerable data for weather forecasting and research. *See also* geostationary satellite; polar-orbiting satellite.

satellite sounding A SOUNDING, such as an instantaneous vertical profile of temperature and humidity, obtained from instruments on a meteorological satellite. Several thousand soundings are produced every day by meteorological satellites, which are fed into operational NUMERICAL WEATHER PREDICTION models.

SATEM A specific SYNOPTIC CODE used to relay degraded quality vertical temperature and humidity profiles from the NOAA polar-orbiting satellites. These messages are relayed with some delay over the Global Telecommunications System.

SATOB A specific SYNOPTIC CODE used to relay cloud motion vector values derived automatically from the GEOSTATIONARY SATELLITES, such as METEOSAT. They are transmitted with some delay over the Global Telecommunications System.

saturated adiabat A curved line drawn on a THERMODYNAMIC DIAGRAM that traces the path of a saturated air parcel moving through the atmosphere adiabatically. *See* dry adiabat.

saturated adiabatic lapse rate (SALR, moist adiabatic lapse rate, wet adiabatic lapse rate) The change of temperature with height of a saturated air parcel that is either ascending or descending adiabatically. In a rising parcel of air the saturated adiabatic lapse rate is lower than the DRY ADIABATIC LAPSE RATE as a result of the release of latent heat that takes place when some of the moisture held by the parcel condenses. The saturated adiabatic lapse rate varies depending on the amount of moisture contained by the air: the greater the amount, the smaller the adiabatic lapse rate. It is generally in the range 4–9°C km^{-1}. As the air parcel rises and cools and moisture is lost via condensation the lapse rate increases and in the upper troposphere the magnitude of the saturated adiabatic lapse rate approaches that of the dry adiabatic lapse rate. The saturated adiabatic lapse rate is sometimes called a process lapse rate, as it refers to the thermal behavior of an air parcel as opposed to that of the surrounding environmental air.

saturated air (saturation) A condition in which the level of WATER VAPOR in a parcel of air is the maximum possible at the existing temperature and pressure without bringing about condensation. In saturated air the rate at which water molecules enter the air by evaporation exactly balances the rate at which molecules leave by condensation. For practical purposes the capacity of the air to hold water vapor predominantly depends on temperature. The addition of more water vapor or a decrease in temperature will lead to condensation.

saturation *See* saturated air.

saturation deficit The difference between the actual VAPOR PRESSURE of a moist sample of air at a given temperature and the SATURATION VAPOR PRESSURE corresponding to that sample. It is also referred to as the *Vapor Pressure Deficit.*

saturation mixing ratio A measurement of the maximum capacity of air to contain water vapor. It is usually expressed as the grams of water vapor that would be needed to saturate the air in each kilogram of air.

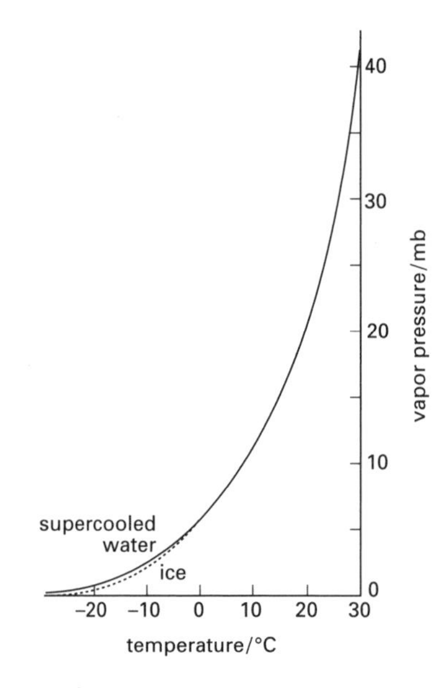

Saturation vapor pressure curve

saturation vapor pressure The maximum possible partial pressure of water vapor in the atmosphere at a given temperature. The higher the air temperature, the greater the saturation vapor pressure, as expressed by the CLAUSIUS–CLAPEYRON RELATION; saturation vapor pressure increases rapidly with temperature and at 32°C (90°F) is around double the value at 21°C (70°F). Saturation vapor pressure is greater over a curved surface (e.g. a water droplet) than over a plane surface, and is greater over pure water than over water with a dissolved solute.

savanna (savannah) Tropical or subtropical open grassland with widely spaced drought-resistant trees and shrubs. It has a pronounced dry season during which brush fires are periodically a natural occurrence. Savanna extends over much of tropical Africa and also occurs in South America, India, Madagascar, and Australia. In the Köppen climate classification savanna is type A.

savanna climate *See* tropical wet-dry climate.

scattering The process by which solar radiation incident on particles (e.g. air molecules, water droplets, pollutants) is absorbed and rapidly emitted in another direction; it may be dispersed in all directions, though not necessarily in equal amounts. The particles have an index of refraction that differs to that of the medium within which they are held. The scattered radiation does not undergo a change of frequency but its phase and polarization may change considerably.The type of scattering depends on the wavelength of radiation and the size of the particles. *See* backscatter; Mie scattering; non-selective scattering; Rayleigh scattering.

scavenging The process by which particulate matter in the atmosphere is captured and removed by precipitation.

scintillation The rapid changes in brightness giving a twinkling effect in the appearance of stars and other celestial bodies. It results from refraction produced by variations of density of air within the atmosphere through which the light rays pass. *Compare* shimmer.

scirocco (sirocco) A local, often dusty, wind that affects the Mediterranean Basin, especially during the spring; it rarely occurs in summer. Associated with the leading edge of a low-pressure system, the scirocco sweeps hot and dry air from the Arabian or Sahara deserts northward. While the northern African scirocco is dry, the wind's passage across the Mediterranean Sea results in the acquisition of moisture in its lowest layers. The scirocco typically arrives on the northern shoreline of the Mediterranean as a warm but damp and oppressive wind. A number of regionally specific names exist for the scirocco, for example KHAMSIN in Egypt, LEVECHE in southeastern Spain, GHIBLI in Libya, and *chili* in Tunisia. The scirocco is thus the generic name for a collection of warm south to southeasterly LOCAL WINDS.

Scotch mist (mizzle) A type of PRECIPITATION that consists of a mass of minute water droplets in a combination of both MIST (or FOG) and DRIZZLE. It usually occurs in upland or mountainous areas where the cloud base is close to ground level and the air mass is saturated but stable.

screen *See* instrument shelter.

scud A type of rapidly moving ragged gray or dark gray low cloud, usually FRACTOSTRATUS, that occurs beneath higher rain cloud (NIMBOSTRATUS or CUMULONIMBUS). Scud clouds frequently occur in association with, or behind, the gust front of a thunderstorm.

SDM *See* statistical-dynamical model.

sea breeze A thermally induced diurnal wind that blows onshore, and is most common in low and mid-latitude locations. It develops as a result of the land warming more rapidly than the sea during the day, which creates a PRESSURE GRADIENT directed from higher pressure over the sea to lower pressure over the land. A sea breeze can reach speeds in excess of 25 knots (29 mph), although values are typically 8–14 knots (9–16 mph). The sea breeze's vertical depth increases throughout the day, sometimes reaching 1.2 km (4000 ft). Its cooling effect also penetrates inland, occasionally being felt at distances of more than 50 km (30 miles) from the coast by 9 pm. A convergence line called the *sea breeze front*, marks the boundary between the incoming sea air and the air over the land; a zone of enhanced CONVECTION. The development of a sea breeze is particularly favored when conditions are calm and a weak offshore GEOSTROPHIC WIND component exists. *Compare* land breeze.

sea fog *See* advection fog.

sea fret *See* fret.

sea ice Ice that forms on the surface of the sea from sea water when temperatures fall below approximately –2°C (28°F). During the process of freezing the salt in the sea water is released and returned to the sea leaving ice that is virtually pure water. Vast areas around Antarctica and over the central Arctic are covered by sea ice, which forms during the winter season. The extent of sea ice has a bearing on climate; it thermally insulates the ocean and decouples it from the surface atmospheric winds, but also reflects solar radiation back to space. The extent of sea ice is largely dependent on the local currents. Fast ice is a type of sea ice that is attached to land.

Seasat 1 A short-lived but pioneering satellite, specifically designed to observe the oceans, which was launched in June 1978. It orbited at a height of 800 km (500 miles) and had an on-board SYNTHETIC APERTURE RADAR, MULTISPECTRAL SCANNER, radar altimeter (*see* radar altimetry), and scatterometer. It gathered significant data but unfortunately lasted only 105 days before malfunctioning.

season A number of months of the year with pronounced climatic characteristics. The methods to determine the limits of each season are not clearly defined. In many middle-latitude countries of the northern hemisphere the year is conventionally divided into four seasons (WINTER, SPRING, SUMMER, and AUTUMN), even if there is little climatic justification for this.

The four seasons are often taken to begin on the winter solstice (December 22/23), the vernal equinox (March 20/21), the summer solstice (June 21/22), and the autumn equinox (September 22/23). In polar latitudes there is a short summer season and long winter with no spring or autumn clearly identifiable. In the tropics, the division is usually based on receipt of precipitation, between the wet and dry seasons, or the MONSOONS. Poleward of the tropics, seasonal variations in temperature produce hot and cold (or cool) seasons.

seasonal prediction Longer-term and mainly experimental forecasts undertaken by some weather services. Such computer-based predictions normally take the form of a global assessment of the anomaly pattern of PRECIPITATION and/or TEMPERATURE for a seasonal period (e.g. June to August).

sea-surface temperature (SST) The temperature of the surface layer of the sea or ocean (approximately 0.5 m deep).

Secondary Ambient Air Quality Standards *See* National Ambient Air Quality Standards.

secondary cold front (secondary front) The leading edge of markedly colder air within a cold AIR MASS behind a primary COLD FRONT. It is usually associated with a TROUGH within the polar air.

secondary depression Generally, a relatively small-scale low-pressure area that forms on the flank of a much larger 'parent' LOW, and is embedded within the large-scale winds that circulate around the larger feature. *See* polar low.

secondary front *See* secondary cold front.

secondary pollutant A pollutant that is not emitted directly into the atmosphere but is formed as a result of chemical changes in the atmosphere, such as photochemical action (*see* photodissociation) and other reactions. Examples are ozone and nitrogen dioxide. *See also* smog.

secular trend In climatology, the slow change, either an increase or a decrease, in the values of a climatic element (e.g. temperature) that takes place over a long period, after fluctuations that occur over a comparatively short period have been eliminated. For example, the trend of increased carbon dioxide concentration in the atmosphere produced by the combustion of fossil fuels.

seiche A stationary wave (i.e. a wave that oscillates without progressing) that occurs in an enclosed or semi-enclosed body of water, such as a lake, bay, or estuary. It is caused, for example, by strong winds, earthquakes, landslides, or changes in atmospheric pressure. The oscillation continues for some time after the force initiating its formation has ceased, occasionally lasting for several days.

sensible heat Energy held by a body as a form of heat that can be monitored or sensed, for example by means of thermometers; it is transferred by CONDUCTION and/or CONVECTION without a change of phase. It is a component of the BOWEN RATIO. *Compare* latent heat.

sensible temperature The temperature as it is felt or experienced by an individual. This can feel different to the temperature actually recorded as a result, for example, of humidity levels, which can make conditions feel hotter (*see* heat index; temperature-humidity index) or wind speed, which can make conditions feel colder (*see* wind chill).

sensor The part of a scanner (e.g. on a satellite) that is sensitive to light or other electromagnetic radiation and measures the RADIANCE from natural objects. A scanner may have multiple sensors to either sense different bands or increase the spatial resolution.

serein The phenomenon of fine rain that falls from an apparently cloudless sky overhead. It can be explained by either the movement of the cloud system away from the observer's overhead position at the time the rain reaches the ground surface, or by dissipation of the cloud following the formation of raindrops.

severe thunderstorm An intense THUNDERSTORM with downdraft winds exceeding 50 knots (58 mph) and/or hailstones with a diameter of at least 20 mm (0.75 in). These thunderstorms are capable of producing heavy showers, flash floods, and tornadoes.

severe weather threat index (SWEAT) A numerical index developed by the US Air Force to indicate the probability of severe weather; it utilizes wind shear, wind speed, and instability.

shade temperature The temperature of air measured by a thermometer mounted in an INSTRUMENT SHELTER. In this location, air circulates freely around the thermometer while it is sheltered from the effects of precipitation, direct solar radiation, and energy emitted from the ground and surrounding objects.

shamal A northwesterly wind that blows across the Tigris and Euphrates plains in Iraq and down into the Persian Gulf. It occurs most often on summer days and is associated with dry and hot conditions, its force sometimes initiating dust storms. It decreases in intensity at night.

sharav A shallow low-pressure area that forms to the LEEWARD of the Atlas Mountains in North Africa; a form of lee cyclogenesis. The controlling upper-level trough intensifies slightly when crossing the mountain range and then induces a low-level cyclone in the lee. Sharav events bring hot and dry weather to the region.

shear *See* wind shear.

sheet lightning A diffuse illumination of a cloud by a LIGHTNING discharge in which the forked discharge form is not visible because of the presence of obfuscating cloud.

shelf cloud *See* arcus.

shelterbelt *See* windbreak.

shimmer The wavering or tremulous effect in images of features produced in the lowest layers of the atmosphere, above a heated surface. It results from the varying atmospheric refraction in the line of sight produced by variations in the temperature, and hence density, of the air. *Compare* scintillation.

ship surface observations Meteorological observations made aboard a weather ship.

short-fuse warning A warning of short-lived severe weather issued by the US National Weather Service (NWS). These include warnings for tornado events, thunderstorms, and flash floods.

short-range weather forecasting (short-term weather forecasting) The prediction of the weather, normally for 48 hours ahead.

shortwave radiation The ELECTROMAGNETIC RADIATION, with a peak in intensity at a wavelength of 0.5 μm, within a range of 0.3 μm to 4 μm, that is received at or near the Earth's surface. *Compare* longwave radiation.

short-wave trough A smaller scale feature that is part of a LONG-WAVE trough. Such smaller troughs often move rapidly through the long-wave pattern, and are frequently associated with major cyclonic developments.

shower A short-lived precipitation (e.g. rain, hail, snow) event that often has a duration of less than 10 minutes and starts and ends abruptly. Showers range in their intensity between light, moderate, heavy, and violent and often change intensity abruptly. They are typically associated with deep convective clouds (e.g. cumulonimbus), which do not usually completely cover the sky; brightness is frequently evident during shower events.

Showers can occur at any time of year, especially in the tropics; in mid-latitude environments they occur with greatest frequency on hot afternoons during summer.

Shurin season A warm season peak in rainfall beginning in late August and lasting through into October in the vicinity of Japan. Much of the rainfall occurs in association with tropical storms that recurve north in the western North Pacific.

Siberian high (Siberian anticyclone, Asian high, Asian anticyclone) An extensive ANTICYCLONE (high) that covers most of middle- and high-latitude Asia during the winter months and disappears in summer. It is a thermal high formed by RADIATIVE COOLING, is characterized by extremely low temperatures, and extends up vertically to some 2–3 km (1–2 miles). The air circulating out of it to the south forms the northeast MONSOON across the Indian subcontinent and adjacent regions.

side-looking airborne radar (SLAR) An airborne radar system used to map morphological features on the Earth's surface in such a way that the images exaggerate the relief, somewhat analogous to photography at low solar illumination angles.

silver thaw *See* glaze.

simoom (simmoom) A local short-lived SCIROCCO-type southerly wind experienced in the deserts of North Africa and Arabia, particularly in spring and summer, and also beyond in central and southern parts of the Mediterranean. In desert regions, the simoom is extremely hot at around 50°C (122°F) and dry, and carries much dust and sand, sometimes reducing visibility to practically zero. It is similar in appearance to a DUST DEVIL or WHIRLWIND. In Turkey, this wind is referred to locally as the *samuel.*

singularity A period of distinctive weather or a weather phenomenon that has a tendency to recur in the majority of years on or about the same time. For example, the January thaw of New England, which occurs on or around January 20–23 in most years.

In a development of the BUCHAN SPELL, singularities were formerly used to seek to identify some regularity or pattern in the climate of Western Europe. Verification of the objectivity in such a scheme is difficult and controversial.

sink In air pollution, the surface or vegetation receiving pollutants removed from the atmosphere. It is the final phase in the atmospheric process of source, transport, transformation, and sink. For example, soil and trees act as natural sinks for carbon from the atmosphere.

sinking *See* downwelling.

sirocco *See* scirocco.

SI units (Système International d'Unités) The internationally adopted system of units used for scientific purposes. See Tables.

skew-*T* log-*P* diagram (modified emagram) A type of THERMODYNAMIC DIAGRAM widely used in the US, in which the dry adiabats make an angle of about 45° with the ISOBARS and isopleths of SATURATION MIXING RATIO are almost vertical and straight. Modified in 1947 from the original emagram, in order to create straight horizontal isobars, the skew-*T* log-*P* diagram is similar in appearance but not identical to a TEPHIGRAM. *See also* Stüve diagram.

sky cover The amount, in OKTAS, of the sky that is covered by different types of cloud.

sky radiation *See* diffuse radiation.

SLAR *See* side-looking airborne radar.

slash and burn The practice of clearing land, typically in tropical rainforest, by cutting down the vegetation and clearing it by burning. The cleared plots of land are cultivated for a number of seasons before being abandoned and a new site cleared.

TABLE 1: BASE AND SUPPLEMENTARY SI UNITS

Physical quantity	*Name of SI unit*	*Symbol for SI unit*
length	meter	m
mass	kilogram(me)	kg
time	second	s
electric current	ampere	A
thermodynamic temperature	kelvin	K
luminous intensity	candela	cd
amount of substance	mole	mol
*plane angle	radian	rad
*solid angle	steradian	sr

*supplementary units

TABLE 2: DERIVED SI UNITS WITH SPECIAL NAMES

Physical quantity	*Name of SI unit*	*Symbol for SI unit*
frequency	hertz	Hz
energy	joule	J
force	newton	N
power	watt	W
pressure	pascal	Pa
electric charge	coulomb	C
electric potential difference	volt	V
electric resistance	ohm	Ω
electric conductance	siemens	S
electric capacitance	farad	F
magnetic flux	weber	Wb
inductance	henry	H
magnetic flux density	tesla	T
luminous flux	lumen	lm
illuminance (illumination)	lux	lx
absorbed dose	gray	Gy
activity	becquerel	Bq
dose equivalent	sievert	Sv

TABLE 3: DECIMAL MULTIPLES AND SUBMULTIPLES USED WITH SI UNITS

Submultiple	*Prefix*	*Symbol*	*Multiple*	*Prefix*	*Symbol*
10^{-1}	deci-	d	10^{1}	deca-	da
10^{-2}	centi-	c	10^{2}	hecto-	h
10^{-3}	milli-	m	10^{3}	kilo-	k
10^{-6}	micro-	μ	10^{6}	mega-	M
10^{-9}	nano-	n	10^{9}	giga-	G
10^{-12}	pico-	p	10^{12}	tera-	T
10^{-15}	femto-	f	10^{15}	peta-	P
10^{-18}	atto-	a	10^{18}	exa-	E
10^{-21}	zepto-	z	10^{21}	zetta-	Z
10^{-24}	yocto-	y	10^{24}	yotta-	Y

sleet 1. In the US, PRECIPITATION in the form of ice pellets of less than 5 mm (0.2 in) diameter. These grains are either frozen RAINDROPS or SNOWFLAKES that have partially melted and then refrozen after passage through a cold layer of air at below freezing temperature close to the Earth's surface.
2. In the UK, precipitation that takes the form of a mixture of rain and snow, or partially melted falling snow.

sling psychrometer (whirling psychrometer) A type of PSYCHROMETER used to measure relative humidity that is designed to ventilate the wet-bulb thermometer and thus speed up evaporation; the dry-bulb and wet-bulb thermometers are both mounted on a type of frame. The wick of the wet-bulb thermometer is dipped in distilled water and the psychrometer is then whirled around by the handle before readings are taken.

slope convection Generally large-scale motion that is transporting heat, for example, in a shallow sloping manner rather than the more localized upright CONVECTION made visible by CUMULIFORM clouds. FRONTAL CYCLONES are examples of sloping convection in the sense that the warm, moist air that streams poleward through such a feature is acting as a massive gently inclined convective current.

slope wind *See* along-slope wind system; anabatic wind; katabatic wind.

slush A colloquial name for soiled snow and ice that has become softened and saturated with water from rain and/or warmer temperatures. It often clogs roads and sidewalks during the thaw that follows a snowfall.

small hail *See* graupel.

smog A type of polluted atmosphere containing smoke and/or a high concentration of aerosols and primary pollutants due to adverse weather conditions. The term was coined by H. A. Des Voex in 1905 to describe 'smoke-impregnated fog' but is now also used for concentrations of pollutants within the atmosphere above urban areas where fog may not be present. *Industrial smog* results from fog in combination with smoke and sulfur dioxide released in the burning of coal. This was responsible for the severe 'killer smogs' that occurred in the UK, during the mid-20th century, such as the London fog of 1952, which led to the passing of the UK's Clean Air Act of 1956. *Photochemical smog* is a hazy atmosphere resulting from a cocktail of mostly gaseous pollutants that have an oxidizing action, occurring especially in urban areas experiencing high solar-radiation values (originally renowned as a feature in Los Angeles, but now common throughout the urbanized world). The commonest oxidant is OZONE formed in the environs of the BOUNDARY LAYER by solar radiation dissociating molecules of oxides of nitrogen (NOx) and volatile organic compounds (VOCs), emitted chiefly by vehicle exhausts but also by industrial power plants. Ozone is an irritant to humans and also causes harm to plant tissue. In addition to ozone, other secondary pollutants include many hazardous chemicals.

smoke management In the UK, as a result of 'killer smogs', the Clean Air Act of 1956 was passed. Local authorities were granted powers to authorize the use of smokeless combustion fuels in order to promote smoke management. These powers were increased in 1968. More efficient fuel burners were promoted until use of North Sea gas eased the problem. *See* smog.

Snell's law The law of refraction stating that the ratio of the sine of the angle of refraction (r) to the sine of the angle of incidence (i) is equal to the ratio of the refractive index (n_i) of the material through which the incident ray is passing to that of the medium through which the refracted ray passes (n_r), i.e.

$$\sin r/\sin i = n_i/n_r.$$

The law is named for the Dutch physicist Willebrord Snell (1591–1626).

snow A solid form of PRECIPITATION comprised of ice crystals formed from water vapor in the atmosphere when the temperature is below freezing-point. When the temperature is well below 0°C (32°F), the snow may fall in the form of individual crystals or minute *ice spicules*. As the temperature approaches 0°C (32°F), the ice spicules aggregate to form SNOWFLAKES. The properties of snow (e.g. density) thus vary according to the temperature: dry and powdery at very low temperatures, becoming increasing wet, compact, and structureless as the temperature approaches 0°C (32°F).

The variable density of snowfall makes it difficult to measure accurately; for instance, a normal rain gauge is blocked by falling snow. Conventionally, snowfall is measured by inserting a ruler into a horizontal layer of undrifted snow. A depth of 100–150 mm (4–6 in) of wet compact snow typically produces 25 mm (1 in) of water; a much greater depth of 500–700 mm (20–28 in) of dry powdery snow is needed to produce the same quantity of water. Snow samplers can also be used to provide more accurate measurements of the water content of a snow pack.

snow crystal A minute ice spicule that aggregates with other snow crystals to form a SNOWFLAKE. The shape of a snow crystal is influenced by the temperature and humidity at which it forms. At least 6000 different shapes of crystal have been recognized: these range from hollow columnar prisms that form only in dry air between –25°C (–13°F) and –50°C (–45°F); dendritic crystals that develop in quite moist air at temperatures of –12°C to –16°C (10°F to 3°F); and thin hexagonal needle crystals, which are most likely to form in moist air between 0°C (32°F) and –5°C (23°F).

snowflake An aggregation of SNOW CRYSTALS. The size of a snowflake varies according to the temperature of formation, being generally larger as the temperature approaches freezing-point. *See also* snow.

snow gauge An instrument used for collecting and measuring the depth of snow. Snow gauges collect the solid precipitation and melt it before taking a reading.

snow grains A form of PRECIPITATION comprising small opaque particles of ice. Snow grains are usually flattened and elongated in shape, having diameters of less than 1 mm (0.04 in). Such grains often fall from stratus clouds or from banks of fog that occur on days with subfreezing temperatures.

snow line The *permanent snow line* is the altitude on a hillslope or mountain that marks the lower limit of permanent snow cover. Above the snow line, ACCUMULATION of snow during the winter always equals or exceeds the amount lost by ablation (melting and evaporation) in the summer. The height of the snow line, which is irregular and not always sharply defined, is heavily affected by summer temperatures and therefore by latitude. On tropical mountains it may be as high as 5000 m (16,000 ft), in the European Alps it typically occurs between 2500 m (8000 ft) and 3000 m (10,000 ft), and in S Greenland it occurs at around 600 m (2000 ft), before falling to sea level near the poles. The degree of exposure or shelter, slope angle, and the amount and seasonal distribution of PRECIPITATION also influence the altitude of the snow line. These factors can result in marked variations in the altitude of the snow line over short distances. The *seasonal (ephemeral) snow line* marks the lower limit of seasonal snow cover during winter.

snow pellet *See* graupel.

snow sampler An instrument used to collect samples of snow. It generally consists of a long metal cylinder, closed at one end, which is driven into the snow pack. The water from the melted sample is measured and analyzed.

soft hail *See* graupel.

SOI *See* Southern Oscillation.

soil temperature The temperature of soil measured at a variety of different depths (e.g. 5, 10, 20, 30, and 100 mm depths). These measurements are recorded at many weather stations.

solar constant The amount of energy passing in unit time through a unit surface perpendicular to the Sun's rays at the top of the atmosphere. Current satellite measurements indicate a value of 1367 W m^{-2}. Lower values are recorded by instruments on the surface of the Earth, as a result of atmospheric ATTENUATION.

solar day The interval of time during which the Earth makes one complete revolution on its axis in relation to the Sun. The average length of the solar day is 86,400 seconds.

solar energy *See* solar radiation.

solar flare A solar protuberance of incandescent gas emanating from the Sun's chromosphere, usually above certain types of sunspot, and resulting from an explosive release of electromagnetic radiation and energetic particles. Flares last for a few hours and may be classified on an ascending scale of one to three plus based on the intensity of the emitted light. The radiation and particles affect the Earth's geomagnetic field to produce aurorae and geomagnetic storms.

solarimeter *See* pyranometer.

solar radiation The radiation emitted from the Sun, nearly all of which is in the form of ELECTROMAGNETIC RADIATION. A very small fraction is emitted in the form of particles, known as particulate or corpuscular radiation.

solar wind The electrically charged electrons and protons that stream out into space at about 400 km s^{-1}, produced by the ionization of particles in the high solar atmosphere. Though this corpuscular radiation (plasma) comprises less than a millionth of the total solar energy emitted by the Sun it interacts with the geomagnetic field of the Earth and can produce storms in the Earth's magnetosphere.

solstice **1.**. Either of the two points at which the Sun reaches its greatest or least declination (i.e. is at its greatest angular distance above or below the celestial equator) during a year.
2. The date on which the Sun reaches either of these points. In the northern hemisphere the summer solstice is on about June 21 and the winter solstice on December 21. In the southern hemisphere the dates are reversed.

sonde *See* radiosonde; rocketsonde.

sounding A vertical profile of atmospheric properties, such as DRY-BULB TEMPERATURE, RELATIVE HUMIDITY, and WIND DIRECTION and WIND SPEED produced by instruments attached to a RADIOSONDE balloon or ROCKETSONDE.

sounding balloon *See* balloon sounding.

source region An extensive area associated with an ANTICYCLONE (high) over an ocean or continent from which air spirals out. It gives an AIR MASS its distinctive properties; for example, warm and moist from the SUBTROPICAL ANTICYCLONES and cold and dry from the winter CONTINENTAL ANTICYCLONES. The areas over which the anticyclones lie are of generally even surface and light winds. Once air flows out from the source regions it becomes modified, especially in the lower levels, by exchanges of heat and moisture with the underlying surface. For example, air following a maritime track passing over ocean surfaces will have its moisture content increased in the lower layers.

southeast trade winds (southeast trades) The tropical TRADE WINDS, best marked over the oceans between about 5° and 15° latitude, that blow with significant strength and notable constancy from the southeast. They occur over the southern-hemisphere tropics, flowing toward the Intertropical Convergence Zone (ITCZ), where they converge with their counterpart NORTHEAST TRADE WINDS. They cross the Equator over much of the year to flow into the ITCZ, and on doing so are deflected into southwesterlies.

South Equatorial Current *See* Equatorial Current.

southerly buster An intense cold squally southerly wind that advects polar air across southern and southeastern Australia, especially the coast of New South Wales, in spring and summer. Thirty busters affect this region in an average year, the majority of which occur between October and March. The southerly buster typically follows the BRICKFIELDER, the 180° switch in wind direction producing a sudden fall in temperature of 10–15°C within a few minutes. It usually occurs on the westward limb of a low-pressure trough, which sweeps a COLD FRONT northward from Antarctica; the mountains of the Great Dividing Range along the south coastline tend to block the cold air and channel it round to the east coast. The southerly buster is frequently accompanied by thunderstorms and strong gale-force winds with gusts up to 60 knots (70 mph). In South American countries, such as Argentina and Uruguay, a similar phenomenon is known as the PAMPERO.

southern lights *See* aurora.

Southern Ocean (Antarctic Ocean) The body of water encircling Antarctica. This ocean comprises the southern parts of the PACIFIC, ATLANTIC, and INDIAN OCEANS. Unlike the northern-hemisphere oceans, the Southern Ocean is not divided into geographic sectors by the continental landmasses. The narrowest part of the ocean, only 970 km (600 miles) wide, is the Drake Passage between South America and the Antarctic Peninsula. Currents in the Southern Ocean are dominated by the sinking of water that has been cooled by coastal landmasses. This cold water flows north along the ocean floor (*see* Antarctic bottom water) and is replaced by water flowing south at the surface from the Indian, Pacific, and Atlantic Oceans.

Southern Oscillation A large-scale fluctuation in monthly/seasonal mean-sea-level pressure across part of the low-latitude Pacific Ocean. The oscillation is an occasional 'see-saw' effect exhibited by changes in the average pressure observed in the South Pacific SUBTROPICAL ANTICYCLONE (high) and the low-pressure area across Indonesia and northern Australasia. The *Southern Oscillation Index* (SOI) is mapped by monitoring the difference of monthly average mean-sea-level pressure between Tahiti (in the high) and Darwin (in the low). Tahiti is always higher, but when it is unusually high, the value at Darwin is anomalously low. In addition, when Tahiti's high is weaker than average (a lower-pressure value), Darwin's is higher than average (a higher-than-average pressure value). A time series of Tahiti minus Darwin monthly mean pressure exhibits the oscillation that is closely linked to the EL NIÑO. Most often the difference in pressure is strong enough to drive the SOUTHEAST TRADE WINDS across the South Pacific, but when it is weak, the trade winds weaken or even reverse into westerlies near the Equator. This is linked to the development of an El Niño event. *See also* El Niño Southern Oscillation.

Southern Oscillation Index (SOI) *See* Southern Oscillation.

South Pacific Convergence Zone (SPCZ) A major year-round feature of the tropospheric circulation that is characterized by marked CONVERGENCE and an associated cloud band that stretches from the low-latitude Pacific Ocean, southeastward into the middle-latitude central South Pacific.

SPARC *See* Stratospheric Processes and their Role in Climate.

SPCZ *See* South Pacific Convergence Zone.

specific heat capacity *See* heat capacity.

specific humidity A measurement of atmospheric moisture content (WATER VAPOR), usually expressed as the dimensionless ratio of the mass of water vapor to unit mass (kilogram) of moist surrounding air (air including water vapor); it is almost identical to, but slightly smaller than that of MIXING RATIO. The specific humidity of an air parcel remains constant unless water vapor is added to or taken from the parcel.

speed shear The rate of change of wind velocity with height. This is a function of atmospheric STABILITY and the roughness of the surface. In the neutral stability state, the variation of wind speed (v) with height (Z) can be expressed as: $v = kZr$, where k is a constant and r depends on the roughness of the surface. In this state, speed shear is typically greater over a city than a wooded area, which in turn is greater than that experienced over undulating grassland. Regardless of the surface, speed shear is greatest in UNSTABLE conditions and least pronounced or sometimes non-existent in STABLE conditions. It is rare for the speed shear to exceed 10 knots (11 mph) between 10–40 m (30–130 ft) above the Earth's surface; typically it is of the order of 3 knots (3.5 mph) per 30 m (100 ft).

spissatus A cloud species; CIRRUS that is thick enough to appear gray toward the Sun, it may also obscure the Sun completely. *See also* cloud classification.

splitting storm A THUNDERSTORM that splits into two separate systems. The new systems tend to follow diverging paths, one to the north and the other to the south. In general (in the northern hemisphere) the southern storm is most likely to develop into a SUPERCELL, while the north-moving storm tends to dissipate and weaken.

SPOT (Système Probatoire d'Observation de la Terre) A French commercial satellite with high spatial resolution that images the Earth's surface. It covers the whole Earth once every 26 days, with an orbit similar to those of LANDSAT 4 and 5.

spring In the middle latitudes, the SEASON of the year in which the Sun approaches the summer solstice (June 21/22 in the northern hemisphere), usually referring to the months of March, April, and May in the northern hemisphere and September, October, and November in the southern hemisphere. It is marked by a decrease in the strength of the mid-latitude westerly winds and their associated storms. Air frosts become less frequent and daily totals of incoming solar radiation show a marked rise. In parts of Western Europe DROUGHT may occur as temperatures increase without a commensurate increase in moisture.

sprite A weakly visible reddish tendril-shaped electrical discharge of considerable volume, occurring above a THUNDERSTORM.

squall A sudden and violent gust of wind that can disappear as suddenly as it appeared. It is similar to a gust but has a longer duration. The World Meteorological Organization defines a squall as a sudden increase of wind speed by at least 16 knots (18 mph), rising to a wind speed of 22 knots (25 mph) or more, and lasting for at least one minute. It is commonly accompanied by rain, sleet, and snow.

squall line A line or band of active THUNDERSTORMS that produce strong winds and rain orientated along the line. The band of thunderstorms may only be a couple of kilometers wide resulting in sharp increases in wind speed followed by heavy rainfall.

SSP *See* subsatellite point.

SST *See* sea-surface temperature.

stability The atmospheric state in which an air parcel returns to its original position after an initial perturbation. Consequently, vertical air motion cannot be sustained in a stable atmosphere. It is sometimes referred to as a state of stable equilibrium, leading to stable air. *See* absolute stability.

stable air *See* absolute stability; stability.

stable boundary layer The stable layer of air that develops close to the Earth's surface and spreads upward where there is no tendency for the air to rise. It occurs typically at nighttime and the stability is produced by near-surface cooling under conditions dominated by outgoing radiation, known as radiative cooling, on calm clear nights, leading to the creation of temperature INVERSIONS. The boundary layer may also become stable for long periods over snow and ice surfaces due to the absence of turbulence, or be created by active subsiding air under anticyclones (high-pressure systems).

staccato lightning *See* beaded lightning.

stade (stadial) During an ICE AGE (glacial period), a cooling of conditions and secondary advance of glaciers constituting a subdivision (substage) of a glacial stage. *Compare* interstade.

stadial *See* stade.

standard atmosphere A hypothetical atmosphere chosen to represent the average distribution of pressure, temperature, and density in the real atmosphere. A number of standards are used. That defined by the International Civil Aeronautical Organization (ICAO) assumes a mean-sea-level temperature of 15°C, a standard sea-level pressure of 1013.25 hPa, and a temperature lapse rate of 0.65°C per 100 m up to 11 km (7 miles) in the atmosphere. The US standard atmosphere is identical to the ICAO standard atmosphere up to 32 km (20 miles) and then extends to 1000 km (620 miles).

static lapse rate *See* environmental lapse rate.

stationary front (quasi-stationary front) A narrow elongated frontal region that shows little or no horizontal progression over periods of several hours or occasionally more than a day since its last synoptic position. It is usually associated with extensive cloud and precipitation that can lead to serious flooding or severe snow storms. *See* front.

station model The internationally agreed method of plotting surface observations around the station 'circle'. The principal information represented, either as a number or a symbol, is DRY-BULB TEMPERATURE, DEW-POINT temperature, MEAN-SEA-LEVEL PRESSURE, PRESSURE TENDENCY, VISIBILITY, PRESENT WEATHER, PAST WEATHER, CLOUD TYPE, CLOUD AMOUNT, and CLOUD BASE height, and WIND DIRECTION and WIND SPEED (see diagram).

station plot The complete set of surface observations that are plotted in the form of a STATION MODEL.

statistical-dynamical model (SDM) A computer-based MODEL based on statistical averages of the climatic system. One such model is the *statistical-dynamical downscaling model* in which statistical relationships are developed between large-scale atmospheric parameters (such as sea-level atmospheric pressure) and localized parameters (such as precipitation or temperature).

steam fog A low thin fog that forms when cold air moves across warm water or a warm wet surface. Water evaporates into the rising air, increasing the dew point, and the air above becomes saturated; the water vapor condenses in the cold air to form visible water droplets, which resembles steam rising from the water. ARCTIC SEA SMOKE is a type of steam fog that occurs over open water in the Arctic.

steering The process by which the direction and speed of motion of large-scale pressure systems, such as FRONTAL CYCLONES, ANTICYCLONES, TROPICAL CYCLONES, and THUNDERSTORMS, is controlled. *See* steering current.

steering current (steering flow, steering wind) A circulation feature that is of im-

The station model. Observations from a weather station are plotted automatically or by hand onto the surface weather chart.

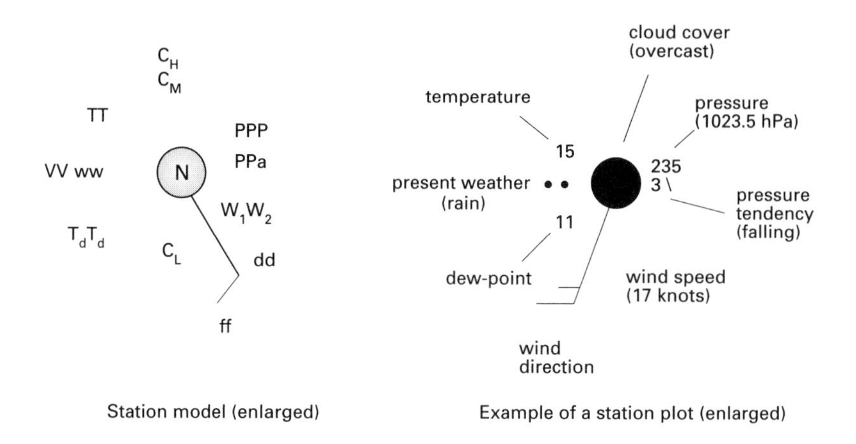

Station model (enlarged) Example of a station plot (enlarged)

TT	Temperature (top left of station) Degrees Celsius (degrees Fahrenheit in the US)	dd	Wind direction Line represents direction wind is blowing from
T_dT_d	Dew-point temperature (bottom left of station) Expressed in same units as dry-bulb temperature	PPP	Mean-sea-level pressure (top and right of station) Units are coded in hPa or mb, given to nearest tenth but without decimal point or leading 9 or 10, e.g. a pressure of 1034.5 hPa is represented by 345
N	Cloud cover (within circle) Assessed to the nearest eighth (okta) or nearest tenth (in the US)	PPa	Pressure tendency (right of station) Change in mean-sea-level pressure in previous 3 hours, represented by a value and a line indicating how changing
VV	Visibility (left of station) A 2-digit code, expressed in units of kilometers or miles		
ww	Present weather conditions (left of station, inside of VV) The weather when the observation is made, e.g. type of precipitation, shown as a symbol	C_H C_M C_L	Cloud Type (C_H and C_M above station, C_L below station) Represented by 9 symbols for each of 3 levels (27 in total)
ff	Wind speed each long barb = 10 knots each short barb = 5 knots ▲ = 50 knots	W_1W_2	Past weather conditions (bottom right of station)

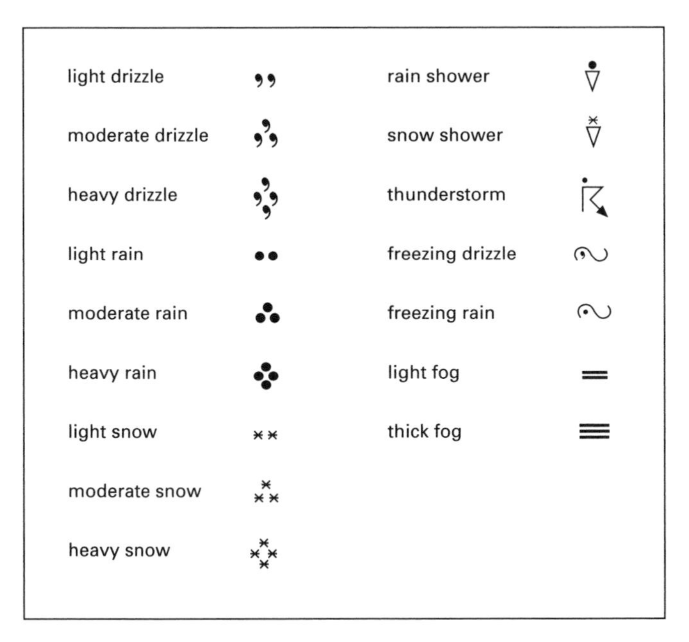

Common weather symbols

clear sky	5/8 covered
covered 1/8 or less, but not zero	6/8 covered
2/8 covered	7/8 covered
3/8 covered	sky completely covered
4/8 covered	sky obscured, e.g by fog

Cloud amount symbols

portance in determining the track of a range of different synoptic and smaller-scale pressure disturbances. TROPICAL CYCLONES, for example, are often embedded in the broad steering current of the TRADE WINDS. However, the wind at one particular level, or in one particular layer, is rarely the complete answer to how such systems are steered. *See also* steering.

Stefan–Boltzmann law The law, which applies only to black bodies, states total radiation in all directions from a black-body radiator is proportional to the fourth power of its thermodynamic (absolute) temperature. Originally formulated in 1879 by the Austrian physicist, Josef Stefan (1835–93), it was derived thermodynamically by another Austrian physicist, Ludwig Boltzmann (1844–1906), in 1889.

Stefan's law *See* Stefan–Boltzmann law.

stepped leader In a LIGHTNING stroke, a short discharge path that initiates the final lightning discharge; it occurs as an advancing region of ionization propagating in luminous steps or sections from the base of the cloud and is often faintly visible. The sequence of steps establishes the channel for the RETURN STROKE, during which the majority of the lightning current flows.

Stevenson screen *See* instrument shelter.

STJ *See* Subtropical Jet Stream.

storm 1. In general, a severe atmospheric phenomenon, for example a gale, thunderstorm, or rainstorm, probably leading to disruption of human life.
2. In SYNOPTIC METEOROLOGY, regions of low pressure carrying increases of wind speed, cloud, and precipitation.
3. Force 10 on the BEAUFORT SCALE.

storm-force wind A damaging wind, corresponding to either force 10 (storm) or force 11 (violent storm) on the BEAUFORT SCALE, with a wind speed of at least 43 knots (49 mph).

Storm Prediction Center An organization in the US that provides forecasts and watches for severe THUNDERSTORMS and TORNADOES across the contiguous US. In addition it monitors heavy rain, heavy snow, and fire weather across the States. Located in Norman, Oklahoma, it is a constituent body of both NOAA's NATIONAL WEATHER SERVICE (NWS) and the NATIONAL CENTERS FOR ENVIRONMENTAL PREDICTION (NCEP).

storm surge A deviation from the normal predicted sea level caused by wind velocity and/or low air pressure over the sea. Storm surges occur with a TROPICAL CYCLONE (HURRICANE or TYPHOON), and are also associated with high winds in advance of a severe storm or low-pressure system. The height of the storm surge is the difference between the observed level of the sea surface and the level that would have occurred in the absence of the storm. Along gently shelving coastlines a storm surge may be many meters high and can cause severe damage and danger to human life, with flooding far inland. Storm surges also cause concern for low-lying cities; the storm-surge threat resulted in the design and erection of the Thames Barrier, near London, UK.

storm tracks (depression tracks) A set of lines that express the route followed by traveling disturbances or cyclonic circulations (low-pressure systems or depressions). They are produced by, for example, plotting a mid-latitude cyclone's central (minimum) pressure every 6, 12, or 24 hours. Mid-latitude cyclones in the North Atlantic most commonly track from southwest to northeast, from off the northeast US and Maritime Provinces of Canada to N West Europe. They tend, on average, to reach their deepest central pressure near Iceland. In the North Pacific, a broadly similar track orientation occurs, running from off NE Asia/Japan toward Alaska, British Columbia, and the Pacific NorthWest. Cyclones tend, on average, to reach their deepest central pressure near the Aleutian Islands.

Strahler climate classification A genetic CLIMATE CLASSIFICATION system proposed by the American geologist and geographer Arthur N. Strahler (1918–) in 1951, with subsequent revisions. The classification is based on the main air masses and primary groups are identified: (1) LOW-LATITUDE CLIMATES, produced by tropical air masses; (2) MID-LATITUDE CLIMATES, produced by tropical and polar air masses; and (3) HIGH-LATITUDE CLIMATES, produced by polar and Arctic air masses. These main groupings are subdivided into 14 climatic types, based on temperature and precipitation variations, together with the undifferentiated highland climates. The classification has been widely used. *See also* Flohn climate classification; Köppen climate classification; Thornthwaite climate classification.

straight-line wind (straight wind) **1.** Any wind that does not show rotation. **2.** A THUNDERSTORM wind, associated with a gust front or a squall line, in which a straight-line wind originates from downdrafts in the MESOSCALE storm system and rapidly spreads out on reaching the ground. It can cause considerable damage to property, for example as experienced in the Mississippi and Ohio Valleys on April 15, 1994. Damage from such winds occurs in a straight line, unlike tornadoes in which the affected area is more circular in shape.

stratiform Of or relating to layered cloud at any level that exhibits little or no vertical development. It includes STRATUS, STRATOCUMULUS, NIMBOSTRATUS, ALTOSTRATUS, and CIRROSTRATUS. *Compare* cumuliform. *See also* cloud classification.

stratiformis A cloud species in which the cloud is spread out as an extensive and uniform horizontal sheet or layer. It applies to cirrocumulus, altocumulus, and stratocumulus. *See also* cloud classification.

strato- A prefix meaning 'layered'.

stratocumulus One of the cloud genera (*see* cloud classification); a type of widespread or patchy LOW CLOUD, often with blue sky visible between the cloud elements, in the form of a sheet that contrasts with STRATUS because its base has a lumpy surface (the cloud elements have a typical angular width of about 5°). It is gray or whiteish in color, formed of water droplets, and often associated with dull weather. It can be associated with light rain although in very cold conditions it can produce light snow or snow pellets.

stratopause The boundary that caps the STRATOSPHERE at approximately 50 km (30 miles) above the Earth's surface and divides it from the MESOSPHERE above. The stratopause is an isothermal layer.

stratosphere The atmospheric layer that lies above the TROPOSPHERE (from which it is separated by the TROPOPAUSE) at an altitude of about 10 km (6 miles); it can extend to around 50 km (30 miles) above the Earth's surface to the MESOSPHERE. The air temperature remains constant up to about 25 km (15 miles) but above this altitude the stratosphere is characterized by increasing temperature with height to about 200 kelvins at the STRATOPAUSE creating a temperature inversion above the troposphere. Temperatures in the stratosphere tend to be high due to the high concentrations of OZONE that absorbs incoming ultraviolet radiation converting solar energy to kinetic energy: the temperature increases with ozone concentration.

The increase in temperature with altitude in the stratosphere has a stabilizing effect on atmospheric conditions, in marked contrast to the troposphere below in which colder air overlies warmer air.

Stratospheric Processes and their Role in Climate (SPARC) A core project of the WORLD CLIMATE RESEARCH PROGRAM (WCRP) that focuses on the interaction of dynamical, chemical, and radiative processes within the stratosphere. The four main areas of study and research are the influence on climate of the stratosphere; processes involved with ozone levels; stratospheric variability; and changes in ultraviolet radiation.

stratus One of the cloud genera (*see* cloud classification); a type of extensive LOW CLOUD in the form of a sheet with a flat base that is usually of a uniform gray tone. Stratus is usually composed of small water droplets and can often be low enough to obscure the upper parts of hills (*see* hill fog) and tall buildings. The lifting of a thick layer of fog can also result in stratus cloud. DRIZZLE and light RAIN occasionally fall from such cloud or ICE PRISMS and SNOW GRAINS in very cold weather.

streak lightning A CLOUD-TO-GROUND DISCHARGE in which the forked appearance is absent and there is a single, often apparently straight, lightning channel. *Compare* forked lightning.

streamline A smoothly curved line that is everywhere parallel to the local direction of the wind vector. A streamline analysis is thus a 'snapshot' of the atmospheric flow's direction field and can highlight zones of CONFLUENCE and DIFLUENCE, for example.

striations Fine grooves or channels in a cloud formation.

Stüve diagram (pseudoadiabatic chart) A type of THERMODYNAMIC DIAGRAM that uses straight lines to represent PRESSURE, TEMPERATURE, and the DRY ADIABATS. It was developed in about 1927 by G. Stüve and is commonly used in the US. *See also* skew-*T* log-*P* diagram; tephigram.

subarctic Denoting those latitudes that lie directly south of the ARCTIC Circle.

Sub-Arctic Current *See* Aleutian Current.

subcloud layer The layer between CLOUD BASE and the surface, normally referring only to the layer below LOW CLOUD.

subhumid In climatology, the zones between the humid zones, in which there is plentiful precipitation, and the arid zones.

sublimation The process in which a substance changes phase from a solid to a gas with no intermediate liquid phase. In meteorology, it refers generally to water vapor evaporated directly from an ice surface, or the reverse process in which water vapor is deposited directly onto an ice surface, for example in the formation of HOAR FROST.

subsatellite point (SSP) The point on the Earth's surface that is vertically below a satellite. It is where an imaginary straight line that joins the satellite to the center of the Earth intersects the surface.

subsidence The process of sinking undergone by a mass of air, usually through some considerable depth of, for example, the TROPOSPHERE. Air can sink from the upper troposphere down to lower tropospheric levels in ANTICYCLONES (highs). Huge volumes of air move slowly down at a rate of perhaps 1–2 centimeters per second. As the air subsides it is compressed and warms adiabatically so that it is often extremely dry and cloud-free. The warming can culminate in a SUBSIDENCE INVERSION at about 1 km (0.6 mile) above the surface. In the lower troposphere it is associated with a layer of DIVERGENCE and increased stability, while in the upper troposphere it is linked to CONVERGENCE.

subsidence inversion A tropospheric layer some few hundred meters deep that has a marked increase of DRY-BULB TEMPERATURE but a marked decrease of DEW-POINT temperature (i.e. it is a dry INVERSION) and is produced by the deep SUBSIDENCE and adiabatic warming that occurs in many ANTICYCLONES (highs). Its base is typically about 1 km (0.6 mile) above the Earth's surface; the layer between it and the surface often has CUMULUS or STRATOCUMULUS clouds.

subsun (undersun) An optical phenomenon, the most commonly occurring type of subhorizon HALO, which appears as a white spot of light, vertically elongated. below the horizon. It is produced by the direct reflection of the Sun by clouds containing ice crystals, typically plate crystals in which the faces are horizontal.

subsynoptic low *See* mesolow.

subtropical Denoting the latitudinal zones that lie between the tropics and the temperate zones. Although this usage is imprecise the word is often taken to refer to those zones poleward of the Tropics of Cancer (northern hemisphere) and Capricorn (southern hemisphere) extending to about 35–40°N and S. Subtropical is used in a number of climate classifications to designate regions in which no month has a mean temperature below 6°C (43°F).

subtropical anticyclone (subtropical high) An extensive semi-permanent ANTICYCLONE that is found over the subtropical oceans (i.e. between 35° and 40° latitude). The subtropical anticyclones are present throughout the year, shifting poleward in the summer and equatorward in the winter. The AZORES HIGH and BERMUDA HIGH over the Atlantic Ocean, and the NORTH PACIFIC HIGH are examples in the northern hermisphere, with the South Pacific and South Indian Ocean highs in the southern hemisphere. The subtropical cyclones are typified by deep sinking motion, dry conditions, and either clear skies or widespread stratiform cloud.

Subtropical Jet Stream (STJ) A westerly upper tropospheric JET STREAM located on the poleward flanks of the HADLEY CELLS in each hemisphere, directly above the SUBTROPICAL ANTICYCLONES (highs) at an elevation of some 13 km (8 miles). It is relatively fixed in latitude from day-to-day but migrates seasonally in association with the north–south shift of the Hadley cells.

suction vortex A small intense vortex found in addition to the central 'parent' vortex of a TORNADO. This type of vortex is embedded in the tornadic circulation and rotates around the parent vortex.

sukhovey (sukhovei) A hot dry SCIROCCO-type wind that affects southern Siberia and Central Asia, especially the steppes of Ukraine and Kazakhstan. Often carrying vast quantities of dust, the sukhovey wind has a desiccating effect, often eliminating cloud cover and allowing intense solar radiation to reach the Earth's surface. It can reduce relative humidity at the ground to very low values both by day and night.

sulfate aerosol Highly reflective particles in the atmosphere generated over continental areas, mainly from the combustion of sulfurous fossil fuels and also from volcanic eruptions. They come ultimately from the release of SULFUR DIOXIDE and sulfur trioxide, which react with sulfur compounds to transform into the minute aerosol particles. Over oceans, sulfate aerosols originate from the DIMETHYLSULFIDE (DMS) produced by phytoplankton. Sulfate aerosol particles have RESIDENCE TIMES of a few days and are therefore not widely well mixed in the atmosphere. They act as CLOUD CONDENSATION NUCLEI.

sulfur dioxide (SO_2) A compound produced primarily by the combustion of sulfurous fossil fuels, but also associated with the natural processes of volcanic eruptions and ocean spray. In the atmosphere it can combine with other sulfur compounds in complex chemical reactions that produce SULFATE AEROSOLS. In moist air it can also combine with water vapor to produce sulfuric acid, and consequently ACID RAIN.

sulfur hexafluoride (SF_6) A colorless gas, a compound of sulfur and fluorine, used as an insulator for circuit breakers and other electrical equipment. It has an extremely low atmospheric concentration enabling it to be used as an atmospheric tracer gas. For example, it has been added to the waste gases leaving power plants so that it can be monitored by instrumented aircraft to assess how far atmospheric pollution from power plants can reach. Sulfur hexafluoride is a powerful GREENHOUSE GAS and, although not emitted in large quantities, it has a lifetime of 3200 years.

sumatra The local name for a squall in the Malacca Strait (between peninsular Malaysia and the island of Sumatra) when a sudden change of wind direction from southerly is accompanied by a rise in wind speed and a high bank of cumulus or cumulonimbus cloud. A sumatra usually occurs at night, commonly between April and November.

summer The SEASON of the year usually designated as June to August in the middle-latitude northern hemisphere and December to February in the southern hemisphere. Temperatures reach a maximum with August often the warmest month. Precipitation is predominantly convectional, with frequent thunderstorms, which may be accompanied by hail, and SUPERCELL storms may develop in continental interiors. Receipt of solar radiation in June can exceed that of the tropics due to increased day length.

Sun dog *See* parhelion.

Sun pillar An optical phenomenon, a type of HALO, in which a luminous streak or pillar appears above and/or below the Sun, especially just before sunset or after sunrise. The pillar can extend to an angular distance of about 20° from the Sun. It results from the reflection of the Sun's light by horizontal ice plates within clouds or ice fog, the flat faces of which are aligned nearly parallel to the ground surface.

sunshine DIRECT RADIATION from the Sun received at the surface of the Earth (as opposed to DIFFUSE RADIATION). The standard measure is of 'bright' sunshine that is strong enough to burn a trace in the daily sunshine card of a SUNSHINE RECORDER. The Sun is too low in the sky to burn the card within about 30 minutes of sunup and sundown.

sunshine recorder An instrument used for measuring the duration of bright sunshine. *See* Campbell–Stokes sunshine recorder.

sunspot A relatively dark cool region of the disk of the Sun with temperatures around 1500 K lower than the surrounding PHOTOSPHERE. This 'umbra' is surrounded by an outer hotter 'penumbra'. The index of 'relative sunspot number', which vary from less than ten in quiet sun periods to over 150 in active years, was introduced in 1848 by Rudolf Wolf (1816–93) of the Zurich Observatory.

Sun-synchronous satellite A satellite that circles the Earth, in near-polar orbit, in such a way that it has an overhead or near-overhead pass above any point on the Earth's surface at or about the same local time each day, throughout the year. In order to achieve this, the plane of its orbit must precess by about 1° every day, to keep pace with the Earth's rotation round the Sun over the year. As a result, the satellite's orbital plane appears to be fixed with respect to the Sun. *See* polar-orbiting satellite. *Compare* geostationary satellite.

Supan's temperature zones An early climate classification in which the climatic regions of the world were based on differences in mean annual temperatures. It was proposed in 1896 by the Austrian climatologist Alexander Supan.

superadiabatic lapse rate A LAPSE RATE that is greater than the DRY ADIABATIC LAPSE RATE of about 10°C km^{-1}. This lapse rate does not occur within the free atmosphere but can occur directly above the surface when it is strongly heated by solar radiation, for example over arid areas in summer. The air has ABSOLUTE INSTABILITY.

supercell A violent rotating thunderstorm with a large rotating updraft known as a MESOCYCLONE. Supercells produce severe weather including hail, heavy rainfall leading to flash floods, severe winds, and frequent lightning; they also produce the most destructive and dangerous tornadoes. The thunderstorm can last up to 6 hours. Supercell formation occurs in environments that have large vertical wind shear and instability.

supercooled cloud Cloud composed of water droplets in which the temperature is below the freezing point of water, 0°C (32°F). Such droplets freeze instantly if they come into contact with ice crystals – this process is critical in the formation of precipitation in most clouds. If freezing nuclei are absent (i.e. in pure air) supercooled water droplets can remain liquid at temperatures down to –40°C (–40°F).

supercooling The cooling of water to below its normal FREEZING POINT of 0°C (32°F) but it remains in the liquid state. The supercooling of water droplets is common in clouds that extend beyond the 0°C isotherm. The supercooled water droplets remain unfrozen if no FREEZING NUCLEI are present; this is possible to temperatures as low as –40°C (–40°F). The inclusion of a seed crystal into supercooled liquid causes it to freeze immediately as the release of the heat of fusion raises the temperature rapidly to freezing point.

superior mirage *See* mirage.

supersaturation A condition in which the atmosphere contains more water vapor than is needed to produce saturation with respect to a flat surface of pure water or ice, and RELATIVE HUMIDITY is greater than 100%. Such a condition may occur when there is no surface upon which condensation can take place (i.e. clean air, in which there is an absence of CLOUD CONDENSATION NUCLEI).

surface-based convection A bubble of air that forms and grows in response to surface heating, moving up through the lower TROPOSPHERE. The height of the bubbles depends partly on the intensity of the heating by the land or sea surface. Dry bubbles are also known as THERMALS, while those with condensed water droplets are seen as CUMULIFORM clouds.

surface boundary layer That part of the PLANETARY BOUNDARY LAYER that lies directly above the ground surface and in which friction effects are generally constant, as opposed to decreasing with height. It extends up from the surface to about 10 m (33 ft) and within this layer insolation and radiational cooling are at their strongest.

surface energy budget *See* energy budget.

surface inversion An INVERSION of temperature in the atmosphere directly above the Earth's surface so that the coolest temperatures are closest to the ground. This can result, for example, from the influx of relatively cold sea air over the ground surface via an onshore breeze. It is also often caused by the overnight radiative cooling of surface air to form a RADIATION INVERSION.

surface weather chart A synoptic chart that portrays one or more aspects of the weather at the Earth's surface for a given geographic area. The most common form is the MEAN-SEA-LEVEL PRESSURE analysis that consists of ISOBARS and also usually FRONTS.

surface weather observations All the visual and instrumental observations that are plotted as the STATION MODEL plus others that are not, such as daily SUNSHINE total.

surface wind The velocity and directional characteristics of the wind flow close to the ground, usually measured at the standard height of 10 m (33 ft). As the airflow experiences frictional retardation in the planetary boundary layer, the surface wind velocity is usually lower than the GEOSTROPHIC WIND or GRADIENT WIND velocities recorded in the free atmosphere.

surge *See* storm surge.

SWEAT index *See* severe weather threat index.

synoptic Denoting the meteorological conditions at a particular point in time and over an extensive area from observations made simultaneously (the word means literally viewed simultaneously).

synoptic chart A weather map for a particular point in time that portrays features having a SYNOPTIC SCALE. Synoptic charts represent features both at MEAN SEA LEVEL and in the upper air.

synoptic climatology The study of CLIMATE over a large related geographic area. In this form of study, synoptic observations are plotted on charts for the purpose of weather analysis and forecasting.

synoptic code The internationally agreed method of representing the range of SYNOPTIC observations made at the surface, from SOUNDINGS, etc. All the data that surround the STATION PLOT are transmitted as groups of five digits in the synoptic code.

synoptic meteorology The branch of METEOROLOGY that is specifically concerned with atmospheric phenomena and systems associated with weather throughout the world. The name originates from the synoptic method, in which observations and atmospheric conditions are plotted on a SYNOPTIC CHART. Data is gathered from the observations from a network of synoptic stations, ground-based radars, balloon-borne radiosondes, aircraft, ships, and meteorological satellites. From this data forecasts of the weather are produced using such methods as numerical weather forecasting, which employs complex computer models.

synoptic scale A particular scale of weather phenomenon; synoptic-scale features are those that are portrayed on a large-scale MEAN-SEA-LEVEL PRESSURE map, such as ANTICYCLONES (HIGHS), LOWS, RIDGES, and TROUGHS. They are defined by analyzing surface weather observations that have a spacing of some few tens of kilometers or more, and so are typically many hundreds or a thousand kilometers across, and last a few days.

synoptic symbols The internationally agreed symbols that are used to define the whole range of PRESENT and PAST WEATHER on a SYNOPTIC CHART.

synoptic weather map *See* synoptic chart.

synthetic aperture radar (SAR) A radar system used in aircraft and on board polar- or lower-angle orbiting satellites, that takes advantage of its rapid speed through space to create a large synthetic aperture (to improve spatial resolution of the image). The size of the synthetic aperture is that created during the period of time that the satellite's radar is emitting a signal down to the surface. Synthetic aperture radar is used to map the roughness of the surface in all weathers so that useful environmental parameters can be deduced for operational purposes.

system In meteorology, some type of weather-producing feature, such as an ANTICYCLONE (high) or LOW.

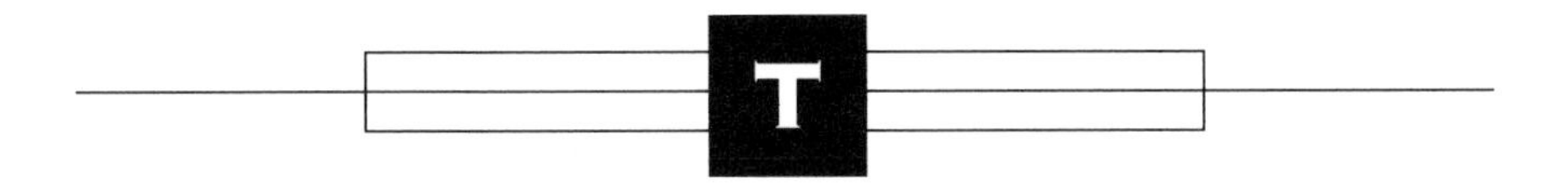

TAF (Terminal Aerodrome Forecast) A forecast transmitted to aviation centers and aircraft to include details of the VISIBILITY, CLOUD HEIGHT, and WEATHER pertinent to the landing conditions at the destination airport.

tail cloud A small feature with a tail-like appearance associated with a TORNADO. It has a horizontal base and consists of fragments of cloud that move from an area of PRECIPITATION toward the WALL CLOUD, which is most often a non-precipitating feature. Where the tail cloud joins the wall cloud there is rapid updraft.

talwind *See* up-valley wind.

teleconnection A connection or linkage between anomalies or variables in the atmosphere and/or ocean over widely spaced geographical areas, frequently on a planetary scale. For example, the teleconnections associated with the EL NIÑO SOUTHERN OSCILLATION (ENSO) and with the NORTH ATLANTIC OSCILLATION (NAO).

temperate climate A climate of the middle latitudes, which is influenced by both tropical and polar air masses. *See also* climate classification.

temperate zone One of the three historic climatic zones, based on temperature and sunshine characteristics, into which the world was divided by the classical Greeks; the *northern temperate zone* lies between the Tropic of Cancer (23°27′ N) and the Arctic Circle (66°32′ N), while the *southern temperate zone* lies between the Tropic of Capricorn (23°27′ S) and the Antarctic Circle (66°32′ S). The temperate zone has distinct summer and winter seasons. *See also* frigid zone; torrid zone.

temperature The degree of hotness of a substance (solid, liquid, or gas), which is related to the average kinetic energy of the atoms or molecules of the substance. It is one of the most important elements in climatology and meteorology. In meteorology, temperature is recorded at stations from thermometers housed within instrument shelters; THERMISTORS are used in automatic weather stations and in upper-level radiosonde measurements. A number of temperature scales are in use: the CELSIUS SCALE is generally used in meteorology although the FAHRENHEIT SCALE is still extensively used in the US; the KELVIN TEMPERATURE is more widely used in scientific studies.

temperature-efficiency index *See* thermal-efficiency index.

temperature gradient *See* lapse rate.

temperature–humidity index (THI) An index based on the combination of temperature and humidity as a measure of the degree of comfort or discomfort experienced by an individual in outdoor conditions in warm weather. Developed by the US National Weather Service, the index gives a single numerical value in the range 70–80. This is an EFFECTIVE TEMPERATURE based on air temperature and humidity, and the index is calculated using the following formula:

$$\text{THI} = 15 + 0.4(T_d + T_w),$$

where T_d and T_w are the dry- and wet-bulb temperatures respectively. Most people are comfortable with an index below 70, and very uncomfortable when it is above 80.

temperature inversion *See* inversion.

temperature inversion layer *See* inversion layer.

temperature range The difference between the maximum and minimum temperatures, or the highest and lowest mean temperatures measured during a specified time period, for example DIURNAL VARIATION or seasonal (annual) variation.

tephigram (te-phi-gram) A type of THERMODYNAMIC DIAGRAM that was developed in 1922 by the British meteorologist Sir William Napier Shaw, who introduced it into the UK Meteorological Office (UKMO). The name is derived from the rectangular Cartesian coordinates used: temperature and entropy. Strictly speaking, the tephigram should be arranged so that temperature lies along the x-axis, and the DRY ADIABATS along the y-axis. In reality, a right-rotated form of the diagram has always been used since the late 1940s. The lines of a tephigram are: ISOBARS (almost horizontal); ISOTHERMS (slope from bottom left to top right); DRY ADIABATS (slope from top left to bottom right); SATURATED ADIABATS (curved lines); SATURATION MIXING RATIO lines (slope from bottom left to top right). The tephigram is the only type of thermodynamic diagram commonly used in the UK. *See also* skew-T log-P diagram; Stüve diagram.

Terminal Aerodrome Forecast *See* TAF.

terrestrial radiation *See* longwave radiation.

thaw The melting of snow or ice at the Earth's surface due to a rise in temperature above 0°C (32°F).

thermal A localized but highly buoyant area of rising air in the atmosphere. Thermals can be generated either by intense heating of the ground surface by the Sun, or by a steep ENVIRONMENTAL LAPSE RATE (ELR), i.e. a superadiabatic temperature profile. A thermal may even reach the CONVECTIVE CONDENSATION LEVEL if its vertical motion is sufficient, leading to CUMULUS cloud formation and possibly precipitation.

thermal advection The horizontal transfer of heat in the atmosphere by the wind. This is distinct from vertical advection, which is usually called convection, and heat that is transferred from one place to another as a result of non-advective processes, such as radiative transfer.

thermal depression *See* heat low.

thermal-efficiency index (temperature-efficiency index) **1.** An index devised by the American climatologist Charles Warren Thornthwaite (1889–1963) for delineating climatic types in his empiric climate classification (*see* Thornthwaite climate classification). The index was used together with an expression for precipitation efficiency to produce the classification.
2. (TE) In ecology, an index that is a measure of the long-range effectiveness of temperature in contributing to plant growth.

thermal equator The zone around the Earth in which the highest temperatures occur, either the mean annual temperature or the temperature at a given time. The mean annual thermal equator occurs at about 5°N, rather than along the actual geographical EQUATOR; one factor for this northward location is the influence of the distribution of the continents. During the course of the year the position of the thermal equator migrates northward to reach about 20°N in the northern hemisphere summer, and southward to around 5°S in the southern hemisphere summer.

thermal infrared The region of the infrared spectrum between about 10.5 and 12.5 μm that is sensed routinely by METEOROLOGICAL SATELLITES. The strength of the radiation emitted to space in this waveband is directly proportional to the temperature of the emitting surface; this is true for cloud, sea, or land surface, for example. This thermal radiation can therefore be converted from a flux measured in watts per square meter at the satellite's sensor, to a temperature of the cloud top or sea surface for each PIXEL. The incessant flow of this radiation into space allows all-day coverage to be achieved, which is very useful operationally.

thermal low *See* heat low.

thermal wind The theoretical wind that blows parallel to the isohypses on a thickness chart, the most commonly drawn chart being that of 1000–500 hPa thickness. The thickness (meters) of any atmospheric layer is proportional to the mean temperature of that layer. The thermal wind on this chart (see diagram) is defined as the VECTOR which quantifies the difference, and hence the degree of WIND SHEAR, between the GEOSTROPHIC WIND at the bottom (1000 hPa) and top (500 hPa) of the atmospheric layer. The thermal wind 'blows' between cooler air (lower thicknesses) to its left and warmer air (higher thicknesses) to its right in the northern hemisphere, and vice versa in the southern hemisphere.

thermistor A resistor in which electrical resistance changes with temperature so that it can be used as the basis for a THERMOMETER. It is made from a mixture of metallic oxides that have been fused together.

thermocline A stable layer of water in an ocean in which vertical mixing is inhibited and the temperature gradient (cooling with depth) is at a maximum (greater than in the warmer layer above it and the cooler layer below). A permanent thermocline can be found in the lower and middle latitudes at a depth of several hundred meters, extending down to over 1000 meters (3280 ft). Seasonal thermoclines occur much closer to the surface during summer months due to the action of solar heating; these are only temporary structures and are destroyed by reduced insolation and increased turbulence during winter.

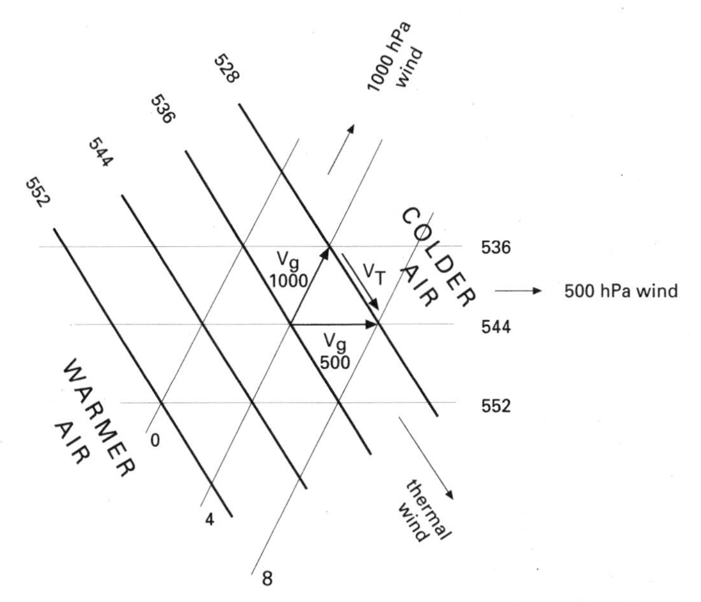

Thermal wind: 1000 hPa and 500 hPa geostrophic wind components *Vg* and the resultant thermal wind V_T blowing parallel to the 1000–500 hPa thickness lines. All heights in 10s of meters.

thermocouple An instrument used to measure temperature. This device, a thermoelectric sensor, consists of two wires composed of dissimilar metals (e.g. copper and iron), joined at each end (junction) to form a circuit. One junction is kept at a constant temperature while the other measures the unknown temperature creating an electromotive force. This emf is related to the temperature difference and is therefore a measure of the temperature difference between the two junctions. *See also* thermometer.

thermodynamic diagram (aerological diagram, adiabatic chart) A diagram used in operational meteorology showing the vertical pressure, temperature, and humidity structure of the atmosphere. Five sets of lines appear on a thermodynamic diagram: ISOBARS; ISOTHERMS; DRY ADIABATS; SATURATED ADIABATS; and SATURATION MIXING RATIO lines. Variations in the DRY-BULB TEMPERATURE (i.e. the environmental lapse rate) and DEW-POINT temperature with height are plotted on to a thermodynamic diagram. A number of types of diagrams are used, all based on the same principles, but the three in most widespread use are the TEPHIGRAM and, especially in the US, the SKEW-T LOG-P and STÜVE DIAGRAMS. A thermodynamic diagram can be viewed as a 'meteorological ready-reckoner' from which derived variables, such as RELATIVE HUMIDITY and air STABILITY, can be quickly determined without having to result to complex and often time-consuming numerical calculations. *See also* absolute instability; absolute stability.

thermodynamic equation An equation representing the law of conservation of energy. The equation states that an air mass will not change its temperature unless heat is added or removed or through changes in surrounding pressure associated with changes in its vertical position. The equation can take the following form:

$$\Delta T = -(g/C_p)\Delta z + \Delta Q_H/C_p.M_{air},$$

where ΔT is the change in temperature of an air parcel with mass M_{air}; $\Delta Q_H/M_{air}$ represents heat input, which can be caused by latent heat through condensation, radiative heating, and heat from chemical reactions; Δz is the change in height; g is the acceleration of free fall; and C_p is the specific heat capacity for moist air.

thermodynamics The branch of physics concerned with the study of heat and its transformation into other types of energy, and vice-versa.

thermodynamic temperature Formerly called *absolute temperature*, which was measured on an absolute scale of temperature, e.g. the Kelvin scale of temperature. Now, however, thermodynamic temperature is regarded as a basic physical property and is usually measured in units called KELVINS. The concept of a temperature scale is now restricted to the *International Practical Temperature Scale*, which is defined by 16 fixed points, so that this scale conforms as closely as possible to the thermodynamic temperature.

thermogram *See* thermograph.

thermograph A self-recording THERMOMETER. It consists of a bimetallic coil, which distorts with changes in ambient temperature; an arm attached to the coil holds a pen that records the distortions on a paper chart, the *thermogram*.

thermohaline circulation A circulation or flow within the ocean that is dependent on differences in water density, governed by temperature and salinity. It has a large vertical component of flow and is responsible for the mixing of deep water in the oceans. For example, deep water is carried from the North Atlantic to the Pacific, where it then upwells and is transported back to the downwelling regions in the North Atlantic (*see* Atlantic conveyor). The sinking of water in the north acts to drag warmer water from the south.

thermometer An instrument used for measuring temperature. The most commonly used type in meteorology is the *liquid-in-glass thermometer*; the liquid used is often mercury but as the freezing point for this element is –38.9°C (–38°F) alcohol, which has a freezing point of −114.4°C (−174°F), is also used, especially for low-temperature measurement and in the MINIMUM THERMOMETER. The liquid, which expands with a rise in temperature, is enclosed in a sealed glass tube; the height of the column measures the temperature, which is read on a graduated scale to the nearest 0.1°C or 0.1°F. Electrical instruments or sensors are becoming more widely used and are of necessity used in AUTOMATIC WEATHER STATIONS, RADIOSONDES, and other automated locations. The sensors include THERMISTORS, THERMOCOUPLES, and resistance temperature detectors (RTD), which work on the basis of the resistance changes of certain metals, such as platinum and copper, to provide accurate measurements. *See also* black-bulb thermometer; dry-bulb thermometer; grass minimum thermometer; maximum and minimum thermometer; maximum thermometer; mercury-in-glass thermometer; wet-bulb thermometer.

thermopile An instrument used for measuring radiation. It consists of a number of THERMOCOUPLES joined in series.

thermosphere The layer of the atmosphere lying above the MESOSPHERE (from which it is separated by the MESOPAUSE) and extending from about 85 km (50 miles) above the Earth's surface. In this layer, temperatures increase with altitude due to the absorption of solar radiation by molecular oxygen.

theta-e *See* equivalent potential temperature.

theta-e ridge An axis along which relatively high values of EQUIVALENT POTENTIAL TEMPERATURE (theta-e) occur. Intense rainfall frequently occurs near to or just ahead of a theta-e ridge.

THI *See* temperature humidity index.

thickness The geopotential height difference between two specified ISOBARIC SURFACES, for example the 1000–500 hPa thickness. The density of air decreases as temperature increases and as a result the thickness between two pressure levels alters with temperature. As the air becomes warmer it expands and therefore the distance between two pressure levels must increase to accommodate the same mass.

Thiessen polygon method A method used in meteorology to assign areal significance to point rainfall values. A polygon is drawn around a precise point within a scatter of points. Polygons are generated from a set of points mathematically defined by the perpendicular bisectors of the lines between all points. For example, Thiessen polygons can be drawn around a network of rain gauges. Each rain gauge will be surrounded by a polygon, drawn so that the boundary is closest to the gauge contained within it than to any other gauge. The area within the polygon is considered to display uniform characteristics, i.e. the area within the polygon, regardless of its size, is assigned the rainfall amount collected by the gauge within it.

Thornthwaite climate classification An empiric climate classification proposed originally in 1931 and subsequently revised in 1948 by the American climatologist, Charles Warren Thornthwaite (1889–1963). The original (1931) classification was based on the PRECIPITATION-EFFICIENCY INDEX and the THERMAL-EFFICIENCY INDEX as a method to specify the climatic effects on vegetation. Five climatic regions with associated vegetation were identified according to the index: > 128 wet (rain forest); 64–127, humid (forest); 32–63, subhumid (grassland); 16–31, semi-arid (steppe); and < 16, arid (desert). Thornthwaite's second classification (1948) is based on an index of POTENTIAL EVAPOTRANSPIRATION (PE) and the moisture budget. In this system the thermal efficiency is derived from the PE value. The system has been widely applied but is complex with a large number of climatic regions and has not achieved the popularity of the KÖPPEN CLIMATE CLASSIFICATION. *See also* Flohn climate classification; Strahler climate classification.

Thornthwaite's index of potential evapotranspiration An index to define POTENTIAL EVAPOTRANSPIRATION that was devised by the American climatologist Charles Warren Thornthwaite in 1948. It was calculated from measurements of air temperature and day length and used in Thornthwaite's second (1948) climate classification. *See* Thornthwaite climate classification.

three-cell model A basic conceptualization of the average large-scale circulation features that occur in the north–south vertical plane in each hemisphere. The three cells are the HADLEY CELL, FERREL CELL, and the POLAR CELL. *See also* general circulation.

threshold wind speed The critical wind speed required to entrain material (e.g. dust) into the atmosphere. Particles having diameters of about 40 μm are most easily entrained into the lower atmosphere; such particles possess the lowest threshold wind speeds. Larger particles are difficult to lift because of their weight; so too are very small particles because they have a greater surface area per unit volume and cohesive forces tend to be greater for smaller particles.

throughfall The PRECIPITATION that passes through a cover of vegetation to fall to the Earth's surface.

thunder The audible acoustic shock wave that follows a LIGHTNING flash. It is caused by the sudden heating and expansion of air along the lightning channel, which propagates outward. Thunder travels at the speed of sound (300 m s^{-1}) so it is heard after a lightning flash has been seen; it is seldom heard at distances beyond 16 km (10 miles). The rumbling of thunder results from atmospheric refraction of the sound produced; a sharp crack is more commonly heard directly beneath a thundercloud.

thunderstorm A severe storm produced by a CUMULONIMBUS cloud in which precipitation (especially heavy rain and/or hail) and updrafts are accompanied by THUNDER and LIGHTNING. A thunderstorm has a typical duration of about one hour, during which electrification processes are sufficiently rapid for many lightning discharges to occur. Thunderstorms are of considerable vertical scale, typically extending from the surface to the tropopause, with an upper ANVIL shape comprising ice crystals sheared in upper level winds. Updrafts and downdrafts can have sharp wind shear associated with them. Electrification probably occurs as a result of collisions between rising ice crystals and falling soft hail (GRAUPEL) particles, carrying negative charge downward in the cloud and positive charge upward into the anvil region. Electrification continues until the electric fields are sufficiently intense to break down the insulating properties of air, when a LEADER is propagated followed by a lightning discharge. *See also* air-mass thunderstorm; multicell storm; supercell.

tidal wave *See* tsunami.

tide The periodic rise and fall of sea level within the Earth's oceans, which results from the relative gravitational attraction of the Moon, and to a lesser extent the Sun, and the Earth. *Spring tides* occur when the gravitational pull of both the Sun and Moon coincide (around new and full Moon) and produce the maximum tidal range between high and low water. *Neap tides* occur when the Sun and Moon are apparently at right angles to each other (first and last quarters). The pattern of *semi-diurnal tides*, in which two high waters and two low waters occur during a tidal day of 24 hrs and 50 minutes is the most common; *diurnal tides* are one high and one low water during a tidal day.

tierra The series of climatic zones within South America, Central America, and Mexico that are defined by altitude. The *tierra helada* is the zone of mountain peaks; the *tierra fria* is the upland climatic zone above 1800 m (6000 ft) in which frost occurs (it is also the upper limit for tree growth); the *tierra templada*, extends between 900 and 1800 m (3000–6000 ft), and the *tierra caliente* is the coastal zone extending from the coast up to 900 m (3000 ft).

tilted storm A type of thunderstorm or cloud tower that is tilted, indicating vertical wind shear and conditions suitable for severe storm intensification.

timberline (treeline) The limit, at high latitudes or high altitudes, beyond which trees do not grow.

timescale *See* Coordinated Universal Time; Greenwich Mean Time; Universal Time; Zulu Time.

tipping-bucket rain gauge A recording RAIN GAUGE in which the delivery tube consists of a bucket divided into two equal compartments, each designed to collect a specific amount of rain. The bucket is balanced on a seesaw so that as one compartment fills to a pre-determined amount, usually 0.3 mm (0.01 in), it tips emptying the water, and the other compartment is positioned in its place to collect the water. Each time a tip occurs an electronic signal is sent to a recorder giving a continuous record of rainfall against time. Some versions of the gauge incorporate a heating element to melt solid precipitation, such as snow. *See also* weighing rain gauge.

TIROS (Television and Infrared Observation Satellite) The first operational meteorological satellite series developed in the US. The first of the series, TIROS–1, was launched on April 1, 1960 to open the era of METEOROLOGICAL SATELLITE observations. It carried two TV cameras, one with high and one with low resolution. The term TIROS is still used occasionally as synonymous with the NOAA polar-orbiting operational satellites (POES).

TIROS operational vertical sounder (TOVS) An instrument consisting of a system of three sensors: the microwave sounding unit (MSU), the High resolution Infrared Radiation Sounder (HIRS), and the Stratopheric Sounding Unit (SSU). It is capable of instantaneously sensing the changing strength of emitted radiation from different levels within the atmosphere. It does this by measuring these signals in a large number of channels or BANDS.

TOGA *See* Tropical Ocean-Global Atmosphere Program.

topoclimatology The study of the influence of topography on CLIMATE.

topography The landforms and features of the surface of the Earth, including relief, vegetation, and human features, such as urban areas and communications. *See also* topoclimatology.

tornadic vortex signature (TVS) A significant distribution of radial velocity components mapped by a DOPPLER RADAR that indicates the possible development of a tornado. The pattern is one in which a strong component away from the radar is juxtaposed horizontally, over a short distance, with a component in which there is strong flow toward the radar. The horizontal WIND SHEAR is extremely large and CYCLONIC. In the US, tornado warnings are issued by the National Weather Service (NWS) based on the strength of the tornadic vortex signature.

tornado A violent rotating column of air extending between a convective thunderstorm cloud and the ground, averaging tens to hundreds of meters in diameter, with extreme vertical updrafts and low pressures in the center. Tornadoes are frequently able to levitate heavy objects and cause widespread devastation in the US, which has a greater incidence of tornadoes than anywhere else in the world; around 1000 annually. Tornadoes form in conditions similar to thunderstorms, and indeed are usually seen beneath thunderclouds, reaching near to the ground. Wind speeds in excess of 500 km per hour have been identified in US tornadoes using Doppler radar and the central pressure can be very low. In the UK around 30 tornadoes occur annually though they are seldom reported;

they rarely, if ever, reach more than F3 strength. *See* Fujita Tornado Intensity Scale. *See also* funnel cloud; wall cloud.

tornado family Conditions for tornado development cover a wide area with numerous thunder cells, it is not unusual for groups or families of tornadoes to occur within a small area or timescale, such as tornado alley in the US.

torrid zone One of the three historic climatic zones, based on temperature and sunshine characteristics, into which the world was originally divided by the classical Greeks; it is the low-latitude zone bounded by the northern and southern limit of the Sun's vertical rays, i.e. the Tropics of Cancer (23°27′ N) and the Tropic of Capricorn (23°27′ S). In this zone temperatures are warm throughout the year and the noon Sun is always high; day and night are nearly equal.

total column moisture *See* precipitable water.

TOVS *See* TIROS operational vertical sounder.

towering cumulus *See* congestus.

trace gas A constituent gas of the atmosphere that is present in extremely small concentration. It does not imply unimportance, however, as trace components, such as OZONE and METHANE, for example, are both very important GREENHOUSE GASES. Other trace gases include CARBON DIOXIDE (on which plant life depends), the oxides of nitrogen, and ammonia.

trade-wind inversion A horizontally very widespread temperature INVERSION associated with the base of deeply subsided air within the Equatorward flank of the SUBTROPICAL ANTICYCLONES (highs). It is an oceanic phenomenon, lying typically some 600 m (2000 ft) above the ocean surface within the high, but increasing in depth toward the EQUATORIAL TROUGH, where it may be as high as 2 km (1.2 miles). Its elevation also increases westward across the subtropical oceans as the heating from the surface increases along its track. This means that the trade-wind CUMULUS clouds increase in depth in association with the gradually upward-sloping inversion. The trade-wind inversion is also dry, which means that within the inversion layer of temperature increasing with height, the HUMIDITY decreases very dramatically upward.

trade-wind littoral climate *See* tropical monsoon and trade-wind littoral climates.

trade winds (trades, tropical easterlies) The winds that blow from the Equatorward margin of the subtropical anticyclones (around 30–40°N and S) to converge in the INTERTROPICAL CONVERGENCE ZONE: the northeast trades in the northern hemisphere; the southeast trades in the southern hemisphere. Unlike many other wind systems, which show great variability over time, the direction and velocity of the trade winds are remarkably constant, especially over the oceans, being of the order of 14 knots (16 mph). The trade winds are strongest in winter, with their latitudinal extent varying by season. Heavy rainfall can result when the trades encounter a topographic barrier, especially on the western side of oceans nearer the Equator where the *trade wind temperature inversion* is weaker. Consequently, the northern coast of many Caribbean islands (e.g. Jamaica) is considerably wetter than the southern coast. Conversely, places located further away from the Equator and on the eastern side of ocean basins within the trade-wind belt (e.g. the Canary Islands) experience very arid conditions because of this temperature inversion. TROPICAL CYCLONES develop in the tropical easterlies. *See also* antitrades.

trajectory The track taken by a parcel of air as it moves within the atmosphere over a given period. The tracks are nearly always three-dimensional, but are often mapped on a two-dimensional weather map, such as at mean sea level.

tramontana A LOCAL WIND experienced in the Mediterranean, especially in Italy where the name is frequently applied to a wind that blows off the country's west coast. In Italy, the tramontana is a cool dry north or northeasterly wind that blows from the Alpine range to the north. A northerly wind that affects the Catalan coast of Spain in winter is also called the tramontana.

translucidus A cloud variety of extensive sheets, layers, or patches translucent enough for the Sun or Moon to be visible through it. It is applied to altocumulus, altostratus, stratocumulus, and stratus. *See also* cloud classification.

transpiration The loss of water vapor through evaporation from the surface of a plant. Liquid water contained in the soil is extracted by plant roots, passed upward through the plant, and discharged as water vapor to the atmosphere. The rate of transpiration during the day is about the same as that from an open water surface in the same meteorological conditions, but is almost zero during night hours.

transverse bands Features that have been noted on satellite cloud images; they are linear bands of clouds organized at right angles to the prevailing flow and may indicate extreme turbulence at upper levels. They have been observed in CIRRUS at upper levels in the vicinity of a POLAR FRONT JETSTREAM.

treeline *See* timberline.

tree-ring analysis *See* dendrochronology.

trend *See* secular trend.

triple point of water The temperature of 0.01°C (273.16 K) at which liquid water, ice, and water vapor coexist in equilibrium, which is used as a fixed point on the thermodynamic temperature scale.

tropical air An air mass that has its source region in the belt of subtropical anticyclones of the middle latitudes. The warm moist maritime tropical air (mT) originates over the oceans, while the hot and dry continental tropical air (cT) has its source over the arid and desert areas of the land masses, such as North Africa.

tropical and subtropical desert climate An arid climate type occurring mainly between 15° and 30° latitude in both north and south hemispheres beneath the descending Hadley cell circulations; it is dominated by subtropical high pressure for most or all of the year. Precipitation is very low and temperatures are high with a high diurnal range. The climate is denoted by BW in the KÖPPEN CLIMATE CLASSIFICATION.

tropical and subtropical steppe climate A semiarid climate that occurs along the margins of the true arid deserts. It has a short wet season associated with the Intertropical Convergence Zone (ITCZ) enabling the growth of grasses, which support the grazing of animals. The Sahel region, which borders the Sahara has a tropical steppe climate. The climate is denoted by BSh in the KÖPPEN CLIMATE CLASSIFICATION.

Tropical Atmosphere-Ocean Array (TAO) *See* Tropical Ocean-Global Atmosphere Program.

tropical climate A general term for the climates of the TROPICS or equatorial zone. Temperatures are high throughout the year with no distinct cool season. Precipitation is also high and in many locations there are distinct wet and dry seasons. *See also* tropical and subtropical steppe climate; tropical monsoon and trade-wind littoral climates; tropical wet-dry climate; wet equatorial climate.

tropical cyclone 1. A general term for all cyclonic circulations that originate within the tropics. *See* cyclone; tropical depression; tropical disturbance; tropical storm.

2. The intense low-pressure system with sustained winds (1-minute average) over 64 knots (73 mph), i.e. above force 12 on the BEAUFORT SCALE, that develops over tropical waters and is known in North and Central America and the Caribbean as a HURRICANE. In the western Pacific region it is known as a TYPHOON, and off Australia and in the Indian Ocean, as a cyclone or tropical cyclone.

Tropical Cyclone Program An integral part of the World Meteorological Organization's WORLD WEATHER WATCH Applications Department. It has the aim of establishing nationally and regionally coordinated systems to ensure the minimization of the loss of life and of damage to property due to TROPICAL CYCLONES. One important thrust of the program is therefore to assist its members in the provision of high-quality predictions of the tracks and intensity of tropical cyclones.

tropical depression A CYCLONE in tropical or subtropical latitudes having one or more closed isobars and in which the maximum sustained surface wind speeds (1 minute average) are 33 knots (38 mph) or less, or less than force 7 on the BEAUFORT SCALE. It may form from an easterly wave or a TROPICAL DISTURBANCE. If it continues to intensify and strengthen if may develop into a TROPICAL STORM, and possibly a hurricane (tropical cyclone or typhoon). Tropical depressions are assigned a number by which they can be recognized and tracked.

tropical disturbance A discrete nonfrontal system of clouds and thunderstorms with organized convection that occurs in the tropics and subtropics. It shows slight circulation but no closed isobars and is in the range of 150–500 km (100–300 miles) in diameter.

The tropical disturbance may develop into a TROPICAL DEPRESSION, and from there intensify into a TROPICAL STORM, or possibly a hurricane (tropical cyclone or typhoon).

tropical easterlies *See* trade winds.

Tropical Easterly Jet Stream A JET STREAM that forms a constituent part of the summer ASIAN MONSOON; it has its ENTRANCE REGION in the upper TROPOSPHERE above Indonesia and its EXIT REGION across sub-Saharan Africa. It is a major seasonal feature of the GENERAL CIRCULATION of the atmosphere.

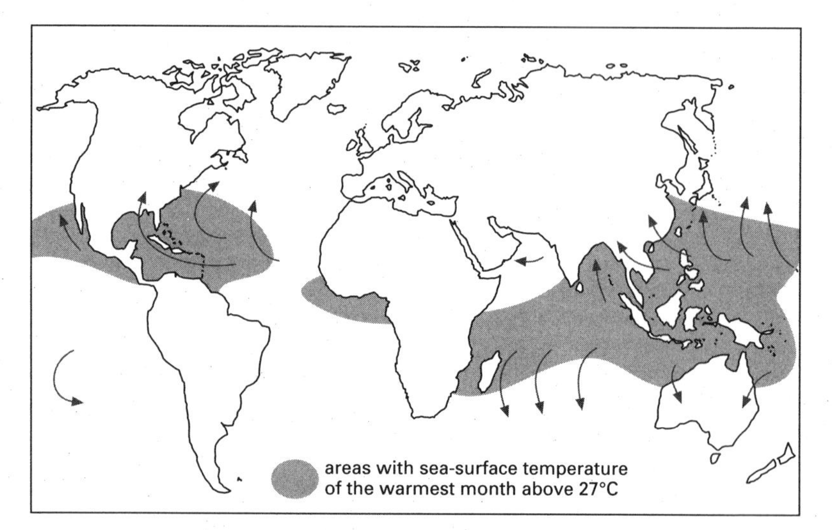

Distribution of tropical cyclones

tropical meteorology The study of the atmospheric phenomena of tropical zones, such as convectional clouds, the trade winds, Hadley cells, Intertropical Convergence Zone (ITCZ), monsoons, tropical depressions, tropical storms, and tropical cyclones.

tropical monsoon and trade-wind littoral climates Climates of the tropical zone with high temperatures, a low annual range of temperatures, abundant precipitation, and a short dry season, which occurs usually during the low-Sun (winter season). The climate type has two different origins. Over southern and southeastern Asia, the largest extent of the type, the ASIAN MONSOON brings precipitation in summer; in winter the outflow from the Siberian anticyclone (high) over Asia brings a drier and cooler season. In the Americas and Africa the TRADE WINDS bring moist air and orographic effects produce abundant rainfall along the coastal zones. Short dry seasons occur with the seasonal migrations in the trade winds. The climates are denoted by Am in the KÖPPEN CLIMATE CLASSIFICATION.

Tropical Ocean-Global Atmosphere Program (TOGA) An international program, a major component of the WORLD CLIMATE RESEARCH PROGRAM (WCRP), aimed at the prediction of climate phenomena with reference to the tropical oceans and their relationship to the global atmosphere. It sought to study the feasibility of modeling the coupled ocean-atmosphere system, and to provide a scientific background for the observation and transmission of data. Lasting from 1985 until 1994, one of its major achievements was the establishment of an ocean-observation system to support seasonal-to-interannual climate studies. This includes the TROPICAL ATMOSPHERE-OCEAN ARRAY (TAO) of nearly 70 moored buoys spanning the equatorial Pacific (which provides measurements of surface winds, sea-surface temperature, upper-ocean temperature and currents, air temperature, and humidity), drifting buoys, observation ships, an island wind profiler network, and satellite data. The data provided by the program contributed to understanding the processes responsible for the EL NIÑO SOUTHERN OSCILLATION (ENSO).

Tropical Prediction Center In the US, one of the nine National Centers for Environmental Prediction (NCEP); it is sited at Florida International University on the outskirts of Miami, Florida. It includes the National Hurricane Center (NHC), the role of which is to provide tropical analysis and forecasts, particularly during the HURRICANE season from May 15 to November 30 for the Caribbean, the Gulf of Mexico, the North Atlantic, and Eastern Pacific. It also prepares and distributes hurricane watches and warnings.

tropical rainy climate One of the major climate groups in the KÖPPEN CLIMATE CLASSIFICATION, designated as A. Temperatures are higher than 18°C (64.4°F) for all months of the year and there is no winter season. The climate has high annual rainfall, which exceeds annual evaporation.

tropical storm A CYCLONE developing in tropical latitudes. It is classified as a tropical storm when the maximum sustained surface wind speed (1 minute average) is between 34 and 63 knots (39–73 mph), force 8 to 11 on the BEAUFORT SCALE. Once it reaches this intensity the tropical storm is given a name by which it can be tracked. The tropical storm may strengthen further to become a TROPICAL CYCLONE, HURRICANE, or TYPHOON (depending on the part of the world in which the tropical cyclone is occurring).

tropical wave *See* easterly wave.

tropical wet-dry climate (tropical savanna climate) A climate type of tropical regions in which there are marked wet and dry seasons and temperature are high but show seasonal variations. Total annual rainfall is around 50–175 cm (20–69 in) and occurs chiefly from convectional rainfall. The migration of the Intertropical Convergence Zone (ITCZ) is responsible for the seasonal shifts in most of the areas. The climate type occurs in Africa and the Americas (latitudes 5–20°N and S) and in Asia (10–30°N). The climate is designated as Aw in the KÖPPEN CLIMATE CLASSIFICATION.

Tropic of Cancer The northern limit on the Earth at which the Sun is directly overhead (which occurs at the solstice on June 21), located on the latitude of 23°27′ N.

Tropic of Capricorn The southern limit on the Earth at which the Sun is directly overhead (which occurs at the solstice on December 21), located on the latitude of 23°27′ S.

tropics (equatorial zone) The zone that lies between the lines of latitude of the TROPIC OF CANCER and TROPIC OF CAPRICORN.

tropopause The narrow boundary that divides the lowest atmospheric layer, the TROPOSPHERE, from the STRATOSPHERE above. The tropopause acts like a lid to convection that occurs in the troposphere; for example, this tends to be the point at which clouds can grow no further. The tropopause is an isothermal layer in which there is no change in temperature with height.

troposphere The atmospheric layer that lies directly over the Earth's surface and below the STRATOSPHERE. Its height varies, being higher at the Equator at around 20 km (12 miles) above the Earth's surface and decreasing toward around 10 km (6 miles) at the poles. The height also varies with the seasons being highest in summer and lowest in winter. The troposphere contains nearly all the Earth's weather phenomena and clouds as well as about 75% of the gaseous mass and 99% of the water vapor of the atmosphere. It is characterized by a decreasing of temperature with height at a rate of approximately 6.5°C/km (the ENVIRONMENTAL LAPSE RATE) until the TROPOPAUSE is reached. The layer is well mixed by vigorous vertical circulations within it related to the intensity of surface heating.

trough A normally elongated feature of low atmospheric pressure on a pressure chart, the axis of which is defined by a sharp 'kink' or angle in the v-shaped isobars. The contour pattern on a chart has the same appearance as the contours of a trough or valley on a land-surface topography map. Troughs often coincide with FRONTS, across which wind direction and therefore isobar alignment change. They are frequently, but not always, associated with unsettled weather, cloudy conditions, and precipitation. *Compare* ridge.

trowal In Canada, the trough of warm air that has been lifted and squeezed up along the upper front of an occluded front. It can be associated with significant weather at the surface, such as may be found with an OCCLUDED FRONT. [From: *tr*ough of *w*arm air *al*oft]

tsunami A potentially destructive wave or series of waves in the ocean produced by sudden submarine movement, such as an underwater earthquake, volcanic explosion, or a landslide (either a landslide into a body of water or a gravity slide of marine sediments). The shallow-water wave or waves, which are about 1 meter high, possess long periods (over 1 hour) and wave lengths (often in excess of 100 km) and can travel at rapid speed (e. g. over 700 km/hr across the Pacific Ocean). When the tsunami approaches land, and water depth decreases, the wave is slowed and grows considerably in height, sometimes in excess of 20 m (66 ft) with the potential to cause serious damage as it travels ashore. Recent tsunamis have included the devastating waves (about 14 m (46 ft) in height) generated by underwater earthquakes that hit the Sissano Lagoon area of the northwest coast of Papua New Guinea on July 17 1998, destroying villages and causing considerable loss of life. The Pacific region has a particularly high incidence of tsunamis associated with the seismically active subduction zone of the Pacific rim that surrounds it. Tsunamis have also been known as *tidal waves* but this term is incorrect. [Japanese: 'harbor wave']

tuba A supplementary cloud feature in the form of a narrow column or inverted cone-like cloud that protrudes from the base of a tall convective cloud, typically CUMULONIMBUS but occasionally cumulus CONGESTUS. It is often the precursor of the FUNNEL CLOUD with which TORNADOES are associated. *See also* cloud classification.

tundra The treeless plains of the northern hemisphere that extend northward from the limit of tree growth (the timberline) across North America (Alaska and northern Canada) and Eurasia to the permanently snow- and ice-covered land of the polar zone. It is generally wet and boggy as a result of the permanently frozen ground – permafrost – which lies below the ground surface preventing drainage, and the vegetation is hardy shrubs, mosses, and lichens.

tundra climate The climate occurring along the Arctic coast of North America and Eurasia and along the coastal fringes of Greenland between 60° and 75°N. It has large annual ranges of temperature and mean annual temperatures are below freezing point of 0°C (32°F). Winters are long and cold and summers short and mild. Precipitation is low overall with dry snow in winter but rises in the brief summer season when cyclonic storms develop along the sea and sea ice margin. In the KÖPPEN CLIMATE CLASSIFICATION it is a type E climate (ET); the mean temperature of the warmest month is above 0°C (32°F) but below 10°C (50°F).

turbidity The reduction in the transparency of the atmosphere to incoming solar radiation caused primarily through ABSORPTION and SCATTERING by particulate matter and aerosols. It excludes scattering by air molecules and by clouds.

turbopause The level, approximately 100 km (60 miles) above the Earth's surface, at which the two main mixing processes in the atmosphere – molecular diffusion and convective motions – are equally important. Below this level turbulent fluid motions dominate and above, molecular diffusion prevails.

turbosphere The part of the atmosphere below the TURBOPAUSE characterized by turbulence in which convective motions play the major role in homogenizing the composition of the atmosphere. *See also* homosphere.

turbulence (turbulent flow) In the atmosphere, random fluctuations that occur in the wind speed and direction as mechanical turbulence when air blows across a rough landscape, creating a complex of EDDY motions. The eddies can produce turbulent flow within a layer extending to a few hundred meters deep. *Thermal turbulence* is created within and around CUMULIFORM clouds, characterized by marked updrafts and downdrafts. This type of turbulence can be troposphere-deep within CUMULONIMBUS clouds. *See also* clear-air turbulence.

turbulent boundary layer *See* mixed layer.

turbulent diffusion (turbulent transport) The transport of heat, water vapor, momentum, and air pollutants throughout the depth of the PLANETARY BOUNDARY LAYER by horizontal and vertical turbulence (the irregular motions embedded in the airflow of the planetary boundary layer). Turbulence is much more effective by several orders of magnitude in transporting quantities through the depth of the boundary layer than the process of molecular diffusivity. *See also* mixed layer.

turbulent flow *See* turbulence.

turbulent transport *See* eddy diffusion.

TVS *See* tornadic vortex signature.

twenty-foot wind The wind speed measured or estimated over a 2- to 10-minute period at a standard 20-foot (6-m) level above a vegetative surface or continuous tree canopy. The twenty-foot wind is used as one input variable into fire danger models in the US.

twilight The time before sunrise (dawn) and after sunset (dusk) when light is increasing or waning and is provided by reflection of sunlight by the upper atmosphere. Specific definitions of twilight are used for the time that twilight begins in the morning or ends in the evening: *civil twilight* begins or ends when the center of the Sun is geometrically 6° below the horizon; *nautical twilight* begins or ends when this distance is 12°; and *astronomical twilight* begins or ends when the distance is 18° below the horizon.

twister A colloquial name, especially in the US, for a TORNADO. It is also used for less severe local events, such as WHIRLWINDS and DUST DEVILS, in which rotational motion occurs.

typhoon The name for an intense TROPICAL CYCLONE that occurs in the western Pacific Ocean. Those typhoons with maximum sustained winds (1 minute average) of over 130 knots may be defined as *super typhoons*. *See also* hurricane. [Chinese *taifun*: 'great wind']

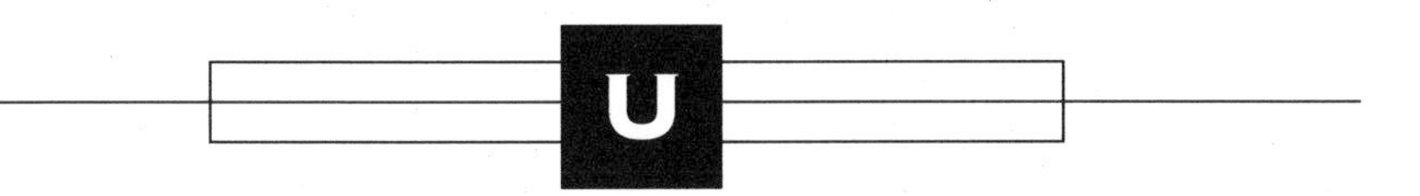

UK Met Office *See* Met Office.

ultraviolet radiation (UV radiation) The part of the electromagnetic spectrum with wavelengths from 1–400 nm in the region below the VISIBLE RADIATION. There are several methods used to subdivide ultraviolet radiation. The most widely known, that based on the biological effects it produces, has three divisions: *UVA* (320–400 nm), which is significant in the production of photochemical smog, *UVB* (290–320 nm), and *UVC* (200–290 nm) which is strongly absorbed by ozone and other gases in the middle stratosphere. Only 1% of solar radiation is within the UVB division, most of which is absorbed by ozone in the atmosphere. The small proportion that reaches the surface can cause damage at the molecular level, for example to DNA, which can affect animals and plants. A number of organizations now produce forecasts of UV levels, such as the *UV Index* issued by the US National Weather Service.

In physics, ultraviolet radiation is subdivided into regions: near (400–300 nm), middle (300–200 nm), far (200–100 nm), and extreme (less than 100 nm).

uncinus A cloud species of CIRRUS in which the high cloud is shaped like a comma that terminates in a hook or a tuft; the upper parts do not show rounded protuberances. *See also* cloud classification.

undersun *See* subsun.

undulatus A type or variety of cloud that occurs in patches, sheets, or layers and displays an undulating form. Sometimes more than one system of undulations is visible. It is a variety of CIRROCUMULUS, CIRROSTRATUS, ALTOCUMULUS, ALTOSTRATUS, STRATUS, and STRATOCUMULUS. *See also* cloud classification.

United Nations Conference on Environment and Development *See* United Nations Framework Convention on Climate Change.

United Nations Framework Convention on Climate Change (UNFCCC) An agreement drawn up in 1992 and opened for signature at the *United Nations Conference on Environment and Development* (UNCED, the *Earth Summit*) in Rio de Janeiro, Brazil, that set out a framework for action to control or cut greenhouse gas emissions to prevent significant anthropogenically forced climate change. It entered into force in 1994. The first meeting of the Conference to the Parties (COP 1) took place in Berlin, Germany, in 1995, where the Berlin mandate was agreed. A protocol to the convention was drawn up in 1997 at the third Conference to the Parties (COP 3) in Kyoto, Japan (*see* Kyoto Protocol).

universal gas constant *See* ideal gas laws.

Universal Time (UT) The name, introduced in 1928, for GREENWICH MEAN TIME (GMT) when used for precise scientific purposes. Universal Time is a general designation for timescales based on the rotation of the Earth and there are several variations. *UT1* is related to polar motion and is proportional to the rotation of Earth in space. *See also* Coordinated Universal Time.

unstable air *See* absolute instability.

unstable equilibrium An atmospheric state in which a displaced air parcel will continue to accelerate vertically after an initial perturbation, leading to rapid vertical air motion and cloud formation. This state occurs when the ENVIRONMENTAL LAPSE RATE (ELR) is greater than the DRY ADIABATIC LAPSE RATE (DALR). *See also* absolute instability; lapse rate; stability.

UN World Meteorological Organization *See* World Meteorological Organization.

updraft The relatively localized ascending current that is a critical constituent of CUMULIFORM clouds. Updrafts vary significantly in magnitude; the updraft that supplies or forms a shallow CUMULUS cloud may have a speed of a few tens of centimeters per second, whereas that forming the heart of a vigorous CUMULONIMBUS cloud may be as strong as 50 m s^{-1} (e.g. in severe weather over the springtime High Plains in the US). The ascending currents that produce STRATIFORM clouds across a very much larger area are sometimes referred to as an updraft.

upper-air station A site at which synoptic observations of the upper-air levels are made. RADIOSONDES are released twice a day from upper-air stations, at 00:00 UTC and 12:00 UTC, to provide observations of ATMOSPHERIC PRESSURE, DRY-BULB TEMPERATURE, RELATIVE HUMIDITY, and WIND DIRECTION and WIND SPEED. In addition, at 06:00 UTC and 18:00 UTC, upper winds are observed by tracking a balloon-borne radar target. The balloons usually ascend into the lower STRATOSPHERE before bursting. Upper-air stations are much more widely scattered than surface-weather stations, and are thus used to define the large-scale smooth features of the troposphere above the friction layer and the lower stratosphere.

upper-air weather chart A map that depicts one or more aspects of the atmosphere at a particular level between the lower TROPOSPHERE and the lower STRATOSPHERE. A common form is the height analysis of, for example, the 500 hPa surface, which takes the form of a field of contours that portray the LONG WAVES. Additionally, charts can be for winds at, for example, 250 hPa, or temperature at 700 hPa. Operational weather services usually recognize standard levels for which charts are always produced: these include 850, 700, 500, 400, 300, 250, and 200 hPa.

upper-level system Mainly synoptic-scale features in the upper TROPOSPHERE that may not necessarily be manifested at the surface. Examples are upper troughs and JET STREAMS.

upslope flow The movement of warm air up the sideslopes of a valley, which is induced by differential heating rates. It occurs during the daytime, the flow being typically shallow in its vertical extent, unlike *upslope winds*, which envelop the entire mountain range and are synoptically forced. *See also* anabatic wind.

upslope fog A type of FOG formed on the upwind or upslope side of a hill or mountain when moist STABLE air undergoing OROGRAPHIC LIFT cools sufficiently, by adiabatic expansion, to produce condensation of fog droplets at ground level.

up-valley wind (talwind) A weak wind that blows up a valley, mostly notably on a warm summer's afternoon. As air is heated during the day, it becomes less dense and so rises, the topographic constriction of the valley sideslopes forcing the air up the valley. It generally occurs at the same time as the upslope ANABATIC WIND.

upwelling A process by which water from deeper levels in the ocean or sea is transported upward to the surface to replace the surface flow that is taking place away from an area. This occurs, for example, along coasts when warm surface waters are driven away from the shore by winds generating offshore currents; these warm waters are replaced by cold denser water brought up from moderate depths. The deeper water is often rich in nutrients, which encourage phytoplankton growth, and many of the world's major fisheries are located in such areas of upwelling (e.g. off the coasts of Peru in South America and Oregon, US).

urban climate A modification of the regional near-surface climate caused by the replacement of components of the biosphere, such as vegetation, by urban fabric, such as buildings, roads, and paved zones. This changes the nature of the heat transfer from the surface by altering its heat capacity. In turn, it changes the atmospheric properties of the SURFACE BOUNDARY LAYER, particularly the temperature, giving rise to the urban HEAT ISLAND effect. Such urban effects on the near-surface climate have significantly increased since about 1950 as urban centers have expanded and cool air ventilation from rural areas has diminished.

UT *See* Universal Time.

UTC *See* Coordinated Universal Time.

UV radiation *See* ultraviolet radiation.

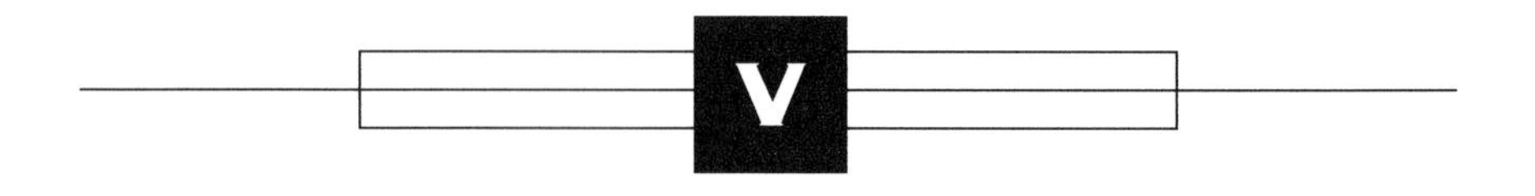

valley breeze *See* valley wind.

valley exit jet A strong down-valley air current that blows from a valley above its intersection with an adjacent plain.

valley wind (valley breeze) The flow of air predominantly up a valley during the day. Such winds are best developed in calm and clear conditions, and hence on days with a high diurnal temperature range. The dynamics of the valley wind are strongly controlled by the topography. *See also* along-valley wind system; anabatic wind; down-valley wind; funneling; katabatic wind; mountain wind.

Van Allen radiation belts Two bands or zones of high-energy charged particles trapped within the Earth's magnetic field; they are concentrated at 3000 km (2000 miles) and 16,000 km (10,000 miles) above the Earth's surface. The belts are named after the American physicist James Alfred Van Allen (1914–), who discovered them in 1958 as a result of experiments carried by space probes and artificial satellites.

vane *See* wind vane.

vane-oriented propeller anemometer *See* propeller anemometer.

vapor concentration *See* absolute humidity.

vapor pressure In meteorology, the component of the total atmospheric pressure that is exerted by water vapor. It is one way of measuring the humidity of the air. At a given temperature, an increase of vapor pressure in the air corresponds to an increase in the humidity of the air. *See also* saturation vapor pressure.

vardar (vardarac) A cold dry northerly wind that blows from Serbia down the valleys of Macedonia, such as the Vardar, bringing polar air into the Aegean Sea region. Like the MISTRAL in its character, the vardar is particularly felt around the Greek city of Thessaloniki. It occurs largely in winter.

vector A force possessing both magnitude and direction. Whereas a *scalar quantity* only possesses magnitude (for example, the wind speed is a scalar quantity), the wind velocity is a vector quantity because both direction and speed are specified.

vector wind *See* thermal wind.

veering (veered) A clockwise change in the wind direction over time. For example, if a southeasterly (SE) wind (135°) blowing over a town at 10.00 pm has been replaced by a southwesterly (SW) wind (225°) by 6.00 am the following morning, then the wind is said to have veered during this 8-hour period. Veering is the opposite change to BACKING.

velum An ACCESSORY CLOUD; a fine extensive sheet of cloud that lies just above or joined to one or several CUMULIFORM clouds that often have grown up through it. It is most often observed with CUMULUS and CUMULONIMBUS clouds. *See also* cloud classification.

vendavale (vendaval) A strong LOCAL WIND associated with a low-pressure system that blows from the southwest and affects the Straits of Gibraltar and the east coast of Spain, chiefly during the winter. Given the wind's Atlantic source, the vendavale wind can bring considerable quantities of rainfall to the region, as well as very squally conditions.

ventilation index (ventilation factor) An index that represents the ability of an airflow to flush out pollutants from within a known volume of polluted air. It is defined as the product of the mean wind speed (in meters per second) through the mixed layer and the depth of the mixed layer itself. In the US the index is forecast by some National Weather Service offices.

Venturi effect The increase in the velocity of a fluid or gas that results from a constriction of flow. In meteorology, the effect is observed when land-surface elements constrict the airflow and hence increase the observed wind velocities of an area. Examples include urban 'canyons' (i.e. streets with high-rise buildings on either side, for example, downtown Chicago and New York) and valleys in mountainous terrain. A low-level temperature inversion can also induce a Venturi effect by constricting and so accelerating the air beneath. This phenomenon is sometimes observed in SEA-BREEZE systems.

veranillo The secondary minimum in precipitation in mid- and late summer in most of Central (Middle) America and the Caribbean. This seasonal feature occurs between the two rainfall maxima in May–June and again in September through November. The cause of this lessening in the rains is associated with the slight rise in pressure over the region, possibly linked with an extension southwestward of the North Atlantic anticyclone into the Gulf of Mexico, before low pressure and tropical storms dominate the latter part of the year.

verglas A thin covering of transparent ice on exposed rock surfaces that forms from the freezing of rainfall or the refreezing of water from melted snow.

vernal *See* spring.

vertebratus A variety of cloud, usually CIRRUS, in a pattern in which the elements appear like vertebrae, ribs, or a fish skeleton. *See also* cloud classification.

vertical instability *See* absolute instability.

vertical motion The component of atmospheric flow directed either up or down. Its intensity, within a low for example, depends directly on the magnitude of the low-level CONVERGENCE and the superimposed upper DIVERGENCE. Gently sloping ascent is tied in with CYCLONIC systems, while SUBSIDENCE occurs with ANTICYCLONES (highs). In meteorology, the vertical motion cannot easily be observed directly but is calculated by the way in which divergence changes sign and magnitude with height in the TROPOSPHERE over a region.

vertical stability *See* absolute stability; stability.

Vienna Convention for the Protection of the Ozone Layer A convention held in Vienna, in 1985, in which nations agreed to take appropriate measure to protect the environment and human health against the adverse affects from human activities that modify or are likely to modify the ozone layer. It sought to encourage research, cooperation between nations, and the exchange of information. The urgency for action was stressed when evidence of severe depletion of the ozone layer above Antarctica was published during 1985; in 1987 the MONTREAL PROTOCOL was drawn up.

virga (fallstreak) A shaft of solid or liquid precipitation that leaves a cloud's base but does not reach the surface because the air in the subcloud layer is dry enough to completely evaporate the raindrops or snowflakes. Virga can be vertical or inclined. In CLOUD CLASSIFICATION virga is classed as a supplementary feature.

visibility In weather observations, the greatest (horizontal) distance over which an object of specified characteristics can be seen clearly by the naked eye. If, when a visibility observation is made, it varies spatially, then the smallest distance must be reported. For the SYNOPTIC CODE, visibility up to and including that of 5 km is reported every 100 m. Upward from 6 km, it is reported every 1 km until 30 km, after which it increases in jumps of 5 km.

Visibility Protection Program In the US, a stated aim of the Clean Air Acts for the protection of good visibility within designated regions of the US, especially the National Parks.

visible radiation (photosynthetically active radiation, PAR) That part of the electromagnetic spectrum that has wavelengths between 0.4 and 0.7 μm, to which the human eye is sensitive. The range of colors across this band is violet, indigo, blue, green, yellow, orange, and red.

visual spectrum *See* visible radiation.

volatile organic compound (VOC) One of a number of organic compounds that readily evaporate into the atmosphere at normal temperatures and pressure. Some contribute to the formation of photochemical smog; others are harmful air pollutants.

volcanic dust Microscopic dry particles emitted principally during volcanic eruptions (as is volcanic ash) that can be injected into the stratosphere. The most important dust chemically is that with a sulfate basis.

vortex flow The air motion linked to low-pressure areas. It ranges from the circulation into a TORNADO on the smallest scale up to that associated with planetary-scale features, such as the CIRCUMPOLAR VORTEX. Sometimes, however, vortex flow is used to refer to the smaller-scale vortices only.

vorticity A measure of the spin or circular motion possessed by a fluid in motion. It is usually taken to mean the spin about the local vertical axis, i.e. in the horizontal plane. *Planetary vorticity* is the quantification of the spin possessed by a planet, such as the Earth. Air that is stationary with respect to the Earth's surface (i.e. in calm conditions) possesses this vorticity. It depends on the rotation rate of the planet (in radians per second) and the sine of the latitude. It is a maximum at the poles and vanishes at the Equator. *Relative vorticity* is the property of spin possessed by fluids (for example, air or water) that move relative to the Earth's surface, usually taken in meteorology to be that about the local vertical axis. It is quantified by considering the horizontal gradients of the west and south components of the wind and is defined as positive for cyclonic circulation. It can also be conceived of consisting of a contribution from the flow's curvature and its shear. For example, a tightly curved wind circulation within which the wind speed increases to the right, looking along the wind, possesses strong *positive relative vorticity*. Anticyclones on the other hand are characterized by *negative relative vorticity*. The total rotation or vorticity is the sum of relative vorticity and planetary vorticity and is known as the *absolute vorticity* (i.e. absolute vorticity = planetary vorticity + relative vorticity). It is a property of the large-scale flow of the atmosphere that is conserved.

Walker cell A tropical circulation feature that exists in the vertical plane along the Equator. The ascending branch is linked to deep moist CONVECTION and heavy rain, while the descending current is linked to clear skies and dry weather. These two vertical components are connected by westerly or easterly flow in the lower and upper TROPOSPHERE. There are a number of such Walker cells around the Equator. The cells are named for the British mathematician and meteorologist, Sir Gilbert Thomas Walker (1868–1958), who first identified them.

wall cloud An ACCESSORY CLOUD feature related specifically to TORNADOES. An important initial sign that a parent cloud is to develop a tornado is that part of its base will abruptly lower as a rotating mass of cloud, 1.5–6 km (1–4 miles) in diameter, which 'hangs' beneath – as the wall cloud. It often resembles a vertical wall lying underneath an intense updraft of a rapidly growing CONVECTIVE CLOUD, the main CUMULONIMBUS. A funnel cloud or tornado may develop within a few minutes, or up to an hour, following formation of the wall cloud.

warm advection The horizontal transfer of the properties of a warm air mass from lower latitudes to higher latitudes. For some purposes, vertical components of motion may also be important.

warm anticyclone An ANTICYCLONE (high) that is characteristic of the semi-permanent subtropical high-pressure areas. In the northern hemisphere, the warm anticyclones include the AZORES HIGH, BERMUDA HIGH, and NORTH PACIFIC HIGH, while in the southern hemisphere they include the subtropical anticyclones located over the South Pacific, South Indian, and South Atlantic Oceans. Warm anticyclones are deep, extending throughout the troposphere, and intensifying with height. They tend to be slow moving and generally bring periods of stable atmospheric conditions with settled and warm weather. *Compare* cold anticyclone.

warm conveyor belt *See* conveyor belt.

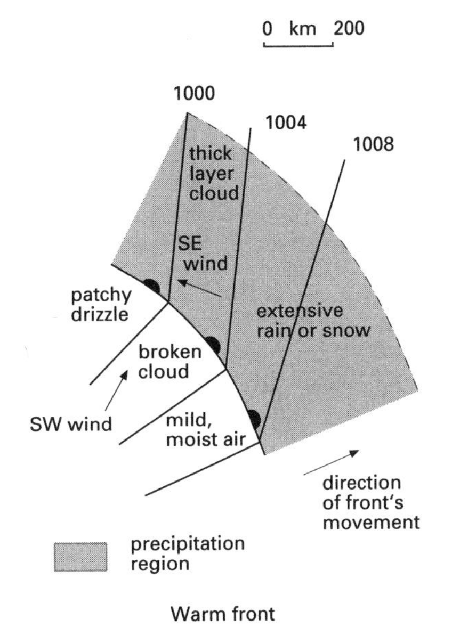

Warm front

warm front The leading edge of an extensive region of advancing warm and, normally, moist air. The front is a shallow sloping zone (the slope typically 1:100 to 1:150) across which temperature and humidity increase on its passage over a site. There is a decline in pressure and often a shift in the wind direction (VEERING in the northern hemisphere). Warm fronts are usually preceded by a belt of widespread PRECIPITATION that lasts for a few hours before the surface front arrives. In the warm front warm air overrides colder denser air underneath, so that it is associated with a classic sequence of clouds. These range from CIRRIFORM cloud some hundreds of kilometers ahead of the surface warm front, gradually lowering to ALTOSTRATUS as the front moves closer to a site, then finally extensive LOW CLOUD and precipitation, typically one to two hundred kilometers ahead of the surface front.

warm occlusion An OCCLUDED FRONT that has extensive cloud and precipitation generally ahead of the surface front. The air ahead of it is colder than the air following it; such occlusions are most common in W Europe, for example, in the winter.

warm rain RAIN generated from clouds in which no ice crystals are present. In such clouds, the BERGERON–FINDEISEN PROCESS of raindrop growth does not operate and the growth of the droplets occurs only by the COLLISION–COALESCENCE PROCESS.

warm sector An extensive region of MARITIME TROPICAL air between the WARM and COLD FRONTS of a LOW or depression. Warm sectors are characterized by warm moist air that is often laden with extensive STRATIFORM cloud. Precipitation often takes the form of widespread drizzle although there can occasionally be heavier bursts of rain from deeper cloud.

warm temperate rainy climates One of the major climate groups in the KÖPPEN CLIMATE CLASSIFICATION, designated as C. Temperatures in the coldest month are between –3°C (26.6°F) and 18°C (64.4°F) and in the warmest month greater than 10°C (50°F).

warning An announcement issued by a weather service, such as the US National Weather Service (NWS), that severe weather has developed, is already occurring, or has been detected on radar. The severe weather poses a hazard or imminent danger, for example a tornado, hurricane, flash flood, or severe thunderstorm.

Wasatch wind A strong easterly CANYON WIND that blows from the mouths of the canyons of the Wasatch Mountains onto the plains of Utah, US.

watch An announcement issued by the US National Weather Service (NWS) indicating a need for preparation and increased awareness that a hazard may be imminent. Watches are produced for tornadoes, thunderstorms, and flash floods.

water The oxide of hydrogen with chemical formula H_20, as well as its liquid form. At atmospheric temperatures and pressures, it can exist in all three phases: solid (ICE), liquid (water), and gaseous (WATER VAPOR).

Water, in its various forms, is of fundamental importance to weather and climate. Water vapor is one of the most important greenhouse gases in the balance of radiative energy and in the HYDROLOGIC CYCLE. Through the processes of evaporation, condensation, and advection water vapor transports energy from lower to higher latitudes, and through cloud formation reflects short-wave solar radiation.

water cycle *See* hydrologic cycle.

water mass A body of water in which properties, such as temperature and salinity, and therefore density, are generally uniform and distinctly different to adjacent water masses.

waterspout A spiraling vertical column of air or vortex of low pressure that occurs over a body of water. *Tornadic waterspouts* form when a tornado passes from land to water (these can be particularly destructive) but the majority form over the water itself. The *fair-weather waterspout* develops at the water surface and in association with warm water temperatures and high humidity within the lower levels of the atmosphere. From the surface it develops upward, visible, at its maximum intensity, from the spray vortex above the water, up through the funnel cloud to the overhead cloud base, reaching a height of around 100 m (300 ft) and moving at speeds of 10–15 knots (9–13 mph). It is generally short-lived with a typical duration of less than 30 minutes. The spout may be subject to windshear but is otherwise often vertical. Waterspouts are most common in the tropics but have also been observed elsewhere, such as over the Mediterranean Sea and the Great Lakes of North America.

water vapor Water in its gaseous form, sometimes also called moisture or aqueous vapor. The distribution of water vapor varies depending on temperature and height, with more than 90% of the water vapor confined to below 500 hPa, usually within the first 5000 m (16,500 ft) above the Earth's surface.

Water vapor plays a fundamental role in the transfer of moisture and latent heat and consequently weather. It enters the atmosphere in the vapor phase of the HYDROLOGIC CYCLE through evaporation and sublimation from water and ice bodies and transpiration by plants. The energy stored as latent heat is transported through advection and released when water vapor condenses or freezes in cloud formation. Water vapor is also the most abundant GREENHOUSE GAS and as such is important in absorbing infrared radiation and warming the Earth's atmosphere. As the temperature of the Earth's atmosphere increases it is capable of holding more water vapor. The warmer temperatures may increase the frequency and magnitude of tropical storms and hurricanes, flooding, and other severe weather. Instruments to carry out routine surface measurements of water vapor include the PSYCHROMETER and DEW-POINT HYGROMETER. The amount of water in the atmosphere is usually measured via a RADIOSONDE, although satellite infrared and microwave sensors are used to obtain total column moisture and low-resolution vertical profiles. A number of definitions are used for the amount of water vapor in air. *See* absolute humidity; dew-point; mixing ratio; relative humidity; specific humidity; vapor pressure.

water vapor channel The wavebands within which scanners on METEOROLOGICAL SATELLITES sense emitted terrestrial radiation. Water vapor absorbs and reradiates electromagnetic radiation at various wavelengths, especially in the infrared of 6.40–7.08 μm. The interception of this radiation by satellites provides a method of remote sensing tropospheric water vapor. The resulting image is of the relative concentration of water vapor in the middle troposphere.

wave In the atmosphere, a generally smooth traveling undulation, in the pressure field, for example, that exhibits a wave-like form. A FRONTAL WAVE is such a feature. A wave can, more widely, describe wave-like patterns on a variety of motion scales, ranging from LEE WAVES to LONG WAVES.

wave cloud A type of cloud that forms in the crest of a LEE WAVE (mountain wave), which is an almost stationary undulation of airstreams at upper levels above or to the lee of a hill or mountain barrier. The lenticular clouds (*see* lenticularis) that form have smooth lens-shaped outlines and can appear to be stationary; they may be individual isolated clouds close to the hill or mountain or form in successions of lines of clouds running parallel to the mountain barrier for some considerable distance downwind.

wave cyclone *See* frontal wave.

wave depression *See* frontal wave.

WCRP *See* World Climate Research Program.

weakening A decrease in the PRESSURE GRADIENT around a low- or high- pressure system over time in conjunction with the decline in the wind strength across a LOW (which is said to be FILLING) or around an ANTICYCLONE (high). *Compare* intensification.

weather The overall condition of the atmosphere at the surface and in the upper air at one particular time. It is normally defined with respect to the data plotted around the STATION MODEL and represented by a variety of maps including UPPER-AIR WEATHER CHARTS and SURFACE WEATHER CHARTS.

weather forecasting The process of predicting future weather conditions, over a wide variety of space scales, from the shortest term nowcast to the medium-range or longer. *See* long-range weather forecasting; medium-range weather forecasting; nowcasting; short-range weather forecasting. *See also* forecast; seasonal prediction.

weather map *See* surface weather chart; synoptic chart; upper-air weather chart.

weather report A surface, upper-air, aircraft, etc., observation normally transmitted to a central collection point by using the SYNOPTIC CODE.

weather satellite *See* meteorological satellite.

weather station An installation designed, equipped, and used regularly to make meteorological observations.

weather symbols *See* synoptic symbols.

weather system A wide range of weather-producing features that occur on a variety of space scales. It includes synoptic-scale features, such as CYCLONES and ANTICYCLONES and smaller, but important, phenomena, such as THUNDERSTORMS and SEA BREEZES.

wedge *See* ridge.

weighing rain gauge A type of automatic recording RAIN GAUGE consisting of a funnel that empties into a bucket mounted on a weighing mechanism. The weight is recorded on a chart mounted on a clock-driven drum, or an electrical signal is sent for data processing. The weight is converted to give an equivalent depth of rain in millimeters or inches. *See also* tipping-bucket rain gauge.

West African Monsoon Trough The most active zone of the southwest moist monsoon air that invades West Africa from the Gulf of Guinea and the neighboring waters of the tropical Atlantic. The trough, which is closely associated with the Intertropical Convergence Zone (ITCZ), moves north from the coastlands of southern West Africa in January, to a location around 18°N in August.

West Australia Current A cold current in the southern Indian Ocean that flows northward along the west coast of Australia as part of the South Indian Current.

west coast desert (cold-water desert) Desert areas occurring along the west coast margins of the hot deserts where upwelling of ocean water is associated with cool currents flowing Equatorward. Such deserts are cooler than expected for their latitude, as a result of air flowing inland off the cool waters; for example, the Peru and Atacama deserts of South America and Namib desert of Africa. The deserts are arid but fogs and heavy dew are common. *See also* tropical and subtropical desert climate.

westerlies The extensive region of westerly winds that are commonplace in the middle latitudes in each hemisphere at the surface. The term usually refers to the average wind direction over the oceans in that latitude belt.

western boundary current *See* boundary current.

West Wind Drift *See* Antarctic Circumpolar Current.

wet adiabatic lapse rate *See* saturated adiabatic lapse rate.

wet-bulb depression The difference between the temperatures recorded by the DRY-BULB and WET-BULB THERMOMETERS of a PSYCHROMETER.

wet-bulb temperature The temperature at which pure water evaporates into a sample of air, dry adiabatically, and at constant pressure, in order to saturate the air.

wet-bulb thermometer A THERMOMETER, the bulb of which is kept moist by a covering of muslin wetted with distilled water. It is used in conjunction with a DRY-BULB THERMOMETER in a PSYCHROMETER to measure the relative humidity of the air. The wet-bulb temperature will be lower than the dry-bulb due to evaporative cooling; the drier the air, the lower the wet-bulb temperature.

wet equatorial climate The climate of the region within about 12° latitude of the Equator, and within the influence of the Intertropical Convergence Zone (ITCZ), in which temperatures are consistently high (30°C; 86°F) and abundant annual precipitation (150–1000 cm) falls. Convergence in weak equatorial lows generates convectional rainfall and thunderstorms, often in late afternoon or early evening. Humidity is high and cloud cover is frequently heavy. Some variations in precipitation occur associated with movements in the location of the ITCZ. In the KÖPPEN CLIMATE CLASSIFICATION it is designated as Af.

wet microburst *See* microburst.

wet snow A form of SNOW that occurs when temperatures are sufficiently high enough to aggregate SNOW CRYSTALS into large SNOWFLAKES by the process of regelation. During regelation, melted ice crystals refreeze under the pressure of the snow pack. Characteristic of more maritime areas in the mid-latitude belt, wet snow has a density of about 0.3 g/cm^3. Unlike DRY SNOW, wet snow does not form a very good material for skiing.

WGNE *See* Working Group on Numerical Experimentation.

whirling psychrometer *See* sling psychrometer.

whirlwind A rapidly revolving column of air, often with a slightly inclined axis, that can sometimes extend to a considerable height into the atmosphere and has at its center a mesoscale low-pressure area. Whirlwinds are common in desert areas (e.g. the DUST DEVIL), the phenomenon being a product of intense solar heating of the ground surface during the day. A whirlwind occurring over water is known as a WATERSPOUT.

white dew *See* dew.

white frost *See* hoar frost.

white-out An optical phenomenon in the form of a uniform white glow that occurs typically in polar regions, and is produced by reflections between a uniform cloud and snow cover. The snow-covered surface appears to merge into the sky, and the horizon and images become blurred or are not visible. This can lead to disorientation and loss of balance on the part of the observer.

Wien's law The radiation law stating that the wavelength of radiation emitted by a body is inversely proportional to its thermodynamic (absolute) temperature; i.e. hotter objects radiate at shorter wavelengths.

willy-nilly (willy-willy) In Western Australia, the local name for a severe TROPICAL CYCLONE; it is also now frequently used for a DUST DEVIL.

wind The horizontal movement of air at any height, ranging from points close to ground level (SURFACE WIND) to levels in the upper atmosphere. A wind is defined in terms of its direction and speed. It is always named for the direction from which it blows (e.g. a northerly wind blows from the north). Wind speed is expressed, by international convention, in KNOTS, but in the US miles per hour (mph) are also commonly quoted; kilometers per hour (km h^{-1}) and meters per second (m s^{-1}) are also used. Wind speed is often classified by means of the BEAUFORT SCALE. Actual wind velocities are measured with an ANEMOMETER; wind direction is obtained with a WIND VANE.

Fundamentally, the surface wind is a result of the difference in pressure, the PRESSURE GRADIENT, between two points on a horizontal plane, which usually arises through unequal heating of the Earth's surface creating differences of temperature and thus density. The rotation of the Earth also has important implications in deflecting wind. The global circulation pattern of winds is known as the GENERAL CIRCULATION of the atmosphere. *See also* anabatic wind; local wind; katabatic wind; thermal wind.

windbreak (shelterbelt) A physical barrier (e.g. row of tall trees) erected to protect crops or buildings from the force of the wind. Such windbreaks are frequently seen in southern France to provide protection from the MISTRAL. Shelterbelts may also be constructed to reduce the risk of soil erosion, and to provide protection against dust storms in summer and blizzards in winter. A windbreak can reduce wind speeds downwind for distances up to 20–30 times the height of the obstruction.

wind chill The effect of wind in the loss of heat from the human body in combination with temperature; an increase in wind speed will carry heat away at a faster rate than on a still day. Several methods are used to calculate wind chill. The *wind chill factor* is a measure of the rate of heat loss produced by the combined effects of wind and low temperature, and is expressed in watts per square meter ($W m^{-2}$). The *wind-chill temperature* is the effective temperature experienced by exposed skin when wind speed is combined with the actual ambient air temperature. A *wind chill index*, for example, that produced by the US National Weather Service (NWS), is used to derive the wind chill temperature from the recorded temperature and wind speed.

wind direction The angle (defined as clockwise from true north) *from* which the air blows. The angles are defined, for example, as 90° for an east wind (i.e. from the east), 180° for a south wind (or southerly), 225° for a southwesterly, and 360° for a north wind. A wind direction of 000 is allotted to calm conditions in the SYNOPTIC CODE. It is sensed by the air flowing over an aerodynamically formed flat plate (a wind vane) designed to always point into the air flow, at a standard height of 10 m (33 ft) above the surface.

wind drift current (drift current) An ocean CURRENT that is induced by wind stresses and in which only the Coriolis and frictional forces are significant.

wind field The spatial distribution of wind on a horizontal plane, this either being at the Earth's surface (mean sea level pressure) or at a standard pressure level (e.g. 850 hPa, 700 hPa, 500 hPa). The locations of high and low pressure can be inferred from the wind field. Such fields may be constructed for varying spatial scales, ranging from that of a thunderstorm to the hemispherical or even global scale.

window *See* atmospheric window.

wind profiler *See* profiler.

wind rose A diagram that portrays information about the wind at a particular location and for a given period of time, for example showing the relation of wind direction to other meteorological variables at a given location, or showing the relative frequencies of wind direction and speed.

wind shadow A region sheltered by an obstruction from the force of the PREVAILING WIND. *See* leeward.

wind shear The rate of change of windflow parameters with height in the atmosphere, these two variables being wind direction (DIRECTIONAL SHEAR) and wind speed (SPEED SHEAR). Wind shear is also occasionally used in the context of WIND FIELDS in a horizontal plane.

wind speed The speed of the wind is expressed most usually in knots (nautical

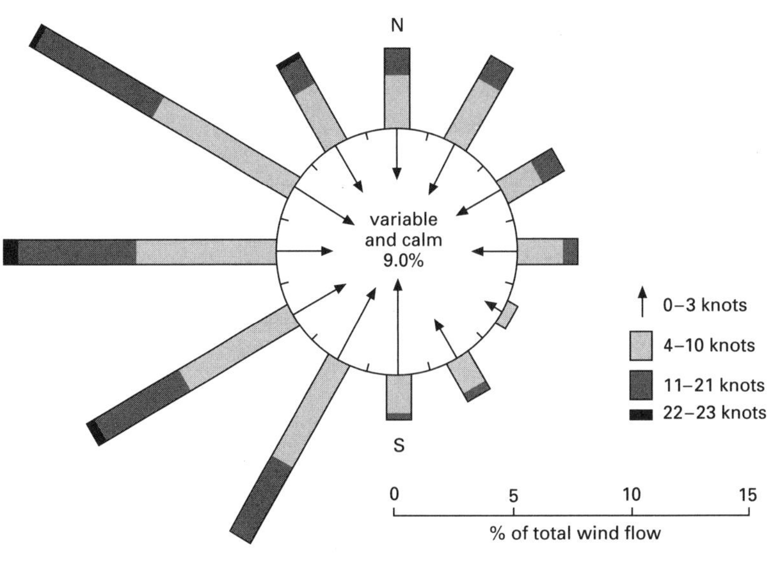

Wind rose

miles an hour) and, more rarely, in meters per second, in the SYNOPTIC CODE. In surface observations it is measured by sensing the rate at which a three-cup ANEMOMETER rotates about a vertical axis at a standard height of 10 m (33 ft) above the surface. The WIND DIRECTION is similarly sensed at this height.

windthrow The uprooting and blowing over of trees by strong winds, especially at gale or hurricane force. The opening up of stands of trees by windthrow can result in changes to the microclimate of forested or wooded areas.

wind vane A device used to measure and record the direction from which the wind is blowing. It consists of a horizontal arm mounted at its center of gravity. The part of the arm with the greatest surface area (usually the fin) provides greater resistance to the wind causing the forward part of the arm to point into the wind. A wind vane is frequently mounted in conjunction with an ANEMOMETER and is used to orient a PROPELLER ANEMOMETER into the wind.

windward The upwind side or side directly influenced by the wind, for example the side of a mountain range that is exposed to the PREVAILING WIND. It is the opposite to the LEEWARD or lee side.

winter The SEASON of the year usually designated as December to February in the middle-latitude northern hemisphere and June to August in the southern hemisphere, characterized by temperatures close to or below freezing and dormant growth in perennial plants. Day length is short and there is often a lack of sunshine. The harshness of winters in the continental interiors of Europe and North America is indicated by the many days of temperatures below 0°C.

WMO *See* World Meteorological Organization.

WOCE *See* World Ocean Circulation Experiment.

Working Group on Numerical Experimentation (WGNE) A group, jointly sponsored by the World Climate Research Programme (WCRP) and the World Meteorological Organization (WMO), that develops atmospheric circulation models for studies of climate and numerical weather predictions.

World Climate Research Program (WCRP) A program established in 1980, jointly sponsored by the World Meteorological Organization (WMO), the International Council for Scientific Unions, and the Intergovernmental Oceanographic Commission. The program aims to develop a fundamental scientific understanding of the climate system and processes, to explore the extent climate can be predicted, and to examine the extent of anthropogenic influence. There are five core projects: The ARCTIC CLIMATE SYSTEM STUDY (ACSYS), CLIMATE VARIABILITY AND PREDICTABILITY PROGRAM (CLIVAR), GLOBAL ENERGY AND WATER CYCLE EXPERIMENT (GEWEX), STRATOSPHERIC PROCESSES AND THEIR ROLE IN CLIMATE (SPARC), and WORLD OCEAN CIRCULATION EXPERIMENT (WOCE).

World Meteorological Center (WMC) One of the three main centers (Melbourne, Moscow, and Washington) within the WORLD WEATHER WATCH (WWW) system for the dissemination of meteorological information on the GLOBAL DATA-PROCESSING SYSTEM (GDPS) and GLOBAL TELECOMMUNICATIONS SYSTEM (GTS).

World Meteorological Organization (WMO) The meteorological agency of the UN, based in Geneva, Switzerland. It commenced operation in 1951, representing 185 member states as the authoritative scientific center covering the state and behavior of the Earth's atmosphere, its weather, and its climate. It plays a supervisory role for all the member states' weather services in terms of proper observational practice, and to ensure the effective operation of the critically important global weather observation network. One of its main aims is the establishment of networks of stations for making meteorological, hydrological, and other observations; in addition it seeks to promote rapid exchange of meteorological information and the standardization of meteorological observations. The WMO operates a number of programs, including the HYDROLOGY AND WATER RESOURCES PROGRAM, the TROPICAL CYCLONE PROGRAM, the WORLD CLIMATE RESEARCH PROGRAM, and WORLD WEATHER WATCH.

World Ocean Circulation Experiment (WOCE) A core project of the WORLD CLIMATE RESEARCH PROGRAM (WCRP), established in 1990, that aims to predict changes in the worlds oceans, such as circulation, volume, and heat storage, that would result from changes in atmospheric climate and net radiation.

World Weather Watch (WWW) One of the WORLD METEOROLOGICAL ORGANIZATION'S (WMO's) major and fundamental scientific and technical programs. Its prime aim is to ensure that its member states receive up-to-the-minute weather observations through the efficient organization of those member states' observation systems. Additionally, the WWW ensures that data is made available through the proper maintenance of telecommunications links from four POLAR-ORBITING and five GEOSTATIONARY SATELLITES, about 7000 ships, 10,000 land stations, and 300 moored and drifting buoys that house automatic weather stations. *See* Global Data-Processing System; Global Observation System; Global Telecommunications System.

WWW *See* World Weather Watch.

Z *See* Zulu Time.

Zaïre Air Boundary *See* Congo Air Boundary.

zenith **1.** That point in the sky vertically above a location on the surface of the Earth, especially with respect to an observer.
2. In general terms, any area of the sky more-or-less directly overhead an observer.

zephyr A name used in the Mediterranean region for a gentle BREEZE that usually blows from the west. It occurs most often around the time of the summer solstice in the northern hemisphere. In Greek mythology, *zephyrus* was the name given to the west wind.

zodiacal light A diffuse zone of luminous light above the horizon that is sometimes visible on the ecliptic before sunrise or after sunset. It is produced as a result of sunlight being scattered by dust particles in the solar system.

zonal flow The west-to-east (positive value) component of the wind direction and speed or the east-to-west (negative value) component. Zonal essentially means parallel to the lines of latitude. *Compare* meridional circulation.

zonal index A measure of the strength of the ZONAL WINDS, usually over a large region for a specific time or short period. An index might be the barometric pressure difference between 40° and 60°N over the northeast Atlantic. A large difference implies a high zonal index, with strong westerlies or easterlies, for example.

zonal mean An average value that is produced by considering individual station values of a particular property (e.g. July mean temperature at the Earth's surface) that are located within a latitude strip, such as 40° to 50°N. The zonal mean of such July average temperatures is the *one* value produced by averaging all the values for that strip. Zonal means cannot, therefore, depict any variation of an atmospheric property in the east–west direction.

zonal winds The component of the observed wind that blows parallel to lines of latitude, either as a westerly or an easterly current.

zonda **1.** A hot and dry FÖHN-type wind experienced in western Argentina. This westerly wind occurs most frequently in spring and can carry considerable quantities of dust. Its properties are a result of the warming of the airflow as it descends the leeward side of the Andean mountain range.
2. A hot and humid southerly wind that carries tropical air from Brazil southward into Argentina and Uruguay. It results in very uncomfortable conditions and is similar to the Australian BRICKFIELDER in character.

Zulu Time (Z) The time zone centered on the 0° meridian of longitude at Greenwich, England, which was allocated the letter 'Z' when used in a military and aviation context; it is now equivalent to COORDINATED UNIVERSAL TIME (UTC), but was formerly equivalent to GREENWICH MEAN TIME (GMT).

Appendix I

Chronology

Major Events in the Development of Meteorology, Climatology, and Atmospheric Physics

c. 525 BC	The Greek philosopher Anaximenes of Miletus (d. c. 528 BC) puts forward the idea that winds, cloud, rain, and hail are formed by thickening of the primary substance, air.
c. 135 BC	The Chinese writer Han Ying first refers to the hexagonal shape of snowflakes in *Moral discourses illustrating the Han text of the "Book of Songs"*.
c. 10 BC	The Greek geographer Strabo of Amaseia (now in Turkey) divides the Earth into frigid, temperate, and torrid zones.
1304	Theodoric of Friebourg (now in Germany) writes *De iride* (On the Rainbow) describing his experiments with bulbs filled with water.
1591	The English mathematician Thomas Harriot (1560–1621) notes that snowflakes have six sides.
c. 1592	The Italian scientist Galileo (1564–1642) invents a simple form of thermometer (the 'thermoscope') based on the expansion of air.
1611	The German astronomer Johannes Kepler (1571–1630) publishes *A New Year's Gift, or On the Six-Cornered Snowflake.*
1621	The Dutch physicist Willebrord Snell (1591–1626) discovers his law of refraction of light (which he did not publish).
1637	The French philosopher René Descartes publishes *Discours de la méthode*, which includes an appendix *Les Météores* containing his ideas about the rainbow and the formation of clouds.
	The refraction law discovered by Snell is independently published by Descartes in another appendix, *La Dioptrique.*
1641	Ferdinand II of Tuscany constructs a thermometer containing liquid.

1643	The Italian physicist Evangelista Torricelli (1608–1647) constructs the first barometer by inverting a tube, closed at one end and containing mercury, in a dish of mercury.
1648	The French mathematician and philosopher Blaise Pascal (1623–1662) investigates Torricelli's barometer. He arranges for his brother-in-law to take an instrument to different heights up a mountain, thereby demonstrating that the height of the column decreased with altitude.
1686	The English astronomer and physicist Edmond Halley (1656–1742) formulates the law relating height to pressure.
	Halley gives a partial explanation of the trade winds, suggesting that hot equatorial air rises and is replaced by cooler air moving in from the tropics.
1687	The French physician Guillaume Amontons (1663–1705) invents a hygrometer.
1695	Amontons constructs an improved barometer.
1702	Amontons constructs a constant-volume air thermometer.
1709	The German physicist Gabriel Daniel Fahrenheit (1686–1736) constructs a thermometer with alcohol as the working fluid.
1714	Fahrenheit uses mercury as the working fluid in thermometers for the first time and defines a temperature scale based on the freezing point of ice and salt (0°) and the temperature of the human body (96°).
1730	The French entomologist and physicist René Réamur (1683–1757) introduces a thermometer using a water/alcohol mixture and devises a temperature scale based on the freezing point of water (0°) and its boiling point (set at 80°).
1735	The German explorer Johann Gmelin discovers the permafrost in Siberia.
	The English meteorologist George Hadley (1685–1768) explains the direction of the trade winds (by reference to the Earth's rotation) in his paper *Concerning the Cause of the General Trade Winds*. This introduces the idea of the *Hadley cell.*
1742	The Swedish astronomer Anders Celsius (1701–1744) devises a temperature scale in which he sets the freezing point of water at 100° and the boiling point of water at 0°.
1743	Jean Pierre Christin inverts the fixed points on Celsius' scale, to produce the scale used today.
1747	The American statesman and inventor Benjamin Franklin

	(1706–1790) begins a series of experiments on electricity. His discovery that charge could be drawn by a pointed metal conductor led to the use of the lightning rod.
1749	Franklin fixes a lightning rod to his house in Philadelphia.
1752	Franklin performs a now famous experiment in which he flies a kite in a thunderstorm to demonstrate that lightning is a form of electricity.
1793	The English chemist John Dalton (1766–1844) publishes *Meteorological Observations and Essays* – an account of his weather observations, which he started in 1787 (and continued for the rest of his life).
1803	The British pharmacist and amateur meteorologist Luke Howard (1722–1846) proposes the first classification of cloud types (still the basis of the present classification).
1804	The French physicist Jean Baptiste Biot (1774–1862) and the French chemist Joseph-Louis Gay-Lussac (1778–1850) make a balloon ascent to a height of three miles to investigate whether the composition of the atmosphere or the Earth's magnetic field changes at this height.
1805	The British admiral Sir Francis Beaufort (1774–1857) proposes a scale of wind speeds (now known as the *Beaufort scale*).
1820	The English chemist John Frederic Daniell (1790–1845) invents the dew-point hygrometer.
1821	The Swiss geologist Ignatz Venetz (b. 1788) proposes that glaciers formerly occurred throughout Europe.
1823	Daniell publishes *Meteorological Essays and Observations*, in which he puts forward theories of wind movement and the atmosphere.
1827	The French physicist Jean-Baptiste Fourier (1768–1830) suggests that human activity may have effects on the Earth's climate.
1835	The French mathematician Gaspard Coriolis (1792–1843) publishes *Mémoire sur les équations du mouvement relatif des systèmes de corps* (Memoir on the Equations of Relative Motion of Systems of Bodies), in which he describes what is now known as the *Coriolis effect*.
1837	The Swiss-born American naturalist Jean Louis Agassiz (1807–1873) first begins to use the term *Eiszeit* (Ice Age).
1840	Agassiz publishes his findings in *Etudes des glaciers* (Studies on Glaciers).
1854	P. Adie designs the Kew barometer.

1857	The Dutch meteorologist Christopher Buys Ballot (1817–90) proposes a rule relating wind direction to the location of the pressure center (*Buys Ballot's law*).
1869	The Scottish meteorologist Alexander Buchan (1829–1907) attempts to identify recurring periods of similar weather (*Buchan spells*).
1853	James Coffin suggests that there are three distinct wind zones in the northern hemisphere.
1863	The British explorer and anthropologist Francis Galton (1822–1911) publishes *Meteographica* (Weather Mapping), which introduces the term 'anticyclone' and describes modern techniques of weather mapping.
1871	The American meteorologist Cleveland Abbe (1838–1916) is appointed chief meteorologist with the Weather Service (part of the US Army).
1874	The American geologist Thomas Chowder Chamberlin (1843–1928) suggests that there were several Ice Ages separated by nonglacial epochs. He published his ideas in a chapter contributed to *The Great Ice Age* (1874–84) by James Geikie.
	The International Meteorological Committee adopts the Beaufort scale for use in weather telegraphs
1888	Gustavus Hinrichs, an Iowa weather researcher, first uses the word 'derecho' for a widespread type of windstorm.
1890s	The British astronomer E. Walter Maunder (1851–1928) identifies a period of low solar activity (the *Maunder minimum*).
1890	The German geographer and climatologist Eduard Brückner (1862–1927) investigates the cycle of cold and damp weather alternating with warm and dry weather in northwestern Europe (the *Brückner cycle*).
1891	Luis Carranza, President of the Lima Geographical Society, contributes a small article to the Bulletin of the society in which he describes a countercurrent flowing N to S from Paita to Pacasmayo occurring annually after Christmas. He reports that fishermen call this El Niño.
	The US Weather Bureau separates from the Army and becomes an independent organization under the direction of Cleveland Abbe.
1896	The Swedish chemist Svante Arrhenius (1859–1927) points out that increased carbon dioxide in the atmosphere can cause the greenhouse effect. He suggests that past Ice Ages may have been caused by decreased amounts of carbon dioxide.

1900	The Russian-born German climatologist Wladimir Peter Köppen first publishes his classification of climatic types (which he modifies several times before 1936).
1902	The British physicist Oliver Heaviside (1850–1925) and the American Arthur Edwin Kennelly (1861–1939) independently suggest the existence of the ionosphere to explain transatlantic radio transmissions.
	The French meteorologist Léon Philippe Teisserenc de Bort (1855–1913) identifies and names the troposphere and the stratosphere as distinct layers of the atmosphere.
	The Swedish physicist Vagn Walfrid Ekman (1874–1954) produces a mathematical model for the result of winds blowing over the ocean (the *Ekman spiral*).
1904	The Norwegian meteorologist Jacob Aall Bonnevie Bjerknes (1897–1975) publishes one of the first scientific accounts of meteorology in his *Weather Forecasting as a Problem in Mechanics.*
1910	Jacob Bjerknes publishes *Dynamical Meteorology and Hydrography.*
1920	The Serbian astronomer and mathematician Milutin Milankovitch (1879–1958) proposes a theory of long-term cyclic climate change (*Milankovitch cycles*).
1922	The United Nations Framework Convention on Climate Change (UNFCCC) is drawn up.
	The British meteorologist William Napier Shaw introduces a type of thermodynamic diagram (the tephigram).
1923	Sir Gilbert Walker, Director of Observations in India, notes "when pressure is high in the Pacific Ocean it tends to be low in the Indian Ocean from Africa to Australia". He calls the effect the Southern Oscillation.
1925	The British physicist Edward Victor Appleton (1892–1965) discovers a layer in the ionosphere (now called the *Appleton layer*).
1927	G. Stüve introduces a type of thermodynamic diagram (the *Stüve diagram*).
1928	Universal Time is adopted as the name for Greenwich Mean Time when used for scientific purposes.
1931	The American climatologist Charles Warren Thornthwaite (1889–1963) devises a precipitation-efficiency index.

1932 The Swedish-born American meteorologist Carl-Gustaf Arvid Rossby (1898–1957) introduces diagrams for indicating the properties of air masses (now known as *Rossby diagrams*).

1935 The Swedish meteorologist Tor Harold Percival Bergeron (1891–1977) publishes a paper *On the Physics of Clouds and Precipitation*, in which he argues that ice crystals in clouds are involved in the mechanism of rain formation.

1939 Work by Walker Findeisen supports Bergeron's theory (later known as the *Bergeron–Findeisen process*).

1946 The American physicist Vincent Joseph Shaefer (1906–1993) induces rainfall by seeding clouds with pellets of solid carbon dioxide.

1947 The American chemist Willard Frank Libby (1908–1980) invents radiocarbon dating, which is used in paleoclimatology.

1948 The American climatologist Charles Warren Thornthwaite (1889–1963) introduces the concept of moisture index.

1950 H. Flohn proposes a climate classification based on global wind belts and precipitation (the *Flohn classification*).

A team led by the Hungarian-born American mathematician John Von Neumann (1903–1957) produces the first computer-generated weather forecasts using the ENIAC computer.

1951 The World Meteorological Organization (WMO) commences operation.

The American geographer and climatologist Arthur N. Strahler (1918–) introduces a climate classification based on air masses (the *Strahler classification*).

1956 The World Meteorological Organization publishes the *International Cloud Atlas*.

The British government introduces an Act of Parliament to limit smoke emissions with the aim of reducing smog.

1957 The American oceanographer Roger Revelle (1909–91) instigates the start of measurements on the amount of carbon dioxide in the atmosphere.

1958 The American physicist James Van Allen (1914–) discovers two bands of ionized particles in the atmosphere as a result of satellite data (the *Van Allen belts*).

1959 The temperature–humidity index is first introduced by the US Weather Bureau.

1960	The National Oceanic and Atmospheric Administration (NOAA) launches the weather satellite TIROS-1.
1961	The American meteorologist Edward Norton Lorenz (1917–) notices that computer predictions of weather are highly sensitive to small differences in initial conditions. This eventually leads to the development of chaos theory.
1963	Lorenz publishes his ideas in a paper, *Deterministic Nonperiodic Flow.*
	The British climatologist Hubert H. Lamb introduces the dust veil index.
1965	The US government introduces the first Clean Air Act to set national air pollution standards.
1966	The American geographer Werner H. Terjung proposes a climate classification based on comfort.
1969	Jacob Bjerknes in California University, Los Angeles, explains the El Niño Southern Oscillation (ENSO).
1970s	The American engineer Herbert Saffir (1917–) and the meteorologist Robert Simpson (1912–) devise a scale for measuring hurricane strength (the *Saffir–Simpson scale*).
1970	The second US Clean Air Act.
1971	The Japanese-born American meteorologist Tetsuya Fujita (1920–98) with Allen Pearson introduces a scale for measuring tornado intensities (the *Fujita* or *Fujita–Pearson scale*).
1972	Earth Resources Technology Satellite 1 (later called Landsat-1) is launched.
1974	The Global Atmospheric Research Program (GARP) conducts the GARP Atlantic Tropical experiment (GATE) to further understanding of the tropical atmosphere.
	The American chemist Sherwood Rowland (1927–) and Marco Molina suggest that CFCs (chlorofluorocarbons) could cause long-term depletion of the ozone layer.
1975	The first GOES (Geostationary Operational Environmental Satellite) is launched.
	The European Space Agency is formed.
1976	The US National Academy of Sciences publishes a report supporting Rowland's theory concerning CFCs and ozone depletion.

1977	The first Meteosat weather satellite is launched by the European Space Agency.
1978	The use of CFCs is banned in the US.
1979	The British scientist James Ephraim Lovelock (1919–) publishes *Gaia*, expounding the idea that the Earth is a self-regulating ecosystem (the so-called *Gaia hypothesis*).
1980	The World Climate Research Program (WCRP) is established.
1984	The British Antarctic Survey obtains conclusive evidence of a hole in the ozone layer.
1985	The Vienna Convention for the Protection of the Ozone Layer takes place.
	The Tropical Ocean-Global Atmosphere Program (TOGA) is started.
1987	The Montreal Protocol – an international agreement to regulate the use of CFCs – is drawn up.
	The Joint Global Ocean Flux Study (JGOFS) is initiated to study the fluxes of carbon in the oceans.
1988	The Global Energy and Water Cycle Experiment (GEWEX) is launched to observe the hydrologic cycle.
	The German scientist Helmut Heinrich obtains evidence for the marked discharge of large numbers of icebergs from the eastern North American ice sheet during glacial periods (*Heinrich events*).
1989	The Global Atmospheric Watch (GAW) is established to collect data on the atmosphere's physical characteristics.
	Projects start to analyze the Greenland ice sheet. The Greenland Ice Core Project (GRIP) is run by the European Science Foundation. The Greenland Ice Sheet Project 2 (GISP 2) is a US project.
1995	Nations meeting in Berlin agree the *Berlin mandate* to take action on climate change beyond the year 2000, which was specified in the United Nations Framework Convention on Climate Change of 1992.
1997	The Kyoto Protocol is agreed to achieve reductions in greenhouse-gas emissions.
1999	Landsat-7 is launched.
2001	The US refuses to ratify the Kyoto Protocol.

Appendix II

Conversion Tables

LENGTH

	m	*cm*	*in*	*ft*	*yd*
1 meter	1	100	39.3701	3.280 84	1.093 61
1 centimeter	0.01	1	0.393 701	0.032 8084	0.010 9361
1 inch	0.0254	2.54	1	0.083 3333	0.027 7778
1 foot	0.3048	30.48	12	1	0.333 333
1 yard	0.9144	91.44	36	3	1

	km	*mile*	*n mile*
1 kilometer	1	0.621 371	0.539 957
1 mile	1.609 34	1	0.868 976
1 nautical mile	1.852 00	1.150 78	1

1 light year = $9.460\ 70 \times 10^{15}$ meters = $5.878\ 48 \times 10^{12}$ miles
1 astronomical unit = 1.496×10^{11} meters
1 parsec = 3.0857×10^{16} meters = 3.2616 light years

AREA

	m^2	cm^2	in^2	ft^2
1 square meter	1	10^4	1550	10.7639
1 square centimeter	10^{-4}	1	0.155	1.0763×10^{-3}
1 square inch	6.4516×10^{-4}	6.4516	1	6.9444×10^{-3}
1 square foot	9.2903×10^{-2}	929.03	144	1

	m^2	km^2	yd^2	mi^2	*acre*
1 square meter	1	10^{-6}	1.195 99	$3.860\ 19 \times 10^{-7}$	$2.471\ 05 \times 10^{-4}$
1 square kilometer	10^6	1	$1.195\ 99 \times 10^6$	0.386 019	247.105
1 square yard	0.836 127	$8.361\ 27 \times 10^{-7}$	1	$3.228\ 31 \times 10^{-7}$	$2.066\ 12 \times 10^{-4}$
1 square mile	$2.589\ 99 \times 10^6$	2.589 99	3.0976×10^6	1	640
1 acre	$4.046\ 86 \times 10^3$	$4.046\ 86 \times 10^{-3}$	4840	1.5625×10^{-3}	1

1 are = 100 square meters
1 hectare = 10 000 square meters = 2.471 05 acres

VOLUME AND CAPACITY

	m^3	cm^3 (cc)	in^3	gal
1 cubic meter	1	10^6	6.1024×10^4	264.169
1 cubic centimeter	10^{-6}	1	0.061 024	2.6416×10^{-4}
1 cubic inch	1.6387×10^{-5}	16.3871	1	4.3289×10^{-3}
1 gallon (US)	3.7854×10^{-3}	3785.4	231.01	1

1 gallon (US) = 0.832 68 gallon (UK)
1 liter = 1 cubic decimeter = 1000 cubic centimeters

VELOCITY

	$m\ s^{-1}$	$km\ h^{-1}$	$mile\ h^{-1}$	$ft\ s^{-1}$
1 meter per second	1	3.6	2.236 94	3.280 84
1 kilometer per hour	0.277 778	1	0.621 371	0.911 346
1 mile per hour	0.447 04	1.609 344	1	1.466 67
1 foot per second	0.3048	1.097 28	0.681 817	1

1 knot = 1 nautical mile per hour = 0.514 444 meter per second

MASS

	kg	*g*	*lb*	*long ton*
1 kilogram	1	1000	2.204 62	$9.842\ 07 \times 10^{-4}$
1 gram	10^{-3}	1	$2.204\ 62 \times 10^{-3}$	$9.842\ 07 \times 10^{-7}$
1 pound	0.453 592	453.592	1	$4.464\ 29 \times 10^{-4}$
1 long ton	1016.047	$1.016\ 047 \times 10^{6}$	2240	1

DENSITY

	kg m^3	*g cm*^{3}	*lb ft*^{3}	*lb in*^{3}
1 kilogram per cubic meter	1	10^{-3}	0.062 428	$3.612\ 73 \times 10^{-5}$
1 gram per cubic centimeter	1000	1	62.428	$3.612\ 73 \times 10^{-2}$
1 pound per cubic foot	16.0185	0.016 0185	1	$5.787\ 04 \times 10^{-4}$
1 pound per cubic inch	$2.767\ 99 \times 10^{4}$	27.6799	1728	1

1 lb/gal (US) = 0.083 082 kg/dm^3

FORCE

	N	*kg*	*dyne*	*poundal*	*lb*
1 newton	1	0.101 972	10^{5}	7.233 00	0.224 809
1 kilogram force	9.806 65	1	$9.806\ 65 \times 10^{5}$	70.9316	2.204 62
1 dyne	10^{-5}	$1.019\ 72 \times 10^{-6}$	1	$7.233\ 00 \times 10^{-5}$	$2.248\ 09 \times 10^{-6}$
1 poundal	0.138 255	$1.409\ 81 \times 10^{-2}$	$1.382\ 55 \times 10^{4}$	1	0.031 081
1 pound force	4.448 22	0.453 592	$4.448\ 23 \times 10^{5}$	32.174	1

PRESSURE

	Pa(N/m²)	*hPa(mb)*	*in Hg*	*atmos*
1 pascal (newton per sq. meter)	1	100	3.3865×10^{3}	9.8692×10^{-6}
1 hectopascal	10^{2}	1	3.3865×10^{5}	9.8692×10^{-4}
1 in Hg	2.9529×10^{-4}	2.9529×10^{-2}	1	3.3422×10^{-2}
1 atmosphere	$1.013\ 25 \times 10^{5}$	$1.013\ 25 \times 10^{3}$	29.92	1

1 pascal = 1 newton per sq. meter
1 pascal = 10 dynes per sq. centimeter
1 hectopascal = 1 millibar
1 bar = 10^{5} pascals = 0.9869 atmosphere
1 torr = 133.322 pascals = 1/760 atmosphere
1 atmosphere = 760 mmHg = 33.90 ft water (at 0°C)

WORK AND ENERGY

	J	*cal_{IT}*	*kW hr*	*btu_{IT}*
1 joule	1	0.238 846	$2.777\ 78 \times 10^{-7}$	$9.478\ 13 \times 10^{-4}$
1 calorie (IT)	4.1868	1	$1.163\ 00 \times 10^{-6}$	$3.968\ 31 \times 10^{-3}$
1 kilowatt hour	3.6×10^{6}	$8.598\ 45 \times 10^{5}$	1	3412.14
1 British Thermal Unit (IT)	1055.06	251.997	$2.930\ 71 \times 10^{-4}$	1

1 joule = 1 newton meter = 1 watt second = 10^{7} ergs = 0.737 561 ft lb
1 electronvolt = $1.602\ 10 \times 10^{-19}$ joule

Appendix III

Temperature Conversions

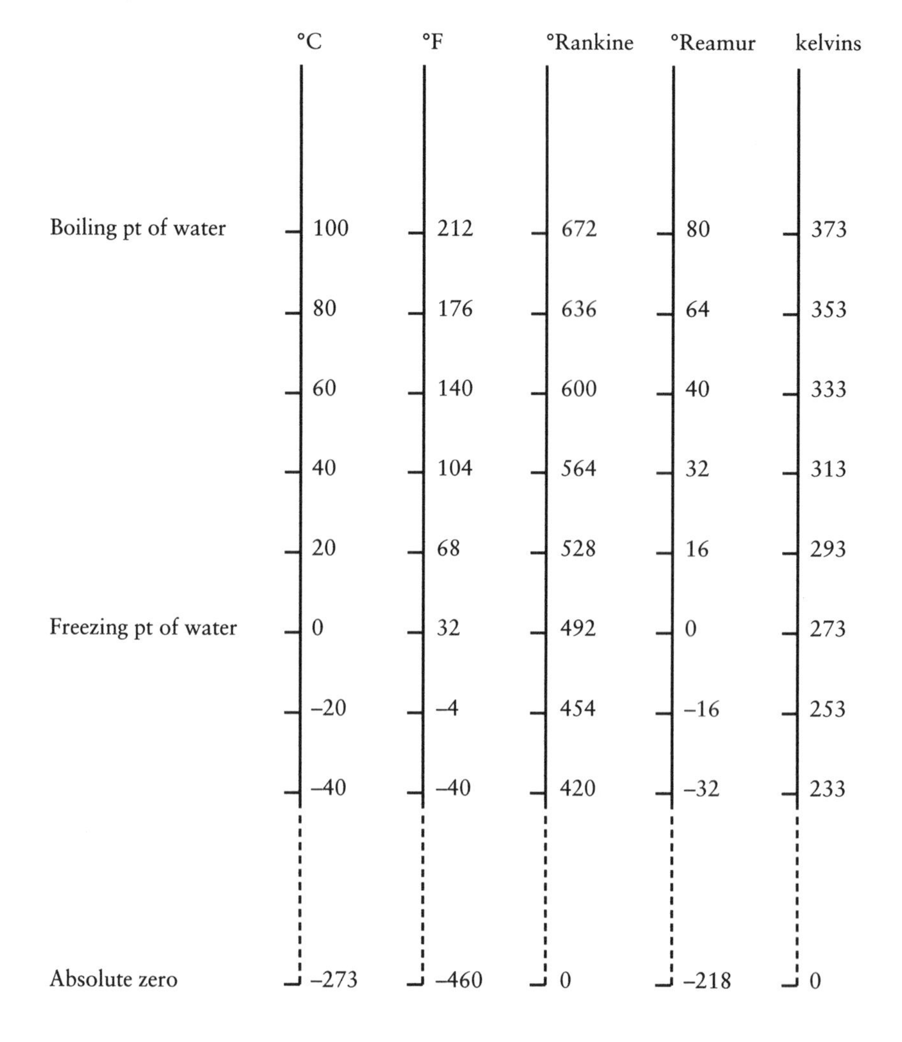

Appendix IV

Web Pages

Weather Sites and Services

Accuweather, Inc.	www.accuweather.com
BBC Weather Centre, UK	www.bbc.co.uk/weather
Meteorological Service of Canada	weatheroffice.ec.gc.ca
National Centers for Environmental Prediction	www.ncep.noaa.gov
National Weather Service (NWS)	www.nws.noaa.gov
Unisys Weather	www.weather.unisys.com
USA Today Weather	www.usatoday.com/weather
The Washington Post Weather	www.washingtonpost.com/wp-srv/weather
The Weather Channel (US)	www.weather.com

Organizations

Bureau of Meteorology, Australia	www.bom.gov.au
Climate Prediction Center	www.cpc.ncep.noaa
Environment Canada	www.ec.gc.ca/envhome
Hadley Centre for climate prediction and research, UK	www.met-office.gov.uk/research/hadleycentre/index
Met Office, UK	www.met-office.gov.uk
National Oceanic and Atmospheric Administration (NOAA)	www.noaa.gov
US Environmental Protection Agency	www.epa.gov
World Meteorological Organization	www.wmo.ch

Education, data, and reference sources

britannica.com	www.britannica.com
Climatic Research Unit, University of East Anglia, Norwich, UK	www.cru.uea.ac.uk

Encyclopedia of the Atmospheric Environment	www.doc.mmu.ac.uk/aric/eae/index
Mount Washington Observatory	www.mountwashington.org
NASA's Earth Observatory	www.earthobservatory.nasa.gov
UK Sci.weather.FAQ	www.booty.demon.co.uk/metinfo.uswfaq
University Corporation for Atmospheric Research	www.edu/ucar/index
WW2010 (weather world 2010 project) University of Illinois	www.2010.atmos.uiuc.edu/(Gh)/home

Meteorological Societies

American Meteorological Society, Boston	www.ametsoc.org/AMS
Royal Meteorological Society, UK	www.royal-met-soc.org.uk
Canadian Meteorological and Oceanographic Society	www.cmos.ca

El Niño

El Niño site, NOAA	www.pmel.noaa.gov/toga-tao/el-nino-story

Satellites

European Space Agency (ESA)	www.esa.int
EUMETSAT	www.eumetsat.de
NASA	www.nasa.gov
NESDIS	www.nesdis.noaa.gov

Severe weather

Storm Prediction Center, NOAA	www.spc.noaa.gov
National Severe Storms Laboratory, NOAA	www.nssl.noaa.gov

Climate change and global warming

Global warming site, US Environmental Protection Agency	www.epa.gov/globalwarming
Climate change site, European Commission	www.europa.eu.int/comm/environment/climate/home_en
US Global Change Research Information Office	www.gcrio.org

Bibliography

Ahrens, C. Donald. *Essentials of Meteorology: an invitation to the atmosphere*. 3rd ed. Pacific Grove, Calif.: Brooks/Cole, 2000.

Ahrens, C. Donald. *Meteorology Today: an introduction to weather, climate and the environment*. 6th ed. Pacific Grove, Calif.: Brooks/Cole, 1999.

Allaby, Michael. *Encyclopedia of Weather and Climate*. New York: Facts On File, 2001.

Bader et al. *Images in Weather Forecasting: a practical guide for interpreting satellite and radar imagery*. Cambridge, U.K.: Cambridge University Press, 1996.

Barry, R. G. & Chorley, R.J. *Atmosphere, Weather & Climate*. 7th ed. London: Routledge, 1998.

Greer, I. W. (ed). *Glossary of Weather and Climate*. Boston, Mass.: American Meteorological Society, 1996.

Houghton, J. T. (ed). *Climate Change 1995: the science of climate change*. Cambridge, U.K.: Cambridge University Press, 1996.

Huschke, R. E. (ed). *Glossary of Meteorology*. 2nd ed. Boston, Mass.: American Meteorological Society, 2001.

Intergovernmental Panel on Climatic Change. *IPCC First Assessment Report*, 1990. *IPCC Second Assessment Report: Climate Change*, 1995.

Lewis, R. P. W. (ed). *Meteorological Glossary*. 6th ed. London: Met Office, Her Majesty's Stationery Office, 1991.

Lye, Keith. *Equatorial Climates*. Austin, Tex.: Raintree Steck-Vaughn Publishers, 1999.

Lye, Keith. *World's Climates*. Austin, Tex.: Raintree Steck-Vaughn Publishers, 1999.

Reynolds, Ross. *Philip's Guide to Weather*. London: George Philip Ltd, 2000.

Schneider, S. H. (ed). *Encyclopedia of Climate and Weather*. New York: Oxford University Press, 1996.

Silverstein, Alvin, & Silverstein, Virginia, & Silverstein Nunn, Laura. *Weather and Climate*. Brookfield, Conn.: Twenty-First Century Books Inc., 1998.

Stevens, William K. *The Change in the Weather*. New York: Dell Publishing Co. Inc., 1999.

Stull, Roland B. *Meteorology Today for Scientists and Engineers*. 2nd ed. Pacific Grove, Calif.: Brooks/Cole, 2000.

Wilson, Francis & Mansfield, F. *The Weather*. EDCP, 1997.

World Meteorological Organization. *International Cloud Atlas*. Vol. I. 1975. *Manual on the observation of clouds and other meteors*. Vol. II. 1987 (plates). Geneva: WMO.